Autodesk® Civil 3D™ 2007
Procedures and Applications

Harry O. Ward, PE

PEARSON

Prentice Hall

Upper Saddle River, New Jersey
Columbus, Ohio

Library of Congress Cataloging-in-Publication Data

Ward, Harry O.
    Autodesk Civil 3D 2007: procedures and applications / Harry O. Ward.
        p. cm.
    ISBN 0-13-171350-7
    1. Civil engineering—Computer programs.   2. Surveying—Computer programs.   3. Autodesk Civil 3D.   I. Title.

TA345.W37 2007
624.0285¢536—dc22   2006044830

Editor in Chief: Vernon R. Anthony
Acquisitions Editor: Jill Jones-Renger
Editorial Assistant: Yvette Schlarman
Production Editor: Louise N. Sette
Production Supervision: Lisa S. Garboski
Design Coordinator: Diane Ernsberger
Art Coordinator: Jaclyn Portisch
Cover Designer: Jason Moore
Production Manager: Deidra M. Schwartz
Marketing Manager: Jimmy Stephens

Autodesk, the Autodesk logo, AutoCAD, AutoCAD®/MAP™, Autodesk® Land Desktop, Civil 3D®, Autodesk® VIZ are registered trademarks of Autodesk, Inc., in the USA and/or other countries. All other brand names, product names, or trademarks belong to their respective holders.

"Virtual Site Design" and "T.I.E. Training" are OutSource Inc. trademarks.

This book was set by Techbooks. It was printed and bound by Bind-Rite Graphics. The cover was printed by Coral Graphic Services, Inc.

Disclaimer:
The publication is designed to provide tutorial information about AutoCAD® and/or other Autodesk computer programs. Every effort has been made to make this publication complete and as accurate as possible. The reader is expressly cautioned to use any and all precautions necessary, and to take appropriate steps to avoid hazards, when engaging in the activities described herein.

Neither the author nor the publisher makes any representations or warranties of any kind, with respect to the materials set forth in this publication, express or implied, including without limitation any warranties of fitness for a particular purpose or merchantability. Nor shall the author or the publisher be liable for any special, consequential or exemplary damages resulting, in whole or in part, directly or indirectly, from the reader's use of, or reliance upon, this material or subsequent revisions of this material.

Pearson Prentice Hall™ is a trademark of Pearson Education, Inc.
Pearson® is a registered trademark of Pearson plc
Prentice Hall® is a registered trademark of Pearson Education, Inc.

Pearson Education Ltd.
Pearson Education Singapore Pte. Ltd.
Pearson Education Canada, Ltd.
Pearson Education—Japan

Pearson Education Australia Pty. Limited
Pearson Education North Asia Ltd.
Pearson Educación de Mexico, S.A. de C.V.
Pearson Education Malaysia Pte. Ltd.

10 9 8 7 6 5 4 3 2 1
ISBN: 0-13-171350-7

This text is dedicated to my son, Joseph, and would not have been possible without the help and support of my wife, Kathy, and my partner, Reiko.

THE NEW AUTODESK DESIGN INSTITUTE PRESS SERIES

Pearson/Prentice Hall has formed an alliance with Autodesk® to develop textbooks and other course materials that address the skills, methodology, and learning pedagogy for the industries that are supported by the Autodesk® Design Institute (ADI) software program that assists educators in teaching technological design.

Features of the Autodesk Design Institute Press Series

JOB SKILLS—Coverage of computer-aided drafting job skills, compiled through research of industry associations, job websites, college course descriptions, The Occupational Information Network database, and *The Job Guide 2004*, has been integrated throughout the ADI Press books.

PROFESSIONAL and **INDUSTRY ASSOCIATION INVOLVEMENT**—These books are written in consultation with and reviewed by professional associations to ensure they meet the needs of industry employers.

AUTODESK LEARNING LICENSES AVAILABLE—Many students ask how they can get a copy of the Civil 3D™ software for their home computer. Through a recent agreement with Autodesk®, Prentice Hall now offers the option of purchasing textbooks with either a 180-day or a 1-year student software license agreement for Civil 3D™. This provides adequate time for a student to complete all the activities in the book. The software is functionally identical to the professional license, but is intended for student or faculty personal use only. It is not for professional use.

For more information about this book and the Autodesk Student Portfolio, contact your local Pearson Prentice Hall sales representative, or contact our National Marketing Manager, Jimmy Stephens, at 1 (800) 228-7854 x3725 or at Jimmy_Stephens@prenhall.com. For the name and number of your sales rep, please contact Prentice Hall Faculty Services at 1 (800) 526-0485.

FEATURES OF *AUTODESK® CIVIL 3D™ 2007*

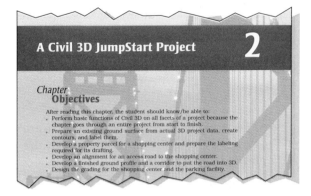

A JumpStart section at the beginning of the book allows users to get up to speed in no time to create Civil 3D drawings.

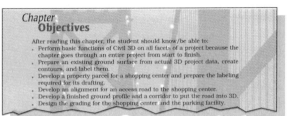

Chapter Introductions with a bulleted list of learning objectives for each chapter provide users with a roadmap of important concepts and practices that will be introduced in the chapter.

Stationing—A unit of measure

computed from the existing ground surface. It is placed inside a Profile View, which is the Civil 3D object that displays profiles. Chapter 6 on Profiles discusses this in more detail, but in summary the Profile View is a Civil 3D concept that contains the grid that profiles are drawn in. The Horizontal Axis of the grid reflects the *stationing* or linear distance along the horizontal alignment and the Vertical Axis represents elevational data. The View will include the existing ground profile with annotation. The settings in the **Toolspace** that control the Profile View are in the Profile View Style, which sets the display and spacing of gridlines, profile titling, and the annotation of stations and elevations within the grid. The Profile View Label Styles control the display of text and annotation. Once you develop the Profile View and the Existing Ground profile, then you develop a finished *grade* profile for our road.

Grade—A change in elevation

Key Terms are bold and italic within the running text and defined in the margin to help students understand and use the language of the computer-aided drafting world.

Surface-Related Point Commands

This next segment delves into procedures for placing Surface-related points into a project drawing.

The next command in this toolbar item is "create new points" based on an alignment's profile information. This can be used for creating 3D stakeout data of the roadway's crown, for example. The command is called Profile Geometry Points. The user is requested to select the alignment and then to select the profile in question. The software then sets the points on the alignment with 3D elevations.

The ability to adjust point elevations based on a surface has also been added. This is a very beneficial routine if points have been set to a surface using the Random Points command, and the surface is modified and updated, then the points that were dependent on that surface can be updated automatically. Change the elevation for a point or a group of points by selecting a location on a surface. This command can be found under the **Points** pulldown >> **Edit... >> Elevations from surface....**

A New to AutoCAD 2007 icon flags features that are new to the 2007 version of the Civil 3D software, creating a quick "study guide" for instructors who need to familiarize themselves with the newest features of the software to prepare for teaching the course.

Many objects in civil engineering are perpendicular to items and this routine simplifies these computations. For instance, side lot lines are usually perpendicular to front or back property lines, or side roads are typically perpendicular to main roads. Because computing a perpendicular is relatively easy, doing so can cut down on construction and surveying stakeout errors.

1. Continue in the same drawing, or open the drawing called My Point Style-Intersections.Dwg.
2. Using the **V** for **View** command, highlight the view called **Points-D**, hit **Set Current**, and then hit **OK**. You see a property parcel surrounding a house. Compute a point perpendicular to the back property line from a point outside the parcel.
3. Using the drop-down arrow to the right of the second icon from the left, choose **Perpendicular**. The prompts follow.
4. Select an arc, line, lot line, or feature line: Select the back property line.
5. Please specify a location for the perpendicular point: **890,1570**
6. Enter a point description <.>: **<Enter>**
7. Specify a point elevation <.>: **<Enter>**

Job Skills link the application of the software to the tasks that students will perform on a daily basis. They provide tips and tricks that are not discussed in other areas of the text.

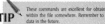

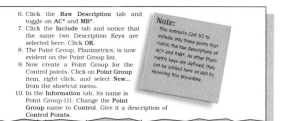

TIP These commands are excellent for obtaining data from other criteria either known or within the file somewhere. Remember to use these as you create parcel and alignment data in the future.

1. Continue on the same drawing, or open the drawing called Chapter-4-Profile.Dwg.
2. Using the **V** for **View** command, highlight the view called **Geom-D**, hit **Restore**, and then hit **OK**. You see an arc and a line.
3. Type **Circle** or pick it from the **Draw** toolbar.
4. Command: _circle Specify center point for circle or [3P/2P/Ttr (tan tan radius)]: **19135, 20050**
5. Specify radius of circle or [Diameter]: Type '**MR** or select the sec-

Tips relate the author's experiences to specific chapter content. These will enhance the student's success in the workplace and provide real-life tips and tricks for the problems.

6. Click the **Raw Description** tab and toggle on **AC*** and **MB***.
7. Click the **Include** tab and notice that the same two Description Keys are selected here. Click **OK**.
8. The Point Group, Planimetrics, is now evident on the Point Group list.
9. Now create a Point Group for the Control points. Click on **Point Group** item, right click, and select **New...** from the shortcut menu.
10. In the **Information** tab, its name is Point Group-(1). Change the **Point Group** name to **Control**. Give it a description of **Control Points**.

Note: This instructs Civil 3D to include only those points that match the raw descriptions of AC* and MB*. As other Planimetric keys are defined, they can be added here as well by revisiting this procedure.

Notes provide additional, related information to the reader outside of the procedure being outlined.

Example 4-1: Find the Angle of a Two-Dimensional Line from the X Axis

To find the angle from the X axis of the two-dimensional line with the same coordinate values as shown in Figure 4-1, follow this process.

Slope = $\Delta Y/\Delta X = (20 - 15)/(10 - 5) = 5/5 = 1$

The angle then is computed as: (ATAN[(ABS)slope]) = ATAN(1) = 45°

Example 4-2: Find the Angle of a Three-Dimensional Line from the X Axis

To find the angle from the X axis of the three-dimensional line with the same coordinate values as shown in Figure 4-1, follow this process.

Slope = $\Delta Z/\sqrt{A^2 + B^2} = (5 - 10)/(7.0710678) = 0.70710678$

The angle then is computed as: (ATAN[(ABS)slope]) = ATAN(0.70710678) = 35.254°

The formula for a line – $Y = mX + b$ was discussed in Chapter 3 under

Project examples throughout the text are one of the primary benefits of this book. Hundreds of examples are procedurized and set up for rapid access to the specific function being addressed. These procedures also act as problem-solving projects. They are actual contextual tasks that provide in-depth solutions for actual problems in civil engineering and surveying. Many other problems at the end of the chapter provide additional opportunity for delving further into Civil 3D's capabilities.

Exercise 4-15: Creating Parcel Styles

This exercise allows some serious practice in establishing styles for parcels. Civil 3D allows for some new features that are often used to add to the aesthetics of parcel drafting such as placing a hatch pattern around the edge of the parcels. You explore this and annotation styles as well in the next series of exercises.

1. Open the file Chapter-4-Parcel.Dwg.
2. Type **V** for **View** and select and restore the view called **Parcel – A**.
3. Then go to the **Settings** tab in the **Toolspace**; expand the items for Parcel.

Hands-On exercises throughout the chapters provide step-by-step walk-through activities for the student, allowing immediate practice and reinforcement of newly learned skills.

Exercise 3-38: Alignment—Import from File Command

The next command, **Import from File**, can be used to import points that were generated based on an alignment and use station, offset values relative to that alignment. The software uses the next available point numbers and prompts for the information needed as it goes.

1. Continue in the same drawing, or open the drawing called My Point Style-Intersections.Dwg.
2. Using the **V** for **View** command, highlight the view called **Points-G**, hit **Set Current**, and then hit **OK**. You see an alignment. Import alignment based points.
3. Using the drop-down arrow to the right of the third icon from the left, choose **Import from File**.
4. A dialog box opens called **Import Alignment Station and Offset File**. Select the file called **Points-import file-sta-off.txt** in the Support files folder for this textbook and hit the **Open** button.
5. The software then prompts you to choose the format for this file. The options are as follows: 1. Station, Offset. 2. Station, Offset,

CD icons in the margins alongside exercises direct students to the Student Data Files on the CD bound into the textbook.

CHAPTER TEST QUESTIONS

Multiple Choice

1. Some uses for Points in civil engineering and surveying comprise:
   a. Property corners for parcels of ownership
   b. Roadway earthworks computations
   c. Developing renderings and animations
   d. None of the above
2. Point groups have the following characteristics:
   a. They have enduring properties that can be easily reviewed or modified, either beforehand or retroactively.
   b. Displaying a points list shows the points included in a point group and it can be updated automatically.
   c. A point group can be locked to so as to avoid

   c. Importing points from external files
   d. All of the above
6. Points can be established when the user has which of the following criteria:
   a. Elevations
   b. A starting point and a distance and a grade
   c. A starting point and a slope
   d. All of the above
7. Points can be:
   a. Defined or set
   b. Edited or modified
   c. Reported on
   d. All of the above
8. Editing of points can be accomplished using

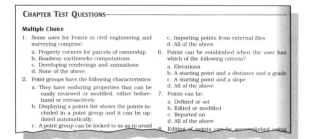

End-of-chapter material, easily located by shading on page edges, includes:

- Chapter Test Questions
  - Multiple Choice
  - True or False
- Student Exercises, Activities and Review Questions

to help students check their own understanding of important chapter concepts.

Contents

Preface

The company I founded, OutSource Inc., is fifteen years old this year, and has trained and worked with thousands of engineers, planners, and surveyors during that time, both nationally and internationally. Those of us at Outsource play a dual role in the industry: we are engineers along with having a relationship with academia. This has likely caused the development of a style of training that combines civil engineering education along with the applied usage of the CADD software. We at Outsource firmly believe that one should understand how the software works and what mathematics the software is built on before using it on a project. This was my philosophy in developing this textbook, and it parallels some of the classes I have taught for many years at Virginia's George Mason University. Those classes consist of lectures—in which I teach theory, formulas, and algorithms—and CADD laboratories—in which I provide lab instruction on application to CADD. This textbook strives to accomplish that same goal.

The two main purposes of this text are to provide (1) basic instruction in what someone would need to know about civil engineering in order to operate this software and (2) procedures and examples of how to operate the software for developing civil engineering project design and drafting. The instruction covers a variety of the disciplines within the field of civil engineering, while outlining the algorithms and formulas for many design functions. It lays down procedurized groundwork for performing design and drafting. Each chapter takes the reader through step-by-step procedures for using this state-of-the-art product, while offering scores of civil engineering–related tips and tricks.

The text will help guide the reader through the paradigm shift occurring in the civil engineering and construction industries as we continue in this new millennium. That shift is directly attributable to one event, that is, the use of three dimensions in designing and constructing projects.

The flow of the text begins in Chapter 1 with an overview of the software and its interface. In Chapter 2 a complete design project is undertaken and brought from inception through completion. It rapidly acquaints the reader with the software's basic functionality. The text then traverses, chapter by chapter, through the complete and advanced usage of each of the software's capabilities. It uses a feature-training methodology in a discipline-oriented organization that mimics design tasks on actual projects.

Each chapter ends with problems, with the answers in the online instructor's manual. These could be used as test questions, additional exercises, or homework assignments.

A few years ago, one of our clients purchased an engineering software package for several thousand dollars from a relatively small software manufacturer. It cost enough to buy a small automobile. The client installed it, had some staff trained, and began using it on a project. Not long after, a frustrated user of the software at the company contacted me to ask whether the software would update the profiles for the company's roads if the engineers modified the horizontal alignment. He also asked whether the surface would update itself if they added breaklines or spot shots to it.

I informed him that the answer was no, that they would need to regenerate the various data sets and re-import them. He said, "That is what I thought, but would you tell that to my boss?" When the principal of the company joined our conversation and I informed him of our discussion, he immediately said, "Then why did they charge me thousands of dollars if I have to do most of the work myself?"

The industry is now in the position in which I can answer his question entirely differently than I did then. The answer is in the civil engineering software solution called Autodesk Civil 3D. I can now provide a resounding "Yes, Civil 3D will update these various components of design and automatically update, regenerate, and import them." The software ripples through the data, effecting change where it is required. Technology has finally reached the level that many civil engineering firm owners and managers thought they were at years ago. Many firms will undergo a paradigm shift in how they work as they implement Civil 3D.

Another anecdote that occurred recently involved a designer computing an earthworks quantity for a large project. He reported to his boss that the developer would require 250,000 cubic yards of fill to build the site. When the boss asked how he knew it would be 250,000 c.y., his response was, "Well, that is what the software said." This type of response always makes a boss apprehensive.

WHY THIS TEXT AND WHY NOW

For years our clients have asked me to write a book on Autodesk's Land Desktop that follows the teaching style at our company. When Civil 3D was released, it provided not only an opportunity to write a textbook on the operation of the software but also an opportunity to document our teaching method consisting of education combined with CADD instruction. The next time a reader of this textbook defends his or her earthworks results to the boss, that reader will be able to say, "Based on the data provided to the computer and my knowledge of how it works, there are 250,000 c.y. of fill required."

Therefore, based on the recent introduction of Civil 3D, its anticipated life span as a solution to engineers worldwide, and the desperate need for improving skills and productivity in the industry, this is the time to introduce a textbook that teaches how to use the software as well as the theories behind the applications. The software and the teaching methods used in this textbook are state of the art and will be around for a long time.

This text is a written version of a set of sophomore-level classes I have been teaching at George Mason University in Virginia since 1997 called Engineering Design and Computations. It combines several teaching concepts that the reader is unlikely to find elsewhere. The first concept is that the theories and computational methods are taught before the student begins any advanced usage of the software. If the readers/users understand the formulas behind the automation, their skills will be much more effective and of higher quality. Readers will come away from this text knowing how to perform computations by hand, by using a calculator, and, of course, by being able to use the Civil 3D software. The second concept is combining feature training with application training. Feature training is a method whereby you learn how the commands work. In application training, you apply the software toward specific problems in civil engineering and are provided with real-life examples of problems that engineers face on a daily basis. This textbook provides both types of training in addition to the education required to use the software effectively and correctly.

Each chapter begins with an introduction of what will be covered, the objectives that are to be attained, and descriptions of the formulas and theories needed to proceed into the chapter. Additionally, there are many types of data that civil engineers deal with and that industry requires them to use. These data types are defined and discussed in detail at the beginning of each chapter that is related to that item. Among these are Points, Point Groups, Curvilinear 2D Geometry (Lines, Arcs, Spirals), Alignments, Profiles, Cross Sections, Corridors, Parcels, DTMs, Surfaces, Contours, Breaklines, Spot Shots, and Text Labels.

Once the foundation for the tasks to be performed has been discussed, problems for the student are defined and procedurized solutions follow. The procedures are supported by actual project data that are prepared for the exercises. Each problem allows the students to perform the work themselves. Then, on subsequent problems, they can begin with a completed file that consists of the data that exemplify the successful completion of the previous functions learned. The example files included with the procedures are set up to allow the users to begin each task cleanly and compare their results with an already completed file. In this fashion, no time is wasted trying to get to the point where you can try the solutions. Simply choose an exercise, open the related AutoCAD file, and try it out. This approach works for users new to the software as well as advanced users who want to skip introductory material and jump directly into advanced functions.

The text includes both lecture material and laboratory material for an Introduction to Engineering class or training classes and is perfectly suited to bridge the lecture and the laboratory. Although the book is of university quality, it will also be an excellent resource for software training centers, technical colleges, and individual users.

IMPORTANT NOTE: *You will notice that data sets are provided for the exercises throughout the text. Generally the exercises follow a precept in that you will be asked to open a specific file. Once you are in the file and before you perform any work, you are advised to **Save the File As** another name so that you can open the virgin file again at any time, without having to reload it from the source. It is suggested that you rename it with the same name with perhaps a suffix of your initials. Once you are finished with a chapter (or at any time in between), there will be another file provided for showing the completed work described in the procedures. It will have the same filename but with a suffix of –**Complete**. It will show the results of the work described in the procedure. This allows you to check your work, or advanced users can jump ahead to accelerate the course.*

SUMMARY DESCRIPTIONS OF THE TEXTBOOK CHAPTERS

Introduction: CADD, the Primary Tool of Our Trade

The text begins with an introduction to the field of civil engineering. Then there is a brief look back at the introduction of CADD to the field of civil engineering and a description of the introduction of Computer Aided Design and Drafting into the industry. The chapter then provides a brief description of the technologies emerging within the field and how Civil 3D begins to fill the needs of these emerging technologies.

Chapter 1: The Components and Interface of Civil 3D

This chapter describes the interface for the software and introduces the reader to some of the state-of-the-art methodologies found in the Civil 3D product.

It also provides an introduction to AutoCAD Map 3D as well as the GIS component of Civil 3D. The new philosophies found within the Civil 3D design application are introduced with discussions of the profoundly new capabilities available to CADD users, including the new interface tools; new objects for use in design; data driven, object style libraries; and new roadway abilities.

Chapter 2: A Civil 3D JumpStart Project

The intent of this chapter is to quickly get the reader acclimated with the basic usage of the Civil 3D software. An introductory project initiates the students into the basic design process. It involves preparing an existing ground surface and then developing a parcel of property for a shopping center and parking lot. It develops an access road alignment, the road's profile, and corridor data for the access road in 3D. Once this is done, the grading for the shopping center and parking lot in 3D is designed and the plans assembled for the road and the shopping center.

Chapter 3: The Simple, but Time Honored Point

Now begins the detailed procedures of the text. The first chapter lays out expectations and discusses the interface and software components. The second chapter shows how to design a entire project very briefly. From here on the text describes the detailed usage of the software on a discipline basis beginning with point usage. Chapter 3 discusses basic math related to points and then introduces Geodetics and the Global Coordinate Systems. Civil 3D is used to develop point aesthetics and build a library for using point descriptions and then develop point groups. Civil 3D **Point Layout** tools are used, and students learn how to compute information using points and then take on advanced points usage and tables.

Chapter 4: Civil 3D, the Modern Curvilinead

This chapter covers linework and geometry. It discusses mathematics and formulas for lines, triangles, arcs, and spirals and introduces different types of angles, bearings, and azimuths.

   Transparent commands in Civil 3D are used in detail and Curvilinear and Rectilinear 2D Geometry (Lines, Arcs, Spirals) is developed. This information is then used to develop Parcels using the **Parcel Layout** toolbar, to develop Alignments using the **Alignment Layout** toolbar, and to learn about Styles and Labels.

Chapter 5: Advanced 3D Surface Modeling

One of the most important parts of Civil 3D has to be the Surface Model because so many computations emanate from surface models. Chapter 5 first discusses the mathematics and theories for Digital Terrain Models and Surfaces and then teaches the TIN process, how ground information is used, and about contour computational methods. We create surfaces in Civil 3D and develop contours and labels. We use breaklines and spot shots and analyze and depict surface data.

Chapter 6: Advanced Profiles and Sections

This chapter covers using Surface Models and how to design vertical alignments. It discusses mathematics, theories, and formulas for vertical tangents and vertical curves and uses Civil 3D for creating profiles, finished grade vertical alignments. Students make use of the **Profile Layout** toolbar to create vertical profiles and then develop Cross Sections and use the **Sample Line Layout** toolbar.

Chapter 7: Advanced Corridor Development

One of the strongest aspects of Civil 3D has to be the corridor design features. The chapter begins with a brief discussion of superelevations for roadways and corridors and the use of subassemblies. Then Civil 3D is used to develop corridor assemblies. Students compute superelevations for a corridor and discuss and develop subgrades for the road.

Chapter 8: Advanced Site Grading and Virtual Site Design™

Site grading is performed for many purposes, but the chief reason is to assure proper drainage across a site following construction. The chapter discusses the building of commercial and subdivision sites in 3D and teaches about quality control methods used in design. Students develop a commercial site in Civil 3D using feature lines and grading objects and then compute earthworks takeoffs and quantities.

Chapter 9: Piping for Storm Sewers and Drainage

This chapter begins with a brief discussion of hydrology, runoff, time of concentration, watersheds, and the rational method and then develops watersheds in Civil 3D. This is followed by a brief discussion of hydraulics and then the development of a pipe network.

Chapter 10: Civil 3D's Visualization Capabilities

Visual communications are a growing requirement in the civil engineering business today, and Civil 3D comes with a photorealistic rendering engine. This chapter contains a discussion and several examples of computer renderings and visualizations from a variety of design projects. Students then use Civil 3D to develop a surface with multiple materials and use Civil 3D for rendering the materials into a photorealistic image.

Chapter 11: Surveying

Surveying is usually where engineering projects commence in earnest. Base data are collected, boundaries are established, and control is set. Once the designers have completed their work, very often the project returns to survey for quality checking, development of stakeout data, or conversion to 3D/GPS Machine Control data. This chapter discusses importing field-collected data, developing traverses for setting control, and balancing those traverses to eliminate (or disperse) any error that may exist.

Appendix A: Data Sharing in Civil 3D

In this appendix several methods for sharing design data with others are discussed. These include inserting a source design file into a another file, using LandXML data sharing, using data shortcuts, and using projects for sharing data.

A unique aspect of this appendix is the discussion of developing data for 3D/GPS Machine Control. This growing area of engineering construction is replacing traditional stakeout activities. Engineers are under increasing pressure to deliver 3D/GPS digital data directly to the contractor for robotic machine control.

Appendix B: Labeling in Civil 3D 2007

In this appendix some of the new labeling methods added to Civil 3D 2007 are discussed.

Appendix C: Software and Hardware Versions

The software version and system specifications section describes the software being used in the text and breaks it down by its components. Hardware specifications are discussed in an imaginative way. The perfect CADD station is specified, a good CADD station that has budget constraints is specified, and the minimum requirements are identified.

Appendix D: Identify the Civil Engineering Industry

Industry components are identified and detail the major fields within the industry of civil engineering along with the tasks that are required on projects falling into these categories. The Civil 3D software is then described and the components found within Civil 3D are introduced.

Appendix E: Supplement to Chapter 3: The Simple, but Time Honored Point

This appendix provides information on additional point commands and is a supplement to the content provided in Chapter 3.

ONLINE INSTRUCTOR'S MANUAL

To access supplementary materials online, instructors need to request an instructor access code. Go to www.prenhall.com, click the **Instructor Resource Center** link, and then click **Register Today** for an instructor access code. Within forty-eight hours after registering, you will receive a confirming e-mail including an instructor access code. Once you have received your code, go to the site and log on for full instructions on downloading the materials you wish

Style Conventions in *Autodesk Civil 3D 2007: Procedures and Applications*

Text Element	Example
Key Terms: Bold and italic on first mention.	*Azimuths*
AutoCAD commands: Bold and capitalized.	**ZOOM EXTENTS.**
Toolbar names, menu items, and dialog box names: Bold and follows capitalization convention in Civil 3D toolbar or pull-down menu.	Unless otherwise specified to use the pull-down, the following convention indicates a menu call from the pull-down menu. **Utilities >> Task Pane.** Next go to the **Toolspace** and choose the **Prospector** tab.
Civil 3D prompts: Command window prompts use a different font (Courier New).	`Specify station:`
Keyboard input: Bold with special keys in brackets. **<Enter>**.	**<300>**

Introduction: CADD, the Primary Tool of Our Trade

A BRIEF INTRODUCTION TO THE FIELD OF CIVIL ENGINEERING

Civil 3D is software for automating solutions to design problems in the field of civil engineering. Therefore, it is of benefit to understand a little about the field itself. Many readers are becoming introduced to civil engineering in the class you are now taking and may have only a partial understanding of the depth and breadth of the field. This introduction serves to briefly describe this; for instance, the corridor design is a highly sophisticated area of the software, as is geometry development and surface modelings. The use of the Civil 3D software spans the field serving some parts very robustly.

Civil 3D—Autodesk Civil 3D™

Civil engineers are often referred to as "jacks of all trades" because the field is so wide and contains so many subsets and disciplines. Civil engineers typically receive training in many of these areas in their undergraduate education and then tend to specialize with a masters or doctoral degree. As one of the oldest branches in the profession of engineering, civil engineering includes the following disciplines, each one a career unto itself:

- Transportation engineering
- Geotechnical engineering
- Structural engineering
- Environmental engineering
- Surveying
- Land development
- Coastal engineering
- Process engineering
- Water resources engineering
- Construction engineering
- Geographic Information Systems (*GIS*) and Geographic Positioning Systems (*GPS*)

GIS—Geographic Information Systems

GPS—Geographic Positioning Systems

Some emerging areas within civil engineering include

- Automated Construction *3D/GPS Machine Control*
- Bioengineering
- Project Management
- Life-Cycle Design
- Real-time Monitoring
- Rehabilitation
- Smart Systems
- Space Structures

3D/GPS Machine Control— Robotic earthmoving

How CADD Became the Primary Tool of the Trade

CADD —Computer Aided Design and Drafting

To understand why Civil 3D is such a breakthrough software application, one must have a familiarity with the software that has been used up until now. To do so, it is important to look back at how *CADD* became the predominant tool of the trade in civil engineering.

Primitive and Manual-Based Tools of the Trade

Prior to thirty-five years ago, the state of the art of civil design generally meant using hard-copy tables, hand-calculated formulas, and simple mechanically operated calculators. PCs did not exist and mainframe computers were extremely rare. Engineers performed trigonometric functions by looking up values in preprinted tables and computed many engineering formulas using a slide rule.

During the 1960s, the largest organizations such as government agencies began acquiring mainframe computers, which consumed huge amounts of square footage. The memory on them was so small that it was common to feed a large set of data for computations into the computer several times before getting complete results. During the 1970s, computers began to appear in engineering organizations and school systems. These computers had either punch cards or paper tape rolls with holes punched in them for data storage. Some surveyors used a tabulation machine called a "Monroe." Usage of these machines tapered off due to the introduction of the hand calculator, which quickly caught on as a main computational tool.

Engineers did technical drawing with ink on linen in the "old days." Mistakes in drafting were not easily tolerated as the linen was quite fragile and not highly conducive to making changes. In order to reproduce the drawings, blueprints were created by passing the original, opaque linens through a light machine. This affected treated paper such that the light could not be emitted through the black ink, thereby leaving a white, etched line on a blue background. Blueprints gave way to ammonia-based Blueline prints when the process was reversed, whereby the light faded a treated paper to white, leaving a blue line where the light could not penetrate through the black ink.

Drafting was performed using drafting tables with drafting arms, ink pens, and pencils. The drafting arms that could be set to any angle allowed the technicians to lean their pens against the straightedges to draw straight lines. The arm could be rotated and locked into that position allowing for rotated lines. By placing right triangles on the arm, the draftsperson could draw orthogonal lines. Other equipment consisted of Koh-i-Noor brand pens, Pelikan brand ink, Kueffel & Esser (K&E) made compasses, mechanical pencils, and so on. The tools of the drafting trade then also included protractors, flexible curves, French curves, railroad curves, straightedges, triangles, and electric erasers. Engineers drew on paper, linen, and mylar.

I consider myself fortunate to have had my career parallel the growth of CADD in my chosen field. After more than twenty years as a professional civil engineer specializing in the use of CADD, I have personally experienced and contributed to the growth of this technology. As far as how I gained this experience, when I was fairly new to the civil engineering business, I was performing design and drafting services using the previously mentioned tools for the City of Houston, Bridge Division in 1978. The economy was very robust with growth in all sectors, and I was soon took employment in the *private sector* with Turner, Collie & Braden, a large firm doing both international and domestic work.

Private Sector —Projects from private sources

AUTOMATED TOOLS OF THE TRADE

Automation was not entirely new to this company as it had been using a mini-computer, with Tektronix computer terminals and homegrown COGO software. Subdivisions were computed by data entry specialists who knew the language and could assemble data files and plot files. They generated these plot files from a single set of batched commands. To make a change to a single property line required that an entire batch file be rerun. To make changes to individual objects without recomputing the entire project would be seen as a highly productive improvement. This was the idea behind interactive graphics and the advent of CADD in civil engineering.

Somewhere in the mid- to late 1970s, M&S Computing began to make a presence in the civil engineering field because of its *CAD* product called IGDS. One could acquire it for about one million dollars for the workstations, mini-computer, plotter, software, and maintenance. These systems tended to be regional acquisitions at that time because part of the United States was suffering from economic troubles. The price of oil and inflation combined to cripple some areas of the country as administrations changed from President Ford, with the "Whip Inflation Now" slogans, to President Carter. Therefore, certain regions, such as those including Texas and California, were doing very well under these conditions at the expense of regions such as the Northeast. Engineering and oil firms in these healthy economic areas could afford the acquisition of million-dollar CAD systems.

CAD —Computer Aided Drafting

INTERACTIVE GRAPHICS-BASED TOOLS OF THE TRADE

Not long after I joined Turner, Collie & Braden, it acquired an M&S Computing CAD system at a cost of close to one million dollars. Because this was such a new technology, most of the company employees were not even sure how it was supposed to be used or where it would enter and depart the workflow. The company acquired two workstations, each with a large digitizing table, a floating menu tablet, and foot pedals to control the height and angle of the table. The system came equipped with dual monitors that could interact with each other; a view from one monitor could be placed on the other monitor so it could be better seen. One monitor tended to have a single large view on it, whereas the other monitor tended to have four smaller views of the subject matter. To this day AutoCAD users still do not work in this way, even though it is highly effective and AutoCAD does support this.

Users of the system could make changes to objects and, when ready, then send them to the plotter, hence the term *interactive*. We take this for granted today but prior methods required recomputation of the entire data set! These were called batch computations.

The tools available in the system at the time were basically commands that drew vector graphics such as lines, arcs, linestrings, shapes, and text. To give the system credit though, it was three-dimensional from the day the company acquired it and it was programmable. These features caused me, in the mid-1980s, to join a small group of progressive people in the MIS (or IT) division of the company who were writing civil engineering and surveying software that interacted with the CAD system. These were pioneering days because very few firms had discipline-oriented CAD software at that time.

When the system arrived on the company's doorstep, an employee from a company called Intergraph came in and placed an Intergraph logo on top of the M&S Computing logo that was on the machine. It turned out that M&S Computing had gone public and was now called Intergraph. The software was called IGDS for Interactive Graphics Design System. By the mid-1980s our company had a half dozen workstations that ran twenty-four hours a day with three shifts of personnel.

PC-Based CADD Is Introduced and Becomes the Primary Tool of the Trade

During the early to mid-1980s, PCs were being introduced to our company and to the industry and standard were Intel 8086 and 8088 processors, often running from floppy disks. Hard disks cost over $5,000 and had to be heavily justified for purchase. Then the new Intel 286s were introduced and the company was struggling with whether to standardize on Apple or IBM-type PCs.

Around that time, PC alternatives began to arise that allowed use of Intergraph Design files on the PC at significantly less cost than a workstation. I recall a couple of PC-based IGDS programs called C-CAD and Pseudo-Station. Intergraph acquired them both; C-CAD disappeared and Pseudo-Station became MicroStation. Later, legal machinations resulted in the rights to MicroStation reverting back to the Bentley Corporation, where they remain today.

3D —Three dimensions

While PC-based CAD was beginning to achieve credibility, in 1982 a small company called Autodesk was founded, which introduced and shipped AutoCAD. The company went public in 1985 at $11 a share. The software was originally two-dimensional and then became 2-½D whereby elevational data were embedded into objects but they were not actually *3D* objects. AutoCAD popularity grew especially among those companies and individuals that did not want to spend hundreds of thousands or millions of dollars to obtain minicomputer-based CAD. AutoCAD responded to the requests of these users and became very strong in the development of construction plans. It excelled in dimensioning and customization. Autodesk put a subset of the List Processing language into AutoCAD (called LISP), which allowed those with minimal programming knowledge to write routines to AutoCAD for redundant drafting functions. The use of the software grew between disciplines as people adapted the software to their needs using LISP.

Visualization —Renderings

As of the late 1980s, Autodesk's AutoCAD Release 10 was introduced as a three-dimensional CAD program. The processing power on the PCs increased with each new processor announcement, and it did not take long before 3D models were being prepared on PCs and color shading was available. Autodesk launched a product in 1987 called Autoshade for this purpose as an add-on package. Then it developed and introduced Autovision in 1993, which was ultimately added into AutoCAD as part of the Release 14 product. Meanwhile, a company called Discreet developed a *visualization* and modeling package called 3D Studio MAX and introduced it in 1990. Autodesk then acquired that company. The 3D Studio MAX product was split into a two pieces: a professional studio production version called 3D Studio MAX and a version for engineers and architects called 3D Studio VIZ. Each solution caters to its intended market.

Contour —A coplanar linestring

In the mid-1980s Dave C. Arnold, a professional engineer in New Hampshire, started a company called DCA Software, which produced civil engineering software tools that worked inside AutoCAD. The software consisted of LISP routines that helped users place points, generate *contours*, and label linework. It was fairly informal in its early days and possessed scarce documentation. DCA Software persuaded AutoCAD dealers to sell the software as an add-on when they sold AutoCAD to engineering companies. That company grew to become Softdesk, a publicly traded company that wrote *soft*ware for Auto*desk* products. Autodesk and Softdesk agreed to merge in late 1996 and finalized it in 1998.

This merger changed the way Autodesk did business because it began selling the combined products as Desktop Solutions and developed a Desktop for each discipline. For instance, Land Desktop catered to the civil engineering industry. Autodesk also produced Architectural Desktop, a Mechanical Desktop, and Building Services Desktop. Land Desktop was one of the most popular engineering programs in the business at the time of this writing. However, with the successful release of the heavily object-oriented Architectural Desktop, Autodesk was taking a similar development track for the civil engineering community.

Autodesk developed and prereleased a new-generation product called Autodesk Civil 3D in 2003. This software is not a new version of the Land Desktop; it is a new product that makes use of the object-oriented technologies that have been introduced over the years in the Land Desktop software but takes them to a much more robust level. Although Land Desktop has point objects, contour objects, and grading objects, Civil 3D increased the use of these concepts and integrated them into a project scenario.

Civil 3D has also developed a sophisticated library structure that allows design data to be embedded into it that guide the design of engineering elements. The first public release of this product, called Civil 3D 2005, occurred in late 2004. In addition to major feature enhancements for sophisticated *corridor* design, it also includes a version of 3D VIZ called VIZ Render as part of the package. The second release, called Civil 3D 2006, occurred in 2005, followed by Civil 3D 2007.

Corridor—A location for passage

Civil 3D is rapidly becoming the state-of-the-art software for civil engineers and is the heir apparent to Land Desktop. It has a dynamic model concept whereby the objects customizes it to work specifically for the civil industry are data driven. The graphics and annotations are style driven. Many objects are parametrically controlled, another first for the civil industry.

This textbook, *Autodesk Civil 3D: Procedures and Applications*, details everything you need to know to use this software rapidly and expands on that knowledge to walk you step-by-step through advanced usage of the software. As you progress through the book, you will learn the theories, algorithms, and formulas behind the software. Then when the supervisor asks, "How do you know that the result is the correct answer?" you can respond with "Because I understand the data provided to the computer and how the software produces computations."

THE FUTURE OF TECHNOLOGY FOR CIVIL ENGINEERING

Now that you have learned a brief history of CADD in civil engineering in this Introduction, what is the future of CADD? The Autodesk Civil 3D software allows users to prepare for the future of technology in civil engineering. As the industry moves into the third dimension, engineers will be better prepared to work with the technologies that are appearing in the related fields such as surveying and construction. These businesses are moving rapidly into advanced usage of GPS even to the point of establishing GPS networks with postprocessed corrected data, 3D/GPS Machine Control whereby robots perform excavation tasks on the bulldozers and excavators that are pushing dirt, to 3D modeling and computer renderings and animations and their respective uses in civil engineering. Engineers must respond to this technology because it requires 3D surface and project data in order for the robotics to do their job. I have been consulting with engineering firms that want to start moving into 3D design because contractors are persuading developers to choose engineering firms that are more modern. The argument is that it will help the developers save money by allowing the contractors to use their investments in 3D/GPS-Guided machine control equipment. Three-dimensional laser scanning is making its way into engineering and surveying firms at around $100,000 per machine. These machines can capture 3D data of subway tunnels, cityscapes, bridges, structures, and dams. They are used to help analyze movements, deflections under load, and for antiterroristic functions. The robotic-reflectorless total stations used by surveyors capture 3D data more readily than ever before. So you can see that technology is rapidly building momentum in the industry, and Civil 3D will help you meet those challenges. Although the advent of 4D has not been broached, that would be another topic for another time!

The intent of this textbook, is to introduce the reader and participant to the Autodesk Civil 3D product and show how it will perform design and drafting functions. The next also discusses the algorithms and design computations that exist in the civil engineering business and relates them to the operations of Civil 3D.

The Components and Interface of Civil 3D

1

Paradigm Shift—A fundamental change

Chapter Objectives

After reading this chapter, the student should know/be able to:
- Have a better understanding of the field of civil engineering and its relationship with Computer Aided Design and Drafting.
- Be familiar with the applications embedded within the Civil 3D offering.
- Realize that Civil 3D is not a program; rather it is a cross-pollination of several programs integrated into a single engineering solution.
- Understand why Civil 3D is creating a paradigm shift in the civil engineering business.
- Understand the software's new philosophies, interfaces, and capabilities.

INTRODUCTION

Chapter 1 introduces Civil 3D, its philosophies, what it is composed of, and its interface. The philosophies are causing a *paradigm shift* in the civil engineering business because the software is an application with entirely new methods, interfaces, and capabilities. These will be described in concept so that the reader, upon beginning the procedurized tutorials and exercises, will at least have an impression of what to expect the software to accomplish.

This chapter also discusses the contents of the Civil 3D suite of software because many solutions come with its install and its adjunct installs. It describes the applications embedded within Civil 3D's capabilities.

Unlike the help system that accompanies the software, this chapter provides detailed illustrations and explanations of the interface and related icons in a single location. This will ease learning these items and undoubtedly save the reader time and reduce frustration in the use of the software. The icons have an enormous number of subtle variations that the user must be aware of in order to stay in complete control of the project data.

REVIEW OF FUNCTIONS AND FEATURES TO BE USED

The features being introduced in this chapter include what the software offers, an explanation of AutoCAD MAP 3D, an example of geodetics, what the interface tools are, and what the icons mean. This chapter explains how the software is Style and Data Driven, how it uses objects for design, and the fact that it offers DOT-level highway design capabilities. A brief explanation of the Sheet Set Manager comes with the software.

WHAT ARE THE COMPONENTS OF AUTODESK CIVIL 3D?—————

Civil 3D consists of the integration of several programs including AutoCAD as the foundation graphics engine. Although the full set of commands is available, only the **File**, **Edit**, **View**, and **Insert** pull-down menus are shown when the software initially launches, as in Figure 1-1. Augmenting AutoCAD is the MAP 3D software, which adds GIS, important utilities, and drafting cleanup tools to the user's toolkit. The Civil engineering component of the software adds on the heavy civil engineering design software. The VIZ Render completes the offering by allowing for the development of some of the industry's highest-quality visualizations.

Figure 1-1   Civil 3D and Map 3D pull-down menus

NEW
to AutoCAD
2007

The VIZ Render tool was included in the 2006 version but the entire Auto-CAD version has been updated to include major rendering tools. Upon first glance, VIZ users will notice a clear similarity to VIZ. But, in 2007, VIZ Render is no longer included as an add-on, rather it is part of the base package. This provides users with benefits in that the AutoCAD data is now directly render-able, if that is indeed a word. This textbook will briefly describe rendering since that function is deserving of a text in itself.

The VIZ Render is Autodesk Civil 3D's native visualization tool. It uses some of the core technology developed in Autodesk VIZ, including much of the photometric lighting and rendering technology from Autodesk VIZ 4. But, there are many differences between the two programs. VIZ Render has optimized methodology to accept data from Civil 3D more effectively. Its user interface has been simplified. The guiding principle in VIZ Render is that Civil 3D is the model building and organization application. These data are then linked to VIZ Render for image making. Although rendering is strong, the modeling creation, animation capabilities, and many utilities and commands are limited from full-blown VIZ solution.

For the purposes of this text, AutoCAD knowledge is assumed to be a prerequisite, as there are many sources and textbooks available from which AutoCAD can be learned. We will, however, briefly introduce the capabilities of the MAP 3D software. The VIZ Render software will be discussed in Chapter 10.

Pull-down Menus, Toolbars, Pop-downs, Dialog Boxes— Interface items

Some of the conventions to be used in this text include references to **pull-down menus**, icons, **toolbars**, **pop-downs**, and **dialog boxes**. An example of the pull-down menu is as shown in Figure 1-2. In fact, ellipses within the pull-down menu indicate that a dialog box will display to complete the information gathering needed to fulfill the command request.

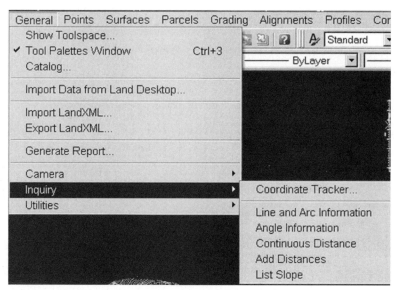

Figure 1-2   Menus and submenus

A toolbar is shown in Figure 1-3 and in this case there are three icons within the toolbar. Toolbars are configurable; that is, the user can reshape them by dragging one corner. Toolbars are also customizable in that commands can be added to or deleted from them.

Figure 1-3   Example toolbar

A pop-down menu is shown in Figure 1-4, and there are usually choices to be made within them.

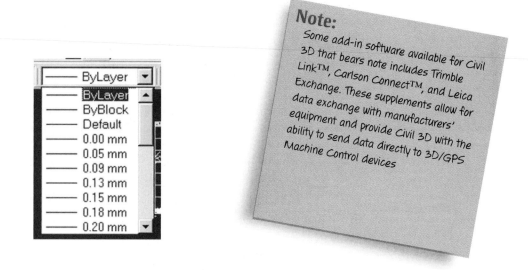

Note:
Some add-in software available for Civil 3D that bears note includes Trimble Link™, Carlson Connect™, and Leica Exchange. These supplements allow for data exchange with manufacturers' equipment and provide Civil 3D with the ability to send data directly to 3D/GPS Machine Control devices

Figure 1-4   Pop-down menu

Dialog boxes are used frequently and might appear similar to that shown in Figure 1-5.

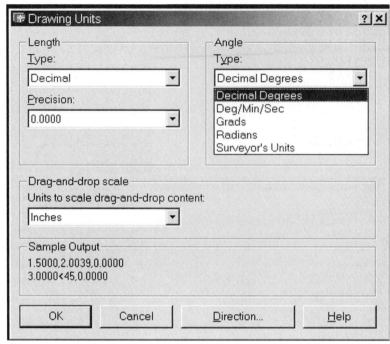

Figure 1-5   Dialog box

INTRODUCTION TO AUTOCAD MAP 3D

The MAP 3D is a required and foundational aspect of Civil 3D, yet it is one of the more misunderstood and underutilized software packages around. Many people do not realize that it was an independent solution sold as an adjunct to AutoCAD for many years. It has been available for over ten years; however, it was not always sold under the name MAP. Years ago, part of it was called the Autodesk Data Extension (ADE). This software, introduced in 1993, possessed many drafting-oriented features that were missing in the main AutoCAD solution.

Together with AutoCAD it made for a great set of tools. Unfortunately, it was not a big seller, although those with very advanced needs probably purchased it. For those who have had a history with AutoCAD products, this was one of their business models, that is, selling "extensions" to AutoCAD. There were at least two other extensions as well: the Autodesk Modeling Extension (AME) and the Autodesk SQL Extension (ASE), where in SQL stands for Structured Query Language used to extend database functionality.

The Autodesk Modeling Extension was included into the base AutoCAD software in Release 14. The Data Extension and SQL Extension, combined into a new product in 1996 called AutoCAD MAP, became fairly popular, especially for civil engineering and surveying users. By combining the graphics tools with the database tools, MAP created an environment that GIS users could benefit from. The ASE toolset allowed users to attach to external databases and create links to the AutoCAD graphical entities, perform sophisticated queries, and generate robust reports and results using AutoCAD.

AutoCAD MAP 3D, introduced in 2004, added surface modeling, surface analysis, and ***point groups*** to the software. These functions are part of the Civil 3D package and as such are removed from the MAP 3D menus. Because of MAP 3D's importance as part of the software, a discussion of its capabilities follows.

Point Groups—A Civil 3D object

AutoCAD MAP 3D augments AutoCAD and enhances the creation, maintenance, and presentation of mapping information. With MAP 3D, users can

- Open and link AutoCAD drawings to associated databases.
- Develop new data or edit data to maps to make them more intelligent.
- Use many unique features to assist in cleaning up "noise" in maps.
- Use a "batch" plotting facility for easy and flexible map plotting.
- Produce basic thematic maps with legends.
- Perform geodetic transformations on graphical objects in a file. This allows for moving or rotating data from one geodetic zone to another.
- Access map data in other formats, such as Oracle, MicroStation, Mapinfo, Arcview, and ArcInfo.
- Link any external documents to objects in their maps.

MAP 3D is the foundation for Civil 3D and helps in developing and managing complex design projects. MAP can increase teamwork efficiency by creating a comprehensive and cost-effective project database. It integrates multiple drawings into one seamless environment providing access, editing, and reporting of drawing, attribute, and related database information within a single AutoCAD session. You can view and edit objects from multiple source drawings in a single work session.

The basis of this product is a robust query engine designed to assist AutoCAD users working with large or complex data sets. MAP lets you query AutoCAD drawings based on any combination of drawing properties, object location, or information stored in related databases. This flexibility speeds up work by letting you focus on just the pertinent information. The query engine enables "super cross-references" because you can access several drawings simultaneously yet limit the reference to precisely the geographic area of the drawing you want. You also have full editing access to all of these queried drawings so edits can be made concurrently.

AutoCAD MAP's multiple-drawing edit capability provides a secure environment for multiple MAP users who may be sharing their data or drawings. Project teams can efficiently and safely share drawings and data. By providing simultaneous access to interrelated drawings, teams can work concurrently from the same source drawings, increasing project team flexibility and efficiency and reducing total process time.

AutoCAD MAP's management tools include routines to automate linking graphic objects to database records, a system to link documents to graphic objects thereby allowing easy access to those documents from your drawings, and an object data structure for storing and accessing data within your AutoCAD drawings.

Features of MAP 3D

AutoCAD MAP 3D allows users to access and modify graphic and nongraphic data stored in multiple AutoCAD drawing (DWG) files. Users perform these modifications from within their current "active" AutoCAD file and the user's environment acts as a single database of information.

One of the most popular features within the package allows multiple users to access central AutoCAD drawings and perform an advanced query for objects and data within those drawings. Once queried, the user can decide where the objects should finally reside; that is, back in the source drawings, with a new drawing, or remain in the current drawing. This awesome power has now been augmented with strong SQL features, import/export capabilities, and geodetic transformations.

The ability to perform graphic data accessing is selective and can occur through multiple drawings whereas nongraphic accessing uses the AutoCAD extended entity data (or object data). The MAP module provides direct access from within AutoCAD to external database management systems (DBMS). Examples of these are Microsoft Access, dBASE, INFORMIX, ORACLE, and PARADOX. Using Extended entity data (EED), a user can also link the AutoCAD data with other applications such as word processing and spreadsheets.

The nongraphic link provides the same command set regardless of the database in use. MAP 3D has drivers for the most popular databases systems. Open DataBase Connectivity (ODBC) is supported for general databases as well.

An Application for MAP 3D

An example of a MAP 3D application is that it can be used to input and/or modify data on a large engineering project in which the data are distributed among many large AutoCAD (DWG) files. With MAP 3D, one can establish connections between entities or symbols in the various drawings with one's current drawing, without carrying the overhead associated with the unwanted data that may exist in those files.

The method of accessing the geographic limits of the data in the drawings is the use of spatial databases. Augmenting this is the ability to access the data via property selection. In other words, one can identify the boundary of the data search and add to that the fact that only the data on the RoadCenterline Layer be brought over to the current working session.

For an engineering project, a key map outlining the project's boundaries may be advantageous, so that users can select data from geographic locations, without becoming overwhelmed by the total volume of data. A little forethought should easily identify anticipated "subboundaries."

Some of the Main Features of MAP 3D

1. Information can be isolated and analyzed by
   - Viewing selected features and statistics
   - Exporting data to external files and accessing EED
2. Data Editing—once the data are extracted they can be
   - Changed or deleted
   - Moved to another drawing or saved back to the original drawing
   - Saved to a new drawing
3. Data Selection Enhancements exist for more effective queries such as
   - Boolean constructs (AND, OR, XOR)
   - Transform data from one coordinate system to another
   - Match features across edges of different drawings
4. MAP 3D has unique editing features to assist in cleaning up mathematical noise in maps. This noise consists of sloppy drafting, errors where entities cross one another or are duplicated, and so on. These features help prepare maps for accurate analysis and the creation of map topologies. They comprise
   - Rubber-sheeting adjustments that compensate for map distortions
   - Boundary clipping
   - Map-edge cutting tools that create clean breaks between linear objects
   - Precisely aligned map-sheet edges
   - Neatly cut and created spaces for annotations
5. The ability exists to import/export other formats including MicroStation (DGN), Mapinfo (MIF), Arcview, Oracle, and ArcInfo file formats.
6. Basic thematic maps with legends can be produced. Create thematic maps while allowing for automatic altering of color, linetype, text, and

other parameters to show correlation in database information. Thematic maps can be based on object properties such as layer and linetype, object data you define in the drawing, or SQL data linked to external databases. As well as limiting objects in the thematic display to a certain area, users can also limit objects based on specific layers or particular block names.

AutoCAD MAP also supports topologies. A Topology in GIS describes the relationship of connecting and adjacent features and adds a level of "intelligence" to the data set. A user can create, modify, and delete topologies; create buffers around points, lines, and polygons; analyze maps with point, line, and polygon overlays using Intersect, Union, Identity, Erase, Clip, and Paste operations; use "shortest path trace" to find the shortest distance between two locations, useful for emergency services; and use "flood trace" to trace out from a point a specific distance in all directions, useful, for example, when analyzing demographic data and comparing alternative retail locations based on driving or walking times. You can also use flood trace to trace out to blocks that have specific attribute, object data, or external database record value, useful, for example, for finding all valves that need to be turned off in a water network.

Security

A newly installed copy of MAP contains no prebuilt securities, except one for the Superuser. The superuser must set these. Login as the superuser by using SUPERUSER as the login name and SUPERUSER as the password (note all caps). To establish security, follow this procedure:

1. Use the EDIT-USER command to set up at least one superuser.
2. Using the EDIT-USER command, the superuser sets the privileges.
3. Using the CONFIGURE command, the superuser enables the FORCELOG variable. After this is set, users must log into MAP with the SET USER command.

The Spatial Database

In a CAD environment, a spatial database is a set of internal and external databases for one or more related drawings. Part of the description defines a CAD drawing. The remainder itemizes nongraphic characteristics such as databases or other documents.

You can think of drawing files as part of a spatial database. MAP evaluates the data that define the graphics and their associated data, and the data linked to it from external databases. A CAD drawing becomes visible only in a drawing editor, such as AutoCAD, which is really a special kind of database manager; it both writes and interprets data that represent graphic elements, called entities.

Thus, a spatial database describes anything that you can represent in part or completely by one or more drawings. For example, a spatial database could describe a treatment plant, a city, an oil refinery, a geographical region, and so forth. Moreover, a spatial database can represent not only shapes but also names, materials, part numbers, suppliers, dates, and the like. *The primary requirement is that all entities share a common coordinate system.*

An AutoCAD spatial database has up to three parts. They are

1. The Graphic database, which is a data file that contains the definitions of all the geometric entities in a drawing such as lines, circles, arcs, and so on.
2. The Extended entity data (EED) or Object data. These data define additional, nongraphic aspects of an entity. Normally unseen, they are part of the drawing file and are stored with individual entities.

3. The External database(s) contains data files that you can read with a database manager. Like the EED database, it also contains nongraphic data. However, it can be linked to entities in one or many drawings.

All of these databases are within the scope of basic AutoCAD. However, the power of MAP 3D lies in that it can take advantage of these databases in several drawings simultaneously.

The power of MAP 3D is its ability to open a working session using a *current drawing*, attaching other AutoCAD files to the working session, performing queries of external information, and importing the results of queries into the current drawing. Once the data are added to, deleted from, or modified, they can then be saved to a new drawing, to the current drawing, or back to the drawings from whence the data came.

Source drawings are AutoCAD drawings, and any data files linked to them, that contain all of a project's data (both graphic and nongraphic) on which you want to run queries. Different source files categorize a project's data in different ways. For example, they can categorize data by location—as for a set of adjacent maps—or by content—for example, by water, gas, electricity, and so on—or by any mix of these and other criteria. When you use the AutoCAD OPEN command, the entire drawing is loaded. All of the drawing's entities, or as many as the current view allows, appear. When MAP "opens" (or attaches) one or more source drawings, no part of them appears. Instead, "attach" means that the contents of those drawings are available for querying and bringing into the current drawing. MAP supports several query strategies.

Queries

Querying, as the central activity of MAP 3D, is the way MAP selects information from one or more drawing files. Several strategies exist for defining query criteria and for handling the entities resulting from the query. Queries range from simple to complex and can be saved for reuse, editing, and supplementing them with AutoLISP expressions. Queries access one or more source drawing files. Usually, MAP 3D queries bring entities directly into the current drawing.

A MAP query has several levels of complexity that are primarily determined by query definition criteria, query mode, and the way queried entities are saved.

Query Definitions

A query definition sets the criteria for selecting entities. The definition can include graphics; Boolean operators such as AND, OR, and XOR; SQL expressions; and AutoLISP expressions. Query definitions use three types of criteria:

1. Location. Selects entities based on their location in the source drawing. The location can be relative to a specific point or feature or within a Window. For example, you can search for entities that lie within a given radius of a specified point or within a given distance on either side of a specified line.
2. Properties. Selects entities based on properties that relate to graphic or Extended entity data (EED) in the drawing file, such as length, layer, material, part number, and so on.
3. SQL. Selects entities based on data in linked SQL database tables, such as supplier, owner, cost, and so on.

Query Access Modes

MAP queries entities in source drawings in one of several query modes including Preview, Draw, and Report modes.

The modes are distinguished by how permanent the query entities in the current and source drawings will be. The Preview and Draw modes display entities in the current drawing with different degrees of completeness, with different consequences for the source drawing, and for different purposes. Report mode exports data to a file without displaying it. While querying, MAP locks the source drawing file so other users cannot access that drawing. After all the entities retrieved from the source drawings appear in the current drawing, all source drawing files unlock so other MAP users can have access. Once data are edited, then a record-locking mechanism kicks in and locks only that entity so that others cannot edit it at the same time. This prevents users from "stepping on each other's toes."

Preview mode

If you execute a Preview mode query, objects that match the query criteria appear on-screen but are not actually copied from the source drawings. Use the **REDRAW** command on the **View** menu to clear the screen. It displays the queried entities on the screen in the current drawing without affecting either the current or the source drawings. These queried entities typically have an entity type of ADEXXXXXX if listed; however, this will disappear on redraw. Show mode displays queried entities in the current drawing only temporarily; however, **ZOOM** and **PAN** will work until a **REDRAW** operation occurs that causes the queried entities to disappear. Use Preview mode for a quick view of a query, to see whether you are on the right track.

Draw mode

If you execute a Draw mode query, objects that match the query criteria are copied into the work session, provided they have not already been copied in by a previous query. The original objects remain unchanged in the source drawings. AutoCAD MAP will not place duplicate copies of an object in the work session. If a previous query copied an object into the work session, the new query will not copy it in again. If you retrieve an object that is on a locked layer, you cannot save changes back to the source drawing. In order to save changes back, you must open the source drawing and unlock the layer before performing the query. Draw mode copies the queried entities into the current drawing, displays the queried entities on-screen, and leaves the source drawings unchanged. Because the queried entities are copied into the current drawing, using **REDRAW** does not make the entities disappear. Once data are modified, MAP locks the entities in the source file. No other MAP user can edit until they are saved back or released. When an object is locked, other users can view it, but they cannot edit it.

Report mode

If you execute a Report mode query, AutoCAD MAP creates the specified text file. Use a text editor, such as Windows Notebook, to view the report file.

Additional tools that make MAP 3D a worthwhile foundation item for Civil 3D include the Drawing Cleanup commands, **TRANSFORM**, **RUBBER SHEETING**, and mass editing commands such as **BOUNDARY BREAK** and **BOUNDARY TRIM**. GIS professionals need these tools to ensure that their linework is accurate and clean, as do engineers and surveyors.

Editing and Graphical Cleanup Tools

The Drawing Cleanup commands correct linear object and node errors to create a clean topology. The commands include these abilities: Delete Duplicate Objects, Erase Short Objects, Break Crossing Objects, Extend Undershoots, Apparent Intersection, Snap Clustered Nodes, Dissolve Pseudo-Nodes, Erase Dangling Objects, Simplify Objects, Zero-Length Objects, and Weed Polylines. The **TRANSFORM** command scales, offsets, and rotates selected objects. The **RUBBER SHEETING** command stretches objects from one set of points to a new set of points and is often used to correct scans that were made from stretched hard copies. The **BOUNDARY BREAK** command cleans map edges by

2D—Two dimensions

cutting lines, **2D** polylines, arcs, and circles that cross a specified edge; and the **BOUNDARY TRIM** command trims objects to a selected or defined boundary.

The MAP menu tools for performing data cleanup can solve problems that have vexed users for years—further evidence of why a land development professional should be using AutoCAD MAP 3D for everyday operations. The following describe several common scenarios that can wreak havoc on engineering projects.

Scenario 1

Breaklines—3D polylines representing breaks in **grade**

Existing Ground—Natural ground

Slop—A change in elevation

On a site where existing surface data are defined by mass points and **breaklines**, the surveyor must process these data and build an **existing ground** surface. Engineers using AutoCAD Land Development Desktop points and faults, on the other hand, might develop the proposed site. In either case, various situations could lead to crossing breaklines or vertical faces. Of course, surface processing does not tolerate vertical faces because the **slope** is infinity, owing to a divide by zero in the calculation.

The software, however, can locate all the invalid crossings and, depending on the type of crossing, will trim, break, or otherwise fix such incidents. Previously, users spent large amounts of time performing these tasks manually.

Scenario 2

The traditional AutoCAD user might trim and erase feverishly to clip data to specific locations or boundaries. Now, the software can break all of the objects so that their properties may be altered and can perform the trimming to boundaries. Users can easily trim data to open or close polylines.

Scenario 3

If a user inadvertently copied AutoCAD drawing data on top of itself, such an error would double not only the file size but also the plot stroking, thereby affecting the quality of the plot. MAP 3D can help users avoid such a nightmare by deleting duplicate objects while preserving the integrity of the original file data.

Each of these scenarios demonstrates how the software can perform sophisticated drawing cleanup operations in a simple and commonsense fashion.

Among the cleanup tools' other features are the following.

- MAP 3D can automatically eliminate the speckling inherent in scanned images and can automatically extend or trim property lines that do not close, according to user-specified tolerances. It can weed out densely populated vertices in polyline data.
- The conversion utilities will redefine lines, arcs, 3D polylines, and circles into polylines.
- After editing, the Mapping commands provide the ability to modify the original entities, create new entities but retain the original ones, or create new entities and delete the original ones.
- Users can perform object selection for editing by using automatic algorithms or by simply selecting objects using traditional AutoCAD methods.
- Rubber sheeting of data is supported in a user-friendly way. Paper stretch could cause inherent errors in data that have been scanned in and converted to vector data. The software's tools will allow users to correct this situation.

Geodetic Transformations

One of the most powerful commands in MAP 3D's repertoire for a surveyor would be the ability to assign a geodetic coordinate system (Global Coordinate System) and transform data from one geodetic coordinate system to another.

In the second example that follows, MAP 3D transforms a file from NAD 83 into NAD 27.

Exercise 1-1: Attach Database and Link to External File

The following exercise has the reader open an AutoCAD file, attach an external database, and link it to the AutoCAD file. Then it shows how to develop text in AutoCAD from the database and explores MAP 3D GIS commands briefly.

1. Launch the Civil 3D software by double-clicking the icon on the desktop.
2. If the **Task Pane** is not displayed, then invoke it by using the **MAP** pull-down menu and selecting **Utilities >> Task Pane**. The system may appear similar to that shown in Figure 1-6. If it is already there, skip this step.

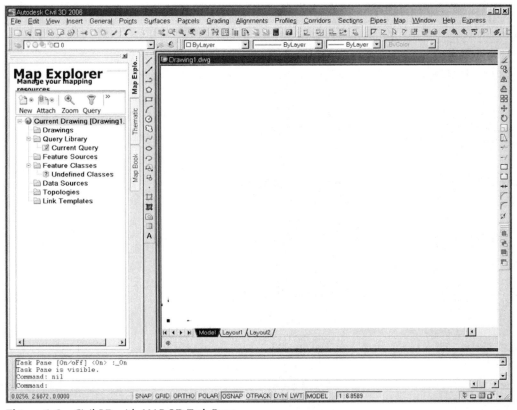

Figure 1-6   Civil 3D with MAP 3D Task Pane

If the Task Pane is not displayed, click on MAP >> Options and turn on Show on Startup. Close out and restart Civil 3D. In the Task Pane click on the pop-down window and several selections are available such as: MAP Explorer, Map Books and Displays Manager. Choose Map Explorer.

3. Now open Chapter-1a.dwg. You will see three *parcels*. Each parcel has an ID inside it that references a Tax Map Reference, which is a unique identifier for the parcel.
4. In the **Task Pane**, right click on **Data Sources** and select **Attach**...
5. Ensure that the Filetype is a Microsoft Access Database (.MDB). Choose the database called Chapter-1-Database.Mdb. Notice that several **Tables** are now available under **Data Sources**. Of interest at this point is the table called **Property**, as shown in Figure 1-7.

NEW
to AutoCAD
2007

Parcel—A plot of land

Student Files   Student Files   Student Files

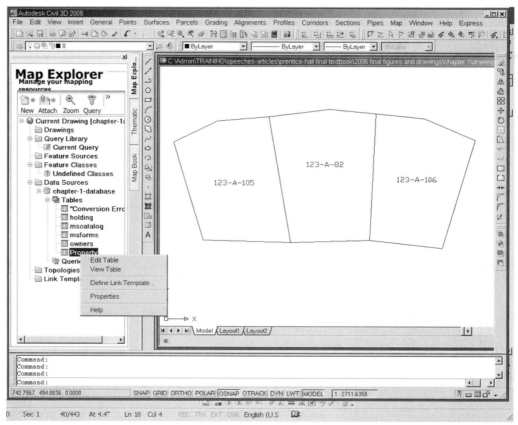

Figure 1-7   Database linking

6. Right click on the **Property** table and select **Edit Table**. The database displays, and notice that by clicking in a cell you can change the value if needed. Note that you do not need Access installed to do this. See Figure 1-8. The column titled Map is the Tax Map Reference. Look down the column and you will see references for 123-A-105, 123-A-106, and 123-A-82. These are the same references in the AutoCAD file. Let us link them together.

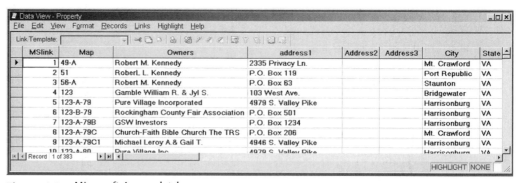

Figure 1-8   Microsoft Access database

7. Right click on the **Property** table and select **Define Link Template...** When the **Define Link Template** dialog box displays, use the template name **Property** and click a check mark ON in the Key column for MAP. That will be the link between the AutoCAD file and the database. Hit **OK**.

8. In the **Task Pane**, expand the **Link Templates** icon and you will see a **Property** template. Right click on the **Property link** template and select **Generate Links**...
9. Set the settings in the **Generate Data Links** dialog box that appears as they are shown in Figure 1-9. Hit **OK**.

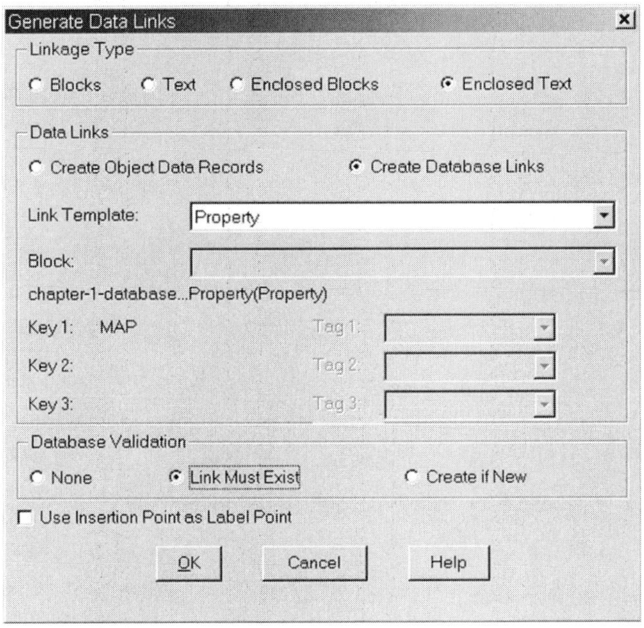

Figure 1-9   Generate Data Links

10. In the prompt, type **A** for **All**. It will respond to inform you that three links were created.
11. Check the links. Under the **MAP** pull-down menu, select **Database >> Link Manager**.
12. It asks you to: Select an object:, so select the property around 123-A-105. Repeat this for 123-A-106 and 123-A-82.
13. Now pick on the 123-A-105 parcel and right click. Select **Label >> Label the Link**..., as shown in Figure 1-10.
14. When the **Select the Database Object** dialog box appears, there is nothing in the window to select (because you are doing this for the first time), so hit **New**.
15. Accept the default as the New label template name and hit **Continue** and a **Text** dialog box will display. This is where you define how you want text to appear when it is imported into AutoCAD from the external database.
16. In the **Label Fields** tab, notice the **Field** pop-down window. From the pop-down choose **Owners** and hit **Add**, and it will place the field into the window below.
17. Now place your cursor in the text window below the Owners and repeat this operation for Address 1, City, State, Zip, and Acres until your screen appears as in the Figure 1-11.
18. Then click on the **Character** tab, type <Ctrl>A to select all of the text, and set the text size to **20.00**. Hit **OK** to exit. Notice that the database information for these fields has been imported into AutoCAD. Using the grip on the text, stretch it to a more readable location.

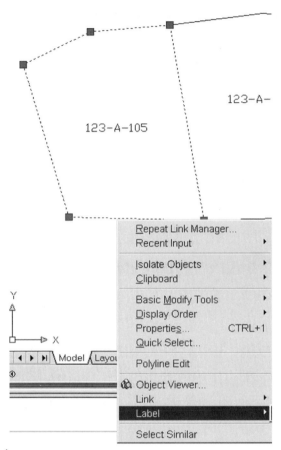

123-A-

123-A-105

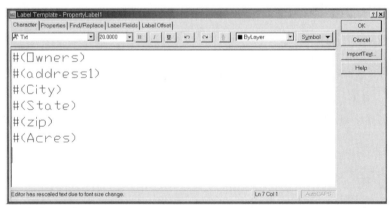

Figure 1-10   Label Links

Figure 1-11   Label Template

TIP

Note that you can add prefixes or suffixes to these fields when they come in to customize the label exactly as you want it.

19. Now click on each of the remaining properties and repeat the sequence. Note that once the Label Template is created, you simply use it and hit **OK**. The drawing should appear as in Figure 1-12.

Now to see that the drawing stays "alive," make a change to the database record and reload the information in AutoCAD to see the label update itself.

Student Files

20. In the same drawing (or you can open Chapter-1a-complete.dwg), right click on the **Property** table and choose **Edit Table**.

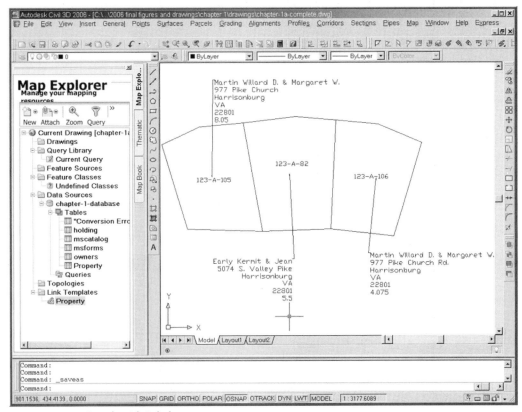

Figure 1-12   Parcels with Labels

21. Go to Row 20 and change the address to **1000 Main Street**; then click Name column when you are finished typing. Then close the database.

If the Task Pane is not displayed, click on MAP >> Options and turn on Show on Startup. Close out and restart Civil 3D. In the Task Pane click on the popdown window and several selections are available such as: Map Explorer, Map Books and Display Manager. Choose Map Explorer.

22. In AutoCAD click on the property for 123-A-105 and right click. Select **Label >> Reload**. Notice that the address now reads 1000 Main Street.

Exercise 1-2: A Geodetic Transformation

This exercise provides an example of transforming an AutoCAD file with one geodetic coordinate system into a different geodetic coordinate system. In it you attach the original file into your working session and then establish your coordinate systems. When you perform the query, MAP 3D transforms the attached file's data into your file with the new coordinate system.

1. Launch the Civil 3D software by double-clicking the icon on the desktop.
2. If the **Task Pane** is not displayed, then invoke it by using the **MAP** pull-down menu and selecting **Utilities >> Task Pane**. If it is already there, skip this step.
3. Now create a new drawing using the **File** pull-down menu and select **New**. Select the **Civil 3D Imperial by Layer** template and click open to create your new file.
4. Now use the **MAP** pull-down menu and select **Drawings >> Define/Modify Drawing Set**.
5. Hit the **Attach** button. Click on the button to **Create/Edit** an alias. For the drive alias, type: **Map Exercises**. For the **Actual Path** location, select the drive or folder for the Civil 3D textbook files. The **Drive Alias**

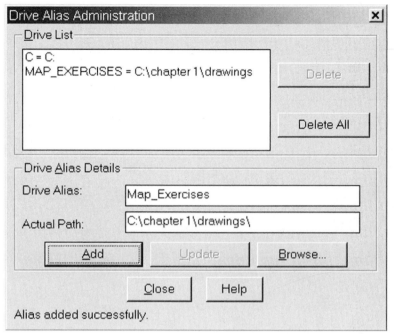

Figure 1-13   Drive Alias

Administration dialog box should appear similar to that of Figure 1-13.

6. Then go to the **Look in** pop-down window, navigate to your location, and select the file called Chapter-1b.dwg. Click **Add** and click **OK**. Then hit **OK** to the **Define/Modify Drawing Set** dialog box to complete the attachment.

7. Then select **Tools >> Assign Global Coordinate System**. This indicates which coordinate system this drawing will be in. When the **Assign Global Coordinate System** dialog box appears, in the Current Drawing area, click the **Select Coordinate System...** button. Choose **USA, Virginia** as the category and then select **NAD 27 Virginia State Planes, North Zone (4501), US Foot** as the Coordinate System.

8. Then in the Source Drawings area, click **Select Drawings...** and select the drawing you attached: Chapter-1b.dwg. Hit the **ADD** button. Click the **Select Coordinate System...** button for the Source Drawings. Choose **USA, Virginia** as the category and then select **NAD 83 Virginia State Planes, North Zone, US Foot** as the Coordinate System. Hit **OK**.

9. The result is shown in Figure 1-14. Hit **OK**.

10. Go to **MAP >> Query >> Define Query**. When the **Define Query** dialog box displays, click the **Location** button, select **ALL**, and hit **OK**. You see the entry in the main query window.

Note:
Nothing has been imported into your working session at this time; it is simply attached for accessing. MAP 3D uses this method to allow for accessing large quantities of data.

Note:
Now that MAP 3D knows what the current drawing is set to and what the source drawing is set to, you can perform a query and extract the data from the source into the active drawing.

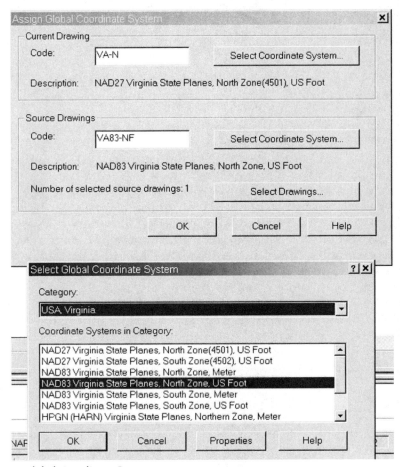

Figure 1-14   Global Coordinate Systems

11. Click **Draw** in the center of the dialog box to physically extract all of the data into the current drawing. Then click **Execute Query**. (Do not hit **OK** at this time.) Hit **OK** to any boxes that appear.

12. When the processing is complete, **Zoom Extents**. If you move your cursor around the data and inspect the coordinate values that appear in the lower left-hand window of AutoCAD, notice that they are in the 2,000,000, 700,000 range. This is the range for NAD 27 in Virginia. If you were to open the Chapter 1-b.dwg file, you would notice that the coordinate range for it is in the 11,000,000, 7,000,000 range. Therefore, the file has been converted. Feel free to save it as some other name.

This exercise allows you to attach a drawing in NAD 83 and extract the data into your current session, which is set to NAD 27. The data physically move millions of feet from where they were to conform to the geodetic coordinate system algorithms. This very sophisticate operation requires some education in this theory before performing it on an actual project. That theory is beyond the scope of this text; however, this example shows how to execute MAP 3D in this function.

CIVIL 3D—A CIVIL ENGINEERING DESIGN APPLICATION WITH NEW PHILOSOPHIES

Civil 3D is a totally new software solution that offers a revolutionary approach to civil engineering design and drafting. It consists of the fusion of several Autodesk solutions including AutoCAD, MAP 3D, and the Civil 3D toolset along with Trimble Link™, Carlson Connect™, and Leica Exchange. All of the traditional AutoCAD commands exist and can be accessed through the usual toolbars, pull-down menus, Custom User Interface, or type-in commands. The AutoCAD Sheet Manager is shown here in a floating palette condition. The MAP 3D commands are accessed via pull-down, type-in, and toolbars. Figure 1-15 that follows shows the MAP 3D pull-down menu and the location of the Civil 3D pull-down menus.

The new philosophies included within this software are data driven design, reactive object technology, "ripple-through effectiveness," state-of-the-art interface enhancements, and an interactive and dynamic library of settings. The data driven design refers to how objects will design themselves based on how the settings data are established. For instance, the horizontal curves for a road can be established such that they always meet a minimum standard allowable curve for the state or county in which they are being designed. Reactive object technology refers to items such as annotation automatically updating itself and reorienting itself when objects change or when the view is altered or rotated. The *ripple-through effect* refers to the impact that modifying an alignment has on the corridor's profiles, sections, and assemblies, whereas they automatically recompute their new values attributable to the modification. The interactive and dynamic library of settings refers to how the objects developed will take on aesthetic and visual characteristics based on how their display settings have been configured. The objects will change their appearance based on any changes to these settings automatically as well.

Ripple-Through Effect—A term describing dynamic modeling

What Civil 3D Introduces

In addition to Civil 3D's fundamentally sound and traditional routines, it introduces profoundly new capabilities to civil engineering users of CADD. These include

1. New interface tools;
2. New objects specifically developed for civil design;
3. Dynamic, model-based design, data driven methods, and the ripple-through-effect;
4. Data driven, object style libraries that control graphics appearances and text;
5. DOT-level highway design capabilities; and
6. Rehabilitation and reconstruction design tools.

Interface Tools

The Civil 3D interface tools are evident in the **Tool Palettes** and **Toolspace**, as the software is launched, as shown in Figure 1-15. Autodesk has gone to great lengths to place more functionality at the user's fingertips, and the invention of the hide-away Toolspace and **Tool Palettes** provides excellent examples. The way that Figure 1-15 is set up is how I recommend that you set up your workspace for production. Notice that the Civil 3D program is not maximized.

An entire set of Survey related tools have been added to the software as evidenced by the Survey pull-down menu.

The following procedure shows you how to accomplish this user workspace. Note that it is not required for the use of the software; it simply places the maximum amount of power at your fingertips without sacrificing design space. The advantage is that you have maximized the working area for design, but have left the **Tool Palettes** and **Toolspace** available. Simply passing the mouse over

NEW to AutoCAD 2007

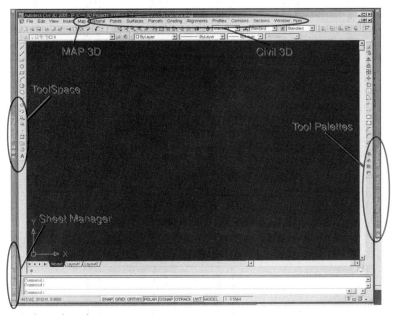

Figure 1-15   Proposed workspace setup

these palettes opens them, and when you finish making a selection from one of them, the palettes automatically close up.

Exercise 1-3: Set Up Production Workspace

1. Launch the Civil 3D software by double-clicking the icon on the desktop.
2. The system may appear similar to that shown in Figure 1-16.

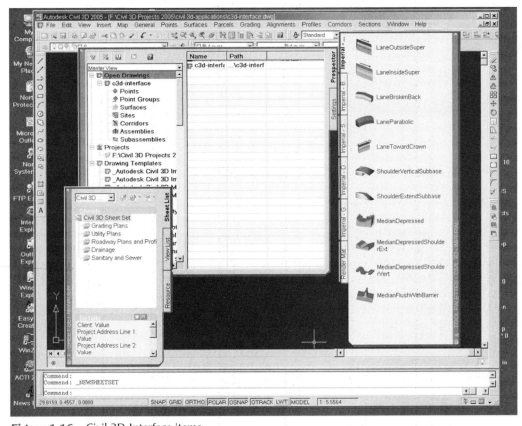

Figure 1-16   Civil 3D Interface items

An Inquiry Toolspace, with instantaneous readouts of many types of data, is now included as part of the interface See Figure 1-17.

Figure 1-17   2007: The new inquiry toolspace

3. If the palettes and **Toolspaces** are not displayed, then from the **General** pull-down menu, select **Show Toolspace...**
4. Right click on the spine of each palette and select **Auto-hide** (Figure 1-18).

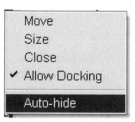

Figure 1-18   Right click menu on palettes

5. When done each palette collapses and floats in the workspace, as shown in Figure 1-19.

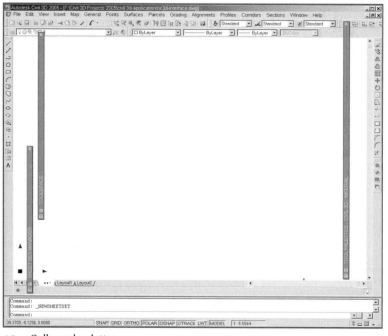

Figure 1-19   Collapsed palettes

6. Size the Civil 3D window to about two-thirds of your desktop size.

7. Drag the **Sheet Set Manager** palette to the far left side of the Windows desktop and drop it. It contracts when you pull your mouse away from it.

8. Drag the **Tool palettes** to the far right side of the desktop and drop it. Drag the **Toolspace** to the far left side of the desktop and drop it. They both contract.

9. Resize the Civil 3D window from the upper left and lower right corners to fit completely between the contracted palettes. You have now maximized your functionality without sacrificing drawing space.

10. To try and set this up without using a procedure similar to this will cause the palettes to "dock," thereby reducing your overall drawing area. See Figure 1-20.

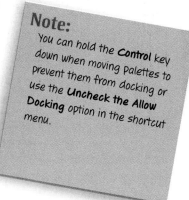

Note: You can hold the **Control** key down when moving palettes to prevent them from docking or use the **Uncheck the Allow Docking** option in the shortcut menu.

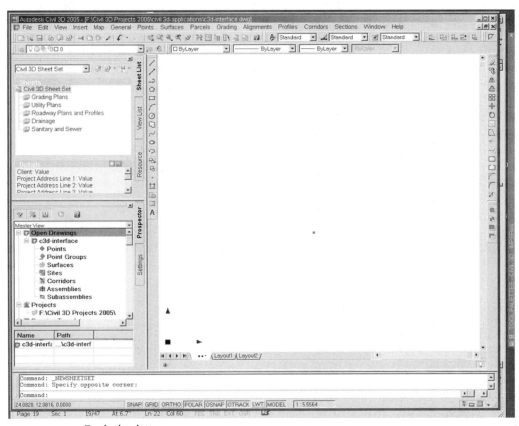

Figure 1-20   Docked palettes

Administrator's Catalog Library Tool

In addition to the interface enhancements discussed previously, Civil 3D also introduces a **Panorama** and a new Administrator's Catalog Library tool. It includes a Render Material Catalog, and Imperial and Metric Subassembly tool catalogs for Corridor design.

Exercise 1-4: Configure Tool Palette

By selecting any of these catalogs, a new window displaying the contents appears.

1. To begin, choose **General >> Catalog ...** and a choice of several catalogs appears. Choose the **Corridor Modeling Imperial Catalog** and the following choices display, as shown in Figure 1-21.

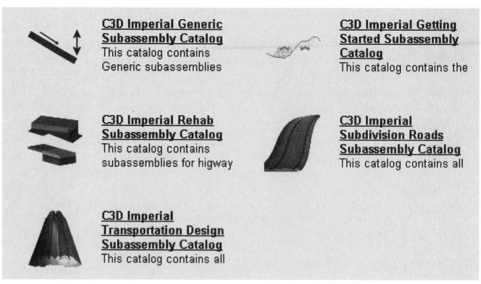

Figure 1-21   Corridor Catalog

A new catalog has been added to 2007 for Channel and Retaining Wall Subassemblies.

2. Then select a subassembly for Generic, Rehab, Transportation, Getting Started, or Subdivision roads. Choose the **Getting Started** option; a window opens showing different subassemblies, as shown in Figure 1-22.

Figure 1-22   Corridor Catalog subassemblies

3. Select the **BasicLane** to import into Civil 3D, then use the "i-drop" facility. Click on the **i** in the blue circle near the **BasicLane**, drag it into Civil 3D, and drop it into the drawing area.
4. Hit **Enter** and place it into the drawing area. It appears as a shaded object with location points on the corners for adding other subassemblies.
5. Now select the **BasicCurbAndGutter** object and drag it using the i-drop facility. The software prompts the following: `Select marker point within assembly or [RETURN for Detached]:` Hit **Return**. The software responds: `Specify location for subassembly:` Select the Center Snap of the top right corner of the pavement. Hit *Cancel* to terminate. See Figure 1-23 and Figure 1-24.

Figure 1-23   Subassembly Pavement section

Figure 1-24   Subassembly Pavement section with Curb

Chapter 7 on corridors covers the development of assemblies in depth. This exercise is just an example of how the catalog works and how to use the i-drop feature.

Panorama Palette

The **Panorama** window is a multipurpose palette with the primary purpose of displaying data such as points, alignment entities, and description keys. The **Panorama** is also a **Tool** palette that provides users with an **Event Viewer** to view messages that are logged during an Autodesk Civil 3D session. For example, if a surface is being created, the user will read in the 3D breaklines that may have been imported into the AutoCAD file.

If there are crossing breaklines, where one breakline physically crosses over or under another, then that breakline is not added into the data set for surface modeling. The **Event Viewer** informs the user of this occurrence. See Figure 1-25.

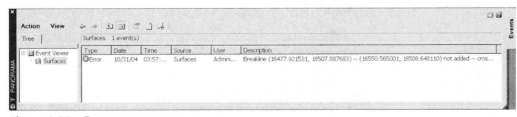

Figure 1-25   Panorama

Toolspace Palette

The Civil 3D environment includes the **Toolspace**, shown in Figure 1-26. The **Toolspace** has two panels, one called a **Prospector** for data and the other for **Settings**. The **Toolspace** allows the user to view project data and status at any time and updates itself dynamically to show the status of all design data within the drawing.

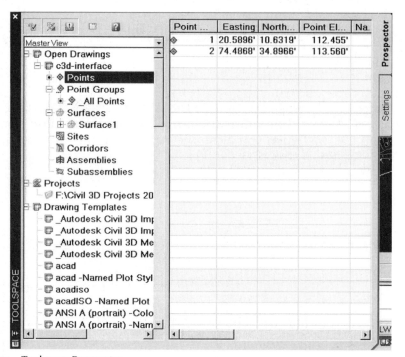

Figure 1-26   Toolspace Prospector

In Figure 1-25, note that Points exist and the two that exist are displayed when the cursor is on the **Points** item. Observe also that the **Surfaces** item is expanded and shows one surface in existence.

Exercise 1-5: Investigate Toolspace

Let us investigate the **Toolspace** a little closer.

1. Make sure that you are in Civil 3D and that your **Toolspace** is displayed. If it is not, select **General ≫ Show Toolspace...**

A Survey Item has been added to the Toolspace. A Survey Tab has been added as well. These allow Surveyors to work with graphical and dynamic objects in the same way as designers can in their part of the Toolspace.

2. Save your drawing to Toolspace.dwg in your project location.
3. There are two essential parts to the **Toolspace**: **Prospector** and **Settings**. The **Prospector** is the *Command Central* for many operations and reviewing tasks.
4. After starting a Civil 3D session, click in the **Toolspace** on the **Prospector** tab.
5. There is a pop-down window in the top left of the palette. Select **Master View** if it is not already selected.
6. Three collections of information can be viewed using the **Master View**: Open Drawings, Projects, and Drawing Templates. Expand the **Open Drawings** item and you see your AutoCAD filename of Toolspace. That file can be expanded to see the potential data types stored within it,

such as Points, Point Groups, Surfaces, Sites, Corridors, Assemblies, and Subassemblies.

7. Right click on **Points** for this example. The **Prospector** displays a short-cut menu pertaining to points. Because this is a new file, the only options are **Create**... and **Refresh**. Hit **Create**... and a **Create Points** toolbar appears, allowing for the creation of points.

8. Take a moment to explore some of the icons in the toolbar. They include **Miscellaneous Manual, Intersection, Alignment, Surface, Interpolation, Slope**, and **Import Points** functions.

9. Choose **Miscellaneous Manual** and, following the prompts, set a point inside your AutoCAD file. Give the point an elevation 112.33 and a description of FG. The point should display where you placed it.

10. Select the point you placed with a left click, then right click and a shortcut menu appears. Choose **Edit Points**... and a **Panorama** palette appears. Notice that you can change the elevation value from 112.33 to 113.22, and when you close the **Panorama**, the point is updated.

11. Close the **Point Creation** toolbar.

These were just some very basic examples of using some of the new interface tools. As the text progresses, you use these tools in detail.

New Objects for Design and Drafting

There is a wide variety of objects to use in designing. Many of these objects were introduced in Land Desktop; however, they are now more fully implemented. But what are objects? Specifically, we are referring to Intelligent Objects, which are Autodesk data types that provide a programming structure for a variety of design items that contain data, functions, and allowable behaviors while interacting with the design and other objects.

A Networks Item and a Figures Item has been added to the Object types. These are found under the Survey item in the Prospector. These items will populate as the surveyor imports or creates traverse networks or linework defined as a figure.

The concept of object relationships is not new to Autodesk users. Land Desktop introduced this in a much more primitive form. Land Desktop contours were objects because they tried to act more like contours than traditional polylines did. What this meant is that the contour label would slide along a contour object when it was moved. It brought the break under the text with it and healed up the location from it which it came. Civil 3D takes this to the next logical level. Contours are no longer objects because they are a part of the bigger picture, but are now treated as a display option for the Surface Object. The Surface Object contains all of the surface data within it, and the user uses style libraries to display the surface in a variety of analytical ways; for example, as contours, as slope shaded analyses, as a rendered object, and so forth. The Surface Object itself is affected by changes to its source data; that is, the points, the breaklines, or other surface data that were used to develop the surface.

The novel and revolutionary aspect of Civil 3D objects is that, because they are style based and dynamic, they react and interact with other Civil 3D objects. Introducing links between design objects that react dynamically minimized the need for many manual tasks, such as ensuring that modifications are enacted between linework and labels, alignments and profiles, and the like. This process creates a ripple-through effect of actions and reactions, and updates the data within the design automatically. Changes in one object can be passed automatically to objects associated to it at the user's option. For instance, if a curve within an alignment is altered, any grading using that alignment as a baseline can be consequently updated. Furthermore, all related alignment stationing, linework labels, and other alignment-related data are likewise updated.

NEW
to AutoCAD
2007

These relationships are quite extensive, and the user should have a feel for which objects update other objects. Some rules of thumb are listed here.

- When Points are modified, they can directly influence Surfaces.
- When Surfaces are modified, they can directly influence Grading Objects, Profiles.
- When Parcels are modified, they can directly influence Grading Objects, Corridors.
- When Alignments are modified, they can directly influence Grading, Corridors, Profiles, and Sections.
- When Grading Objects are modified, they can directly influence Surfaces, Corridors.
- When Subassemblies are modified, they can directly influence Assemblies, Corridors.
- When Assemblies are modified, they can directly influence Corridors.

A relationship exists between design objects, the styles that control their display, and the labels that control their annotation. These styles and labels are also managed as objects within Autodesk Civil 3D.

Objects are the basic working mechanism of the engineering design workflow. The underlying methodology for Civil 3D uses an object-oriented architecture. Therefore, design components such as points, surfaces, and alignments are intelligent and maintain relationships with other objects. The classic example is a horizontal alignment that has associated profiles and cross sections. If the alignment is manipulated, then the profiles and sections linked to that alignment are automatically updated accordingly.

The major object types in Civil 3D are represented as follows.

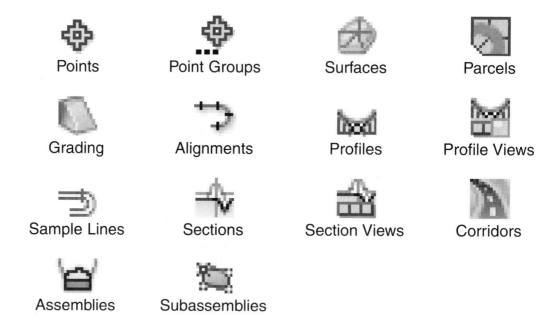

| Points | Point Groups | Surfaces | Parcels |

| Grading | Alignments | Profiles | Profile Views |

| Sample Lines | Sections | Section Views | Corridors |

| Assemblies | Subassemblies |

Civil 3D software is the first version of AutoCAD for civil engineering users in which lines, arcs, and polylines will NOT be the predominant data types for design and drafting. Rather, surface, roadway, parcel, and point objects will be the predominant design objects that are created and modified. Although other civil-oriented software solutions are on the market, none have taken this reactive technology as far as the Civil 3D has. The ability to perform storm sewer drafting and compute watershed locations is also available in Civil 3D.

Dynamic, Model-Based Design, Data Driven Methods, and the Ripple-Through Effect

The concept of a live, dynamic model is new to the civil engineering field and consists of the ability to make changes to integral design components and have those changes update any related project components. The classic example is a horizontal alignment that has associated profiles and cross sections. If the alignment is modified, then the alignment annotation, the profiles, and the sections linked to that alignment are updated automatically. Another example could be parcels with associated distance, bearing, and area labels that are updated automatically when the parcel geometry is altered.

Survey networks and figures are now dynamic parts of the software. This means that changes to setups or backsights can be automatically recalculated in the network object.

Data Driven, Object Style Libraries

Object style libraries are very important to the use of the software in that they control the appearance and design characteristics of the object. When a new object is created, a predefined style can be applied to it to control how it displays. Styles can be modified on the fly and are retroactive. For example, if contours appear red and the user would like them to appear blue, a style modification can occur and be applied to the contours. They will then inherit the property of displaying blue. If the style definition is modified, the changes are immediately applied to all objects using that style, similar to Text Styles or Dimension Styles.

The Styles are managed on the **Toolspace Settings** tab in Figure 1-27. All Civil 3D objects have a Standard style that can be used as is, or new ones can be created as needed. The best approach would be to have someone in the organization predevelop all of the Styles needed as part of the implementation of the software. Styles can be organized and stored in a drawing template (.DWT) file.

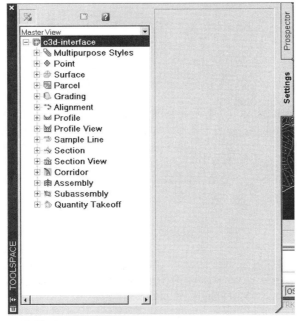

Figure 1-27 Toolspace Settings

A set of Survey settings has been added to the Settings tab. This tab allows the user to create Figure Prefix Libraries for defining linework or figures, Equipment Databases where settings for ambient conditions and other equipment parameters are contained. The Management of the Survey databases occurs here as well.

When developing a Style for an object, the object will be physically placed on a main layer. But component objects can reside on supplemental layers in order to control the display of those components. So an alignment object can reside on the **CL** layer, but the lines can reside on the **CL-Lines** layer while the arcs reside on the **CL-Arc** layer. The user can issue this instruction by using an * to denote using the object's layer as a prefix or suffix. It becomes an interesting visual when the user would like to inspect the arcs within an alignment separate from the linear components.

DOT-Level Highway Design Capabilities

Civil 3D offers Autodesk users high-end highway design tools that allow for multiple baseline designs and multiple regions with each baseline. Baselines are centerlines for adjacent roadway occurrences. Regions are areas within those baselines where different things might be happening, such as *superelevations* or different assemblies.

Superelevations—Banking the road surface on curves

Rehabilitation and Reconstruction Design Tools

Civil 3D provides a highly customized and very specific toolset for designers who are working in the area of roadway rehabilitation and maintenance design. These tools include subassemblies and routines for sidewalk replacement, curb replacement, and several variations on milling and overlaying pavements. The initial installation of Civil 3D may not import these into the **Tool Palettes**, but they can be accessed through the Catalogs and dragged and dropped into a **Tool** palette of your making.

Corridor capabilities have been added and include the addition of more powerful and useful boundaries as well as the ability to add station-based assembly insertions.

A new catalog has also been added for Channel and Retaining Wall Subassemblies.

The following Figure 1-28 and Figure 1-29 indicate the rehabilitation capabilities of Civil 3D.

TIP   In addition to these fine tools, Autodesk has also developed the ability for designers to create designs where the old roadway will be repaired. This is in contrast to prior solutions in which the assumption was that all roads were new roads. An enormous amount of roadway repair is occurring throughout the world, and these abilities allow for milling, overlaying, sidewalk, or curb and gutter replacements on existing roadways.

Job skills will benefit not only from the ability to design DOT-level roadways but from the roadway rehabilitation tools as well. Many regions in the country or in the world do not have a lot of new growth; however, they do perform a large amount of roadway maintenance. These skills will provide new productivity for that staff.

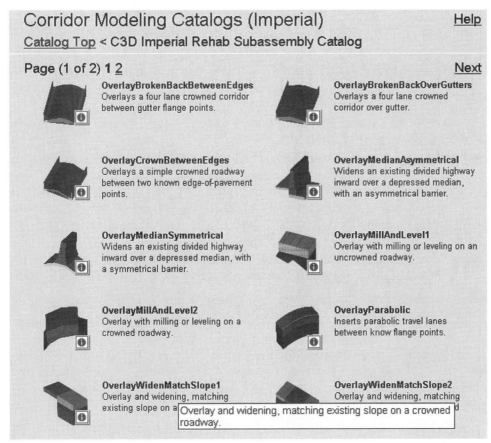

Figure 1-28   Rehabilitation tools, page 1

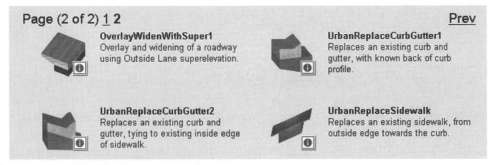

Figure 1-29   Rehabilitation tools, page 2

Understanding Civil 3D Icons

The following information offers the user a description of the icons in Civil 3D. Although Civil 3D Help provides this information, it is scattered around and some find it difficult to locate. This text places all of these items in this single location for easy reference.

The icons at the top of the **Prospector** palette control the display of icons within the **Prospector** panel, display the **Panorama** window, and provide access to **Help**. The icons and their respective meanings are outlined next, as found in Civil 3D Help.

 Toggles the display of **Project Item State** icons in the **Prospector**. The **Project Item State** icon indicates its status with respect to the project, such as whether it is checked in or out. Details for these icons are shown.

An **Open** padlock indicates that the item is checked in.

If the item has an Open padlock with a Check mark on it, the object is checked out to your active drawing.

If the padlock is Locked with a Check mark, the object is checked out to a drawing that is not your active drawing. This could be your drawing or someone else's drawing.

If there is a Locked padlock with No check mark, the object is protected and cannot be modified.

If there is a Pencil with a Red circle through it, the object belongs to a project that could not be found. That project would need to be reconnected. If the project cannot be found and it does exist in a different project path, then right click **Projects** and select **Manage Project Paths ...** to select the correct path for the project.

 Toggles the display of **Drawing Item State** icons in the **Prospector**. The **Drawing Item State** icon indicates the state of an object within a drawing, including whether or not it is locked.

The following icons describe the possible states of information.

No icon means the object is Not referenced by another object, is Not locked, and is Not a reference itself.

The object is **Not** referenced by another object, Is locked, and is Not a reference itself.

 The object **Is** referenced by another object, is Not locked, and is Not a reference itself.

 The object **Is** referenced by another object, Is locked, and is Not a reference itself.

The object is **Not** referenced by another object, is Not locked, but it Is a reference itself.

 The object is **Not** referenced by another object, Is locked, and Is a reference itself.

 The object **Is** referenced by another object, is Not locked, and Is a reference itself.

 The object **Is** referenced by another object, **Is** locked, and **Is** a reference itself.

Toggles the display of **Drawing Item Modifier** icons at the top of the **Prospector**. The **Drawing Item Modifier** icon indicates the status of the item relative to the project or the drawing. A description of the **Drawing Item Modifier** icons is provided here.

No icon means the object is **Not** out of date and Does Not violate constraints, the Local copy of the object **Is** more recent than the Project copy, and the Local copy of the object has Not been edited relative to the Project object.

The object **Is** out of date or **Does** violate constraints, the Project copy of the object Is more recent than the Local copy, and the Local copy of the object Has been edited relative to the Project object.

The object is **Not** out of date and **Does Not** violate constraints, the Project copy of the object **Is** more recent than the Local copy, and the Local copy of the object **Has been** edited relative to the Project object.

The object **Is** out of date or **Does** violate constraints, the Project copy of the object is **Not** more recent than the Local copy, and the Local copy of the object **Has been** edited relative to the Project object.

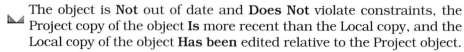 The object is **Not** out of date and **Does Not** violate constraints, the Project copy of the object is **Not** more recent than the Local copy, and the Local copy of the object Has been edited relative to the Project object.

The object **Is** out of date or Does violate constraints, the Project copy of the object **Is** more recent than the Local copy, and the Local copy of the object Has Not been edited relative to the Project object.

The object is **Not** out of date and **Does Not** violate constraints, the Project copy of the object Is more recent than the Local copy, and the Local copy of the object **Has Not been** edited relative to the Project object.

The object **Is** out of date or **Does** violate constraints, the Project copy of the object is **Not** more recent than the Local copy, and the Local copy of the object **Has Not been** edited relative to the Project object.

Toggles the display of the **Panorama** window, if the **Panorama** window contains active vistas.

Indicates that data do exist in this area.

The view in the **Prospector** might appear similar to Figure 1-30. The name of the drawing is EG Drawing. Points do exist, and Point Groups exist. When the tree for a drawing is expanded, there are AutoCAD icons in front of each drawing name. If the icon is blue, it means that the drawing is in the project. If the icon is red, then the drawing is currently open.

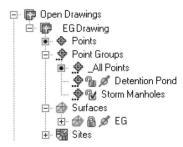

Figure 1-30   Data and Project icons

The **Toolspace** has Windows Explorer style expansion and contraction facilities that are called Trees. They display items in a hierarchical structure. Any item containing other items below it is called a Collection. In Figure 1-29, Open Drawings, EG Drawing, Surfaces, and Point Groups are Collections. A symbol may be shown to the far left of each Collection name and this is called Tree Node. Its intent is to provide information about the display of the items in the Collection. The following list describes these Tree Node symbols:

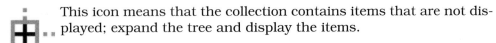

This icon means that the collection contains items that are not displayed; expand the tree and display the items.

This icon means that the items contained in the collection are displayed and the user can collapse the tree and hide the items, if desired.

This icon means that the collection contains items that cannot be displayed in the tree. The user can click on the Collection name to list the items in a **Toolspace** item view, which is another pane displayed beside or below the tree.

This icon means that the Collection's data need to be refreshed. The user can right click on the collection and choose **Refresh**.

This icon means that the item is a reference to an item located elsewhere in the tree. The user can double-click the reference to navigate to the actual item.

This icon means that the Collection contains no items.

NEW
to AutoCAD
2007

New icons have been added for Survey items that incude: Control Points, Non-Control Points, Directions, Setups and Traverses.

Pipe Utilities Grips

Several grips that allow for modifying pipe and components are described here.

The Endpoint Free Grip. Clicking the endpoint free grip in the Plan View moves that endpoint of the pipe to a new point. The midpoint and opposite endpoint of the pipe are maintained during this edit so the length does not change. Curved pipes behave identical to an AutoCAD arc.

The Midpoint Free Grip. Clicking the midpoint free grip moves the whole pipe to a new, specified point. Curved pipes behave identical to an AutoCAD arc.

The Length Constraint Grip. Pipe length can be increased or decreased by gripping the end of the pipe and only that end shortens or lengthens. The direction of the pipe does not change in this mode of operation.

The Midpoint Resize Constraint Grip. This grip allows you to reset the pipe's width based on the pipe sizes available in the parts list and part catalog. When the midpoint resize grip is used on a pipe, a series of parallel lines display next to the pipe whereby each set of parallel lines represents an available pipe width. A tooltip displays the pipe width, and one can snap to any of the parallel lines to select a new pipe width on the fly.

Sheet Set Manager Palette

Another interface item in Civil 3D is the Sheet Set Manager, which was an AutoCAD 2005 enhancement that is included in the Civil 3D product. Sheet sets organize, create, and publish multiple drawing layouts into a single set of plans. Before the Sheet Manager existed, a set of drawings were organized using manual procedures. The individual drawings in the set may have emanated from several different AutoCAD files, and the process of ensuring that they were properly maintained, ordered, and correctly numbered was very time consuming. Enter the Sheet Manager. The CREATE SHEET SET command in the **File** menu opens a Wizard that guides you through the process of creating a sheet set. The user can also right click in the **Sheet Set Manager** window and obtain the menu shown in Figure 1-31.

Once in the Wizard, there are two ways to create a sheet set. You can use a sample sheet set provided by Autodesk or use an existing drawing set of your own. The sample sheet set option works like a template and provides the organization and default settings for the new sheet set. Once the setup is complete, a user can compose individual sheets within the set by importing layouts and placing saved views as needed. The individual sheets created using this method are saved as separate drawing files, in which the original drawing files are linked as referenced drawings. The sheet set is saved as a separate file with the extension of .DST.

If you have already created layouts for the desired drawings and now need to assemble them together into a plan set, you would use the second option. You can specify one or more folders that contain drawing files and the layouts can be imported into the sheet set. In this case, no new drawing files are created. Keep in mind, though, that this option will not fully utilize the

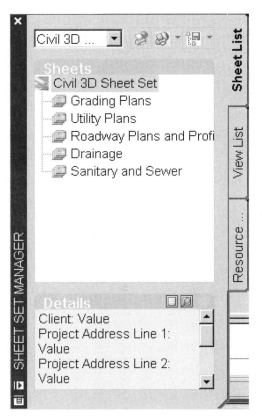

Figure 1-31   Sheet Set Manager menu

capabilities of the sheet set feature to manage and organize such items as sheet numbers, callouts, and such because the title blocks and views have already been defined.

The **Sheet Set Manager** palette performs the management of the sheets. The user can create new sheet sets, open existing ones, list out the sheets, and observe the views within a particular sheet. The sheets can also be arranged into individual groups to assist in navigation and organization. Figure 1-32 shows that a variety of details are available as well as a thumbnail preview if so desired.

Figure 1-32   Sheet Set Manager palette

The sheet file can be opened by double-clicking on it. If the sheet set was created using the **Sample Sheet Set** option, the user must specify the folder location of the AutoCAD files whose views are to be used. These become support drawings and are displayed in the **Sheet Set Manager** dialog box along with their saved views, which can be chosen and placed on a sheet. The sheets and their views are routinely numbered and update dynamically when changes are made to a name or a number. Once the sheet set has been developed, the complete set or just certain selected sheets can be published (or plotted). *Publishing* is a term now used for outputting the data to an external source. Not all plots are to hard copy anymore as some outputs go to the DWF format for viewing in the Express Viewer or on the Internet. Of course, outputting to a plotter is still a normal function for creating hard copies. The Sheet Manager has made plotting drawings individually an obsolete endeavor. Other enhancements include the **eTransmit** option, which allows for the creation of a transmittal package formatted in the Zip format, and the **Archive** option, which creates a compressed archive of the entire sheet set.

CHAPTER TEST QUESTIONS

Multiple Choice

1. The Civil 3D user must know which of the following in order to use Civil 3D for typical subdivision design?

   a. AutoCAD
   b. Geometry concepts
   c. Surface or DTM concepts
   d. All of the above

2. A user must understand which Civil 3D interface items in order to negotiate through the data effectively?

   a. **Tear-off** palettes
   b. **Toolspace**
   c. Slide Sorter
   d. **Kiosk** tools

3. Which of the components allows users to access all of the Properties of project data?

   a. **Settings** tab
   b. **Preview** window
   c. Right clicking on the **Surfaces** item
   d. All of the above

4. Some of the add-in software available for Civil 3D includes which of the following?

   a. Trimble Link
   b. Carlson Connect
   c. Leica X-Change
   d. All of the above

5. MAP 3D can connect and provide read/write access to which of the following?

   a. Microstation files
   b. Photoshop files
   c. Microsoft Access
   d. None of the above

6. One user can use MAP 3D to many AutoCAD drawings simultaneously. How many users can attach to the same AutoCAD drawing simultaneously?

   a. One user
   b. None
   c. An unlimited number
   d. Two users

7. Which components consist of the AutoCAD spacial database for MAP 3D?

   a. Graphic database
   b. Extended entity data (EED) or Object data
   c. External database(s)
   d. All of the above

8. Drafting technicians should know how to use MAP 3D in conjunction with Civil 3D in which of the following ways?

   a. Attaching multiple drawings to a working session
   b. Performing drawing cleanup operations
   c. Importing/exporting data to other manufacturers' programs
   d. All of the above

9. Drawing Cleanup tools allow for which of the following?

   a. Deleting Duplicate Objects, Erasing Short Objects, and Breaking Crossing Objects
   b. Raster and pixel-based photo editing
   c. OLE cleanup and editing
   d. Altering audio sounds from MP3 and WAV files

10. The **Panorama** performs which of the following functions?

    a. Road design
    b. Site grading
    c. Utility piping layout
    d. None of the above

True or False

1. True or False: GIS capabilities are not part of the Civil 3D solution.

2. True or False: All Civil 3D data can be reformatted into 3D/GPS Machine Control data.

3. True or False: 3D surfaces can be formatted to 3D/GPS Machine Control data.

4. True or False: MAP 3D can link AutoCAD graphics to external databases such as Microsoft Access and Oracle.

5. True or False: MAP 3D can help a surveyor compute a geodetic transformation from NAD 27 to NAD 83.

6. True or False: Civil 3D cannot use the Auto-CAD Sheet Set Manager because it can open only one drawing at a time.

7. True or False: Every icon in the Civil 3D **Toolspace** will invoke an engineering command.

8. True or False: Some icons appear in the **Prospector** when project objects are out of date.

9. True or False: The **Prospector** shows when project items are populated with data.

10. True or False: Users can access project data in various ways through the **Prospector**.

STUDENT EXERCISES, ACTIVITIES, AND REVIEW QUESTIONS

1. What are the major object types Civil 3D uses? Why would we use them?

2. What would be some examples of uses for object technology? How does this compare to parametric technology?

3. What is GIS? Explain what it is used for and how it differs from CAD.

4. What is GPS? Explain what it is used for and how the future of engineering and surveying will be changed by the advances in this area.

5. What is a spatial database as related to GIS?

6. Perform research using this text or other resources and develop a few paragraphs on what a geodetic transformation is.

7. What software components come in the Civil 3D delivery? In other words, what is it comprised of?

8. Do you think the cost of designing engineering projects is more than, less than, or the same as it was thirty years ago? Explain your answer.

9. Explain what alignments are used for; provide examples of use and the benefits they provide us with as designers.

10. Explain what Corridor Profiles are used for; provide examples of use and the benefits they provide us with as designers.

11. If someone were provided a series of stations, offsets, and elevations relative to an alignment, explain what this means and what usefulness it might present. What can be accomplished with this information?

12. Write a few paragraphs on what visualizations (or renderings) can be used for in the civil engineering industry.

13. If engineers work in two dimensions, what limitations do they have compared with someone working in three dimensions? Explain your response.

14. How can laser-guided machine control be used in conjunction with GPS-guided machine control to automate earthmoving tasks?

A Civil 3D JumpStart Project

2

Chapter Objectives

After reading this chapter, the student should know/be able to:
- Perform basic functions of Civil 3D on all facets of a project because the chapter goes through an entire project from start to finish.
- Prepare an existing ground surface from actual 3D project data, create contours, and label them.
- Develop a property parcel for a shopping center and prepare the labeling required for its drafting.
- Develop an alignment for an access road to the shopping center.
- Develop a finished ground profile and a corridor to put the road into 3D.
- Design the grading for the shopping center and the parking facility.

INTRODUCTION

We commence immediately with a design project. The details and advanced usage of the commands will be reserved until subsequent chapters; however, this chapter takes you through an entire project from start to finish, with only a minimal amount of settings manipulations. The reader prepares an existing ground surface from actual 3D project data, creates contours, and labels them. You develop a property parcel for a shopping center and prepare the labeling required for its drafting. Then you develop an alignment for an access road to the shopping center, and a *finished ground* profile and a corridor to put the road into 3D. The next step is to design the grading for the shopping center and the parking facility.

Finished Ground—Proposed ground

REVIEW OF FUNCTIONS AND FEATURES TO BE USED

This chapter uses engineering functions to create a terrain or surface model and show the reader how to use this "Object," display it in different ways, and label it. The project creates a Parcel object from primitive geometry entities and displays it to show directional and distance labels as well as parcel areas. It also develops an alignment from primitive geometry and defines is as an object. Styles will be control its appearance as well as that of the other objects. You will create a *profile view* with an *existing ground profile* and then develop a finished ground profile for the access road again. Assemblies are created from subassemblies and a corridor object is designed using assemblies. Then grading objects are developed for the site of the shopping center.

Profile View—a Civil 3D object

Existing Ground Profile— The **existing ground** vertical alignment

Job skills will benefit from the ability to see rapidly what Civil 3D can do from a broad-brush approach. This chapter and its accompanying procedures go from beginning with existing ground data to a completed, albeit, simple, design project.

PROJECT DESIGN APPLICATION

Part One: Project Design Application

To commence you create an Existing Ground surface to work with. You also set some basic settings to make the surface react and display the way you wish. Your first task is to open a drawing with raw data in it and inspect the contents of the data.

Note:
These data were prepared without the benefit of a Civil 3D template and as a result have little to no settings predeveloped within them. Here some of the important ones will be set. For actual projects it is strongly advised that the product be configured to adhere to your organization's CADD standards.

Exercise 2-1: Open Drawing and Inspect Data

1. Launch Civil 3D and when it opens close the Drawing.dwg that initially opens.
2. Select **File** from the pull-down menu, then **Open**, and choose the Chapter-2a.dwg in your student folder.
3. Notice that there are breaklines and mass points in the file, which together represent the existing ground conditions.
4. On inspection of the data notice that there is a road running east to west. Zoom into the middle of file and you will see a cul-de-sac heading toward the south from the road.
5. Click on a breakline somewhere in the file and notice the grips. Place your cursor in one of the grips, but do not click it. Simply look at the lower left corner of the screen and observe that the X, Y, Z coordinates are shown for the grip. Notice that the Z coordinate or elevation has a value representing the elevation of the surface at that location.

Note:
The mass points are actually AutoCAD blocks floating at the correct elevation. An aerial photogrammetist developed the breaklines using stereodigitizing techniques. The three-dimensional breaklines have elevations at each vertex. The mass points are placed in the data set to densify the surface data.

After having observed the data set, you can prepare a Civil 3D surface for Existing Ground. Before doing that, inspect some important surface-related settings. The first one to change is the Precision of the Contour label. Because we are labeling even contours that are multiples of 2, there is no need to have thousandths place precision on the annotation when whole numbers will suffice. The precision is unneeded and the extra decimal places tend to clutter up the drawing. Additionally you need to create a layer for the annotation to reside on.

Exercise 2-2: Develop Settings

1. Go to the **Toolspace**. Click on the **Settings** tab.
2. In the **Master View**, expand **Drawings**. Then expand **Surface**. Then expand **Label Styles**.

3. Choose the **Contour** label style, and note that one exists called **Standard**.

4. Click on the **Standard** label style, right click the mouse, and choose **Edit**.

5. The **Label Style Composer** dialog box displays. Click on the **information** tab and observe the contents of the panel. Click on the **General** tab and observe the contents of the panel.

6. Click the **Layout** tab. Several properties exist in the **Property** column.

7. Click on **Text**; expand text. Choose **Contents** and then refer to the **Values** column. Click on the value for **Contents**. Choose the button with the ellipses . . .

8. This brings up the **Text Component Editor—Label text**.

9. Click into the **Preview** window on the right side of the dialog box, highlight the data there, and erase them so that the window is blank.

10. Click on **Precision** in the Modify column.

11. The value for the Precision is 0.001, which is not appropriate for your contour label because you do not wish to have decimal points on even contours. Click on 0.001 and select **1** from the pop-down window in the **Values** column.

12. Then click the **blue arrow** at the top of the dialog box to populate the **Preview** window with the new data settings.

13. Look on the **Preview** window and you will see some code indicating that new settings exist. Hit **OK**.

14. In the **General** tab, for the **Layer** property, click on the value where it shows Layer zero and notice that a button with ellipses . . . exists. Choose the button and a **Layer** dialog box appears. Choose **New**. In the **Create Layer** dialog box, click on the value for **Layer** property where it says **Layer1**, and type in a new layer name of **C-Topo-Labl**.

15. Click **Apply** and hit **OK**.

There is one more setting to set. When executing the procedure to label contours, several technicalities are established by the software that will provide benefits later on. One of those has to do with a Surface Control Line shown in Figure 2-1. This linework will appear in the center of the contour

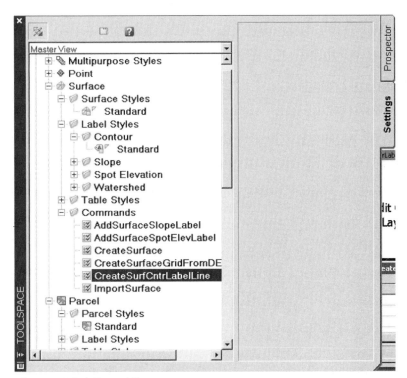

Figure 2-1   Contour Label Control Line

label and can be placed on its own layer so as to be nonintrusive to the user when plotting or negotiating through the drawing. The layer on which this object will fall needs to be set.

Exercise 2-3: Working with Edit Command Settings

1. Go to the **Toolspace**. Click on the **Settings** tab.
2. In the **Master View,** expand **Drawings**. Then expand **Surface**. Then expand the **Commands** item.
3. Click on the item called **CreateSurfCntrLabelLine**. Then right click on the item and choose **Edit Command Settings**. . .
4. A dialog box will display called **Edit Command Settings—CreateSurfCntrLabelLine**. It contains many parameters but the one of interested here is the Default Layer (Figure 2-2).

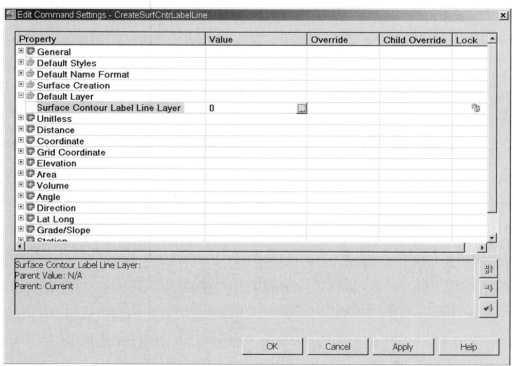

Figure 2-2 Edit Command Settings

NEW
to AutoCAD
2007

5. Within the Edit Command Settings dialog is a new setting called Labeling >> Labeling prompt method. The options are: Command prompt, Suppress prompts, and Dialog. The Command Line means that when a label is being inserted and it contains Referenced Text, a prompt occurs at the command line to select the objects to refer to in the label. The Dialog option inserts a label that also contains Referenced Text; then a Label Properties dialog box is displayed. The Suppress prompts option is used to insert a label that contains Referenced Text; all prompting for objects is suppressed. No objects are specified for Referenced Text fields that may exist in the label style and the label properties can be edited later to specify the objects to which to refer.
6. Select the **Default Layer** under the **Property** column. Click in the **Value** column and hit the button with the ellipses . . .
7. A **Layer Selection** dialog box displays. Click **New** . . .

8. Click in the **Values** column for the Property of Layer and type in **C-Topo-Ctrl-Line.**
9. Click the Value of **Yes** for the **Property Plot** and set it to **No**.
10. Then hit **OK**. Hit **OK** for the **Layer Selection** dialog box. Hit **Apply** and **OK** for the **Edit Command Settings—CreateSurfCntrLabelLine** dialog box.

Next you allocate a surface for the project, develop a surface name, and set the type of surface.

Exercise 2-4: Set Up a Surface

1. Choose **Surfaces** from the pull-down menu.
2. Select **Create surface** . . .
3. The **Create Surface** dialog box appears.
4. Several types of surfaces can be created: a *TIN* surface, a Grid surface, and two types of Volume surfaces.

TIN—Triangulated Irregular Network

> The Volume surfaces are created during earthwork take-off computations. The Grid surface is for when a surface of evenly spaced rectangular data whose elevations are interpolated from the data is desired.

5. We want a **TIN** surface. In the window below is the informational item that includes the name of the surface. Click on the Value column for the Surface name.
6. The Surface name is incremental, so the first surface is Surface1, the second surface is Surface2, and so on. Click in this field and type **Existing Ground** as your name.
7. To the right of the name is the **Layer** field. Pick on the button on the right side of the dialog box for Surface layer. An **Object layer** dialog box appears. Where it says **Base Layer Name**, choose the button on the right for **Layer** and the **Layer Selection** dialog box opens.
8. Hit the button for **New** layer and enter **C-Topo** in the field for **Layer1**. Then where it says **Modifier**, select **Suffix** from the pop-down window.
9. When the field to the right becomes available, type in a dash and an asterisk (-*). This will create a layer with the name of the surface as a suffix to the layer name. You should get a layer called **C-Topo-Existing Ground**, as can be seen in the **Preview** window.
10. Type **Existing Ground** for the **Description** also.
11. Leave all else as default and click **OK**.
12. Next go to the **Toolspace** and choose the **Prospector** tab.
13. Look under **Open Drawings** for your drawing. Expand **Surfaces** to show the new surface you just created for Existing Ground.

The next step is to identify the 3D surface data and build the surface.

Exercise 2-5: Build Surface

1. Expand the **Surfaces** item. You see an item for **Definition**.
2. Expand **Definition** and you see all of the data types that can go into the preparation of the surface. These include **Breaklines, Contours, DEM files,** and **Drawing Objects.**
3. Our data set contains breaklines and blocks, as discussed earlier. Therefore, use the commands for adding **Breaklines** and use **Drawing Objects** to add the blocks.
4. Begin by clicking on **Breaklines**, then right click on **Breaklines** and select **Add** . . .

5. A dialog box for **Add Breaklines** displays. Add a **Description** of **EG**.
6. Use **Standard** breaklines as the type of breakline but notice that other choices exist in the pop-down list.

> The choices allow for **Proximity breaklines, Wall breaklines,** and **Breaklines from a file.** Proximity breaklines are 2D polylines that have points with elevations at each vertex in the polyline. The point elevations will be used by the software in creating 3D breaklines from the 2D polylines. The Wall breakline option is used to represent vertical components of surfaces. In actuality they are not truly vertical because this would cause issues in the software relating to a divide by zero in the denominator for various slope computations. Therefore, the software allows for Wall breaklines that contain wall height data but the top of the wall is slightly offset from the bottom of the wall so as to avoid the vertical computational issue. The last choice is to bring in breaklines from an external text file.

7. Use **Standard** breaklines, which are the 3D polylines in the file. The last setting is a **Mid-ordinate distance** field, which is available for creating 3D breaklines from 2D polylines that have curves in them. In this case if you use the default of 1', a 3D polyline would be created such that small chords simulate the curve but do not deviate from the arc by more than 1'.
8. Hit **OK** and the command prompt instructs you to: `Select objects.` Place a window around the data, and do not worry if there are points mixed in with the selection set of breaklines because the software filters anything that is not a breakline.
9. A **Panorama** dialog box may appear that indicates that there are two crossing breaklines. Coordinates are displayed so the user can check the severity of the issue. In this case it has been determined that this will not harm our surface integrity, and so continue on. Click the green arrow in the top right of the **Panorama** dialog box to continue. Upon completion of this function, the software creates the surface as can be seen by a yellow boundary.

Crossing breaklines can now be found using the "Zoom to" option in the Panorama.

10. Now in the **Prospector**, choose **Drawing Objects** under **Surface >> Definitions** area. Right click and choose **Add . . .**
11. For the **Object type**, choose **blocks**.
12. Type **EG** as the **Description** field in the dialog box and click **OK**.
13. Hit **OK** and the command prompt instructs you to: `Select objects.` Place a window around the data, and do not worry if breaklines are mixed in with the selection set of points because the software filters out the data it does not need.
14. On completion the yellow surface boundary is updated to include the blocks.

Now that the surface is created, you typically want to view contours representing the surface.

Exercise 2-6: Develop Contours for Surface

1. In the **Toolspace**, choose the **Settings** tab. Choose **Surface >> Surface Styles >> Standard**. Right click on **Standard** and choose **Edit . . .**
2. In the **Edit Surface Styles** dialog box, choose the **Display** tab.
3. With the view direction in the pop-down window set to 2D, click on the lightbulb to turn on the Major and Minor contours and turn off the lightbulb for Boundary. Note the colors are preset and can be changed here if so desired.

4. On leaving the dialog box, the contours in Civil 3D are automatically updated to show the contours as defined in the Style.

5. The next step is to create Contour labels on the Contours. Choose **Surfaces >> Labels >> Add Contour labels** from the pull-down menu.

6. A toolbar for performing this function appears and allows for last-minute or on-the-fly changes to the criteria for labeling. Choose the fourth button from the left and click the **Down arrow**. Choose **Multiple Group Interior**. On doing this, a field opens to the right of the **Down arrow**. Type **500** in this field. This instructs the software that a label is to be placed every 500' along the contour.

7. The command prompt then instructs: `Start point:` in order to identify a start point for the labeling to begin. Place a point at the top of the site and place a second point to the bottom and outside of the site.

8. Hit **<Enter>** to terminate the command. On completion of the command, zoom in and inspect the data and the contour labels every 500' across the site.

9. Turn off the layer created in the settings called **C-Topo-Labl-Line**.

10. Save your file to a name with your initials as the suffix of the name, Chapter-2-a-hw.dwg.

Part Two: Project Design Application

Part Two of the exercise is to develop a property parcel. It consists of a fairly simple four-sided shape that is 500 feet wide by 500 feet long. Please note that the software is certainly not limited to this simplicity; rather, this is an introductory exercise and the plan is to keep it simple in order to make several conceptual points clear.

You are going to open the drawing Chapter 2-b.dwg from the CD, a file that contains the completed surface data from the previous set of exercises. The next objective is to develop a parcel boundary in which you design a commercial shopping center. The first step is to create the primitive geometry for the boundary lines. Following that using the Civil 3D functionality will create Parcel objects from the primitive linework.

Exercise 2-7: Develop Geometry for Parcel Boundary

1. Open the drawing called Chapter 2-b.dwg.

2. Begin by creating a layer called **C-Prop-Bndy**; set the color to **yellow** and set it to be the current layer. You draw a polyline as described next. Select the icon for Polyline in AutoCAD or type **PL<Enter>**. The system responds as shown:

Student Files

```
Command: PLINE
Specify start point: <Polar off> 15600,18540
Current line-width is 0.0000
Specify next point or [Arc/Halfwidth/Length/Undo/Width]:
<Coords off> <Coords on> @500<0
Specify next point or [Arc/Close/Halfwidth/Length/Undo/Width]:
@500<90
Specify next point or [Arc/Close/Halfwidth/Length/Undo/Width]:
@500<180
Specify next point or [Arc/Close/Halfwidth/Length/Undo/Width]: CL
```

3. Next address the settings for the boundary to place the Civil 3D parcel on the **C-Prop-Bndy** layer.

4. From the **Parcels** pull-down menu, choose the **CREATE FROM OBJECTS** command. Civil 3D responds: `Select lines, arcs, or polylines to convert into parcels:` Select the polyline you just drew.

5. The **Create Parcels—From objects** dialog box opens (Figure 2-3). Accept the default Site name and Parcel style.

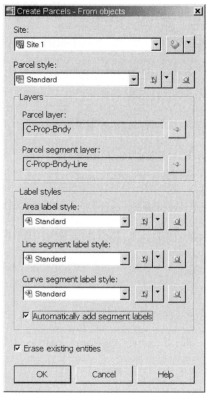

Figure 2-3   Create Parcels—From objects

6. Now select an existing layer for the Parcel layer. In the area for **Layers**, click on the button to the right of **Parcel layer**. This opens the **Object Layer** dialog box. For the parameter called **Base Layer Name:** click on the button to the right for **Layers**. Click on the layer **C-Prop-Bndy**. Hit **OK**.

7. In this example you create a layer to use for the Parcel segment layer. Click on the **Layer** button to the right of **Parcel segment layer** field. This opens the **Object Layer** dialog box. For the parameter called **Base Layer Name:** click on the button to the right for **Layers**. Click on the button for **New** . . . In the Value column for **Layer Name**, type in **C-Prop-Bndy-Line** and then hit the value for Color and select **red**. Hit **OK**.

8. Turn on the **Automatically add segment labels** option at the bottom of the dialog box. Use **Standard** for all of the settings. Ensure that the check mark is **On** for **Erase existing entities**. Hit **OK** to exit and execute.

9. The Parcel object should be created in AutoCAD. Check in the **Toolspace** and click on the **Prospector**. Under **Open Drawings**, expand your current drawing named Chapter-2c.dwg. Expand the **Sites**, and you **Site 1**, which was created when you created the parcel. Expand see **Site 1** and you see **Parcels**; expand **Parcels** to see your parcel, called **Standard _ 100**. If you click on this item, you see the parcel in the right-hand preview window panel. The two accompanying figures, Figure 2-4 and Figure 2-5, show the **Prospector** panel and the AutoCAD data.

10. Close and save your drawing.

Now that there is a parcel boundary for the property, continue on with the design. The next step is to develop an entrance road to the shopping

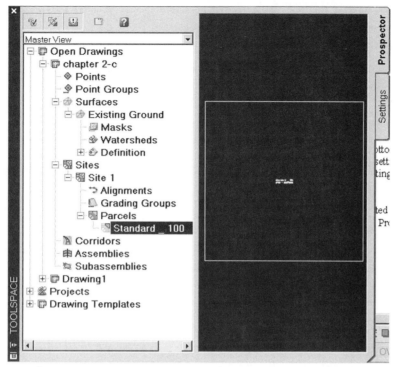

Figure 2-4   Parcel Preview

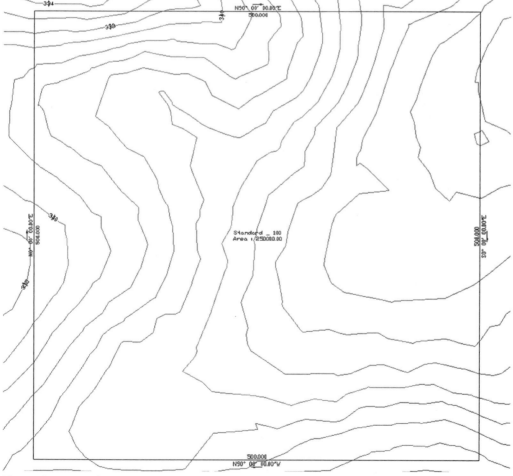

Figure 2-5   Parcel in AutoCAD

mall. You reconfigure the road that runs east to west just south of our parcel boundary. First draw the primitive geometry for the roadway centerline alignment.

Exercise 2-8: Develop a Horizontal Alignment for Access to the Parking Lot

Student Files

1. Open the drawing called Chapter-2c.dwg. It contains the surface you developed and the parcel boundary.
2. Create a layer called **C-Road-Cntr**, set the color to **green**, and set the layer current.
3. Type **L** for **Line** in AutoCAD and when asked for the start point, type **15925,18543 <Enter>**.
4. Then type **@330<S<Enter>**.
5. Then type **@500<N55dE<Enter>**.
6. Then type **@600<E<Enter>**.
7. Next you place curves into the linework using the **FILLET** command by typing **F** for **Fillet**. Type **R** for **Radius** and set the first radius to **100**. Then pick the first two *tangents* that you drew and a 100' radius curve will be drawn. Type **F** for **Fillet** again. Type **R** for **Radius** and set the radius to **500**. Then pick the second and third tangents that you drew and a 500' radius curve will be drawn.

Tangent—A geometry term

8. Type **PE** for **PolyEdit** and when the software asks: Do you want to turn it into one? <Y>, hit **<Enter>**. The software responds with: Enter an option [Close/Join/Width/Edit vertex/Fit/Spline/Decurve/Ltypegen/Undo]:. Type **J<Enter>** and it will ask you to Select the objects. Select the first tangent, the first curve, the second tangent, the second curve, and finally the third tangent. Hit **<Enter>** when done. The geometry for the center of the road is now a polyline.

Alignment—A structured path

9. From the *Alignment* pull-down menu, choose **Create from Polyline** . . . and the software responds with Select polyline to create alignment: Select the polyline for your roadway. On selecting the polyline, the **Create Alignment—From Polyline** dialog box appears. Accept Site 1 for the site, Alignment (<[Next Counter (CP)]>), and the Alignment style of Standard. Now click on the **Layer** button to the right of Alignment Layer. This opens the **Object Layer** dialog box. For the parameter called **Base Layer Name:** click on the button to the right for **Layers**. Click on the button for **New**. . . In the Value column for **Layer**, type in **C-Road-Cntr** and then hit the value for Color and select **green**. Hit **OK** until you exit to the **Create Alignment—From Polyline** dialog box.
10. Use the default for Alignment Label Set of Standard. Click the **down arrow** next to the button to the right of **Alignment Label** set. Click on **Edit Current Selection** (a check mark may already be set here, which is fine). The **Alignment Label Set—Standard** dialog box appears.

Stations—A unit of measure

11. At the top of the dialog box in the **Labels** tab, shown in Figure 2-6, are pop-down windows for the **Type: of label** and the **Major Station Label Style**. Make sure that the **Type** is **Major *Stations*** and that the **Major Station Label Style** is **Standard**. Then, click on the **down arrow** next to the button to the right of the **Major Station Label Style** pop-down menu. Select **Edit Current Selection**. The **Label Style Composer—Standard** dialog box displays. Choose the **General** tab.

    Under the Property column is a Label item. A subitem for the Label is Layer. Select in the Value column for the Layer Property, and see the Layer button with ellipses. . . Select that button to change the layer for labels. Choose the **C-Road-Cntr** layer for the labels. Hit **OK**. Hit **Apply** and **OK** for the **Label Style Composer—Standard** dialog box.

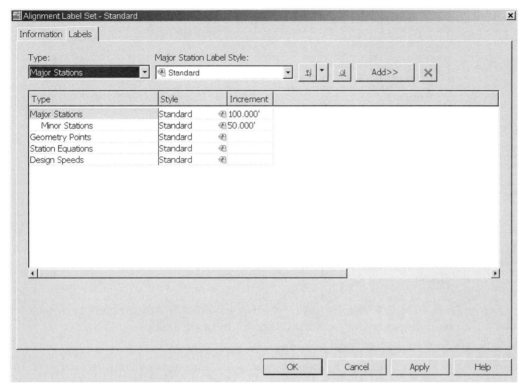

Figure 2-6   Alignment Label Set—Standard

12. Next, change the **Type** to **Minor Stations** and ensure that the **Minor Station Label Style** is **Standard**. Click on the **down arrow** next to the button to the right of the **Minor Station Label Style** pop-down menu. Select **Edit Current Selection**. The **Label Style Composer—Standard** dialog box displays. Choose the **General** tab. Under the **Property** column is a **Label** item. A subitem for the Label is **Layer**. Select in the Value column for the **Layer Property**, and see a button with ellipses. . . Select that button to change the layer for labels. Choose the **C-Road-Cntr** layer for the labels. Hit **OK**. Hit **Apply** and **OK** for the **Label Style Composer—Standard** dialog box.

13. Next, change the **Type** to **Geometry Points** and ensure that the Major Station Label Style is **Standard**. Click on the **down arrow** next to the button to the right of the **Geometry Points Label Style** pop-down menu. Select **Edit Current Selection**. The **Label Style Composer—Standard** dialog box displays. Choose the **General** tab. Under the **Property** column is a **Label** item. A subitem for the Label is **Layer**. Select in the Value column for the **Layer Property**, and see a button with ellipses. . . Select that button to change the layer for labels. Choose the **C-Road-Cntr** layer for the labels. Hit **OK**. Hit **Apply** and **OK** for the **Label Style Composer—Standard** dialog box.

14. Hit **Apply** and **OK** to exit the **Alignment Label Set—Standard** dialog box.

15. In the area for **Conversion Options**, turn off the toggle for **Add curves between tangents**, because you have already added them using the **FILLET** command. Hit **OK**. Chapters 6 and 7 on road design goes into detail about road design works, whereas the intent of this exercise is to prototype a site rapidly. See Figure 2-7.

16. You should see a roadway alignment, fully labeled on the **C-Road-Cntr** layer. Observe in the **Prospector** that Alignment-(1) exists under the

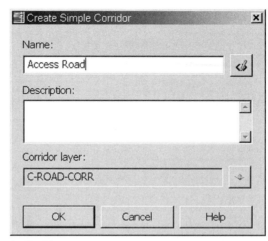

Figure 2-7   Create Simple Corridor

Sites, Site 1, Alignments item. If you click on **Alignment-(1)** a preview of the alignment shows up in the panel to the right.

17. Close and save your drawing.

Now there is a horizontal alignment for the access road into the shopping area. The next task is to develop a vertical profile for the roadway you just created. To accomplish this develop a Profile along the alignment and computed from the Existing Ground surface. It will be placed inside a Profile View, which is the Civil 3D object that displays profiles. Chapter 6 on Profiles discusses this in more detail, but in summary the Profile View is a Civil 3D concept that contains the grid that profiles are drawn in. The Horizontal Axis of the grid reflects the *stationing* or linear distance along the horizontal alignment and the Vertical Axis represents elevational data. The View will include the existing ground profile with annotation. The settings in the **Toolspace** that control the Profile View are in the Profile View Style, which sets the display and spacing of gridlines, profile titling, and the annotation of stations and elevations within the grid. The Profile View Label Styles control the display of text and annotation. Once you develop the Profile View and the Existing Ground profile, then you develop a finished *grade* profile for our road.

Stationing—A unit of measure

Grade—A change in elevation

Profiles—A structured vertical path

Exercise 2-9: Develop a Profile View

1. Open the drawing called Chapter-2d.dwg. It contains the surface you developed, the parcel boundary, and the roadway centerline.
2. From the *Profiles* pull-down menu, click **Create Profile from Surface** and the **Create Profile from Surface** dialog box appears. In the Alignment pop-down, Alignment-(1)(1) should already be there from the alignment creation performed earlier.
3. Click on the **Existing Ground** surface in the window on the right. Click the **Add>>** button to send it to the bottom window for calculation. The station range should be correct from the alignment computations.
4. Now click on the **Draw in Profile View** button.

TIP   If you hit **OK**, the computation is active but has no place to display the results because the Profile View has not been established.

5. The **Create Profile View** dialog box appears. In the **Profile view name** field, type **Access Road**, and from the **Alignment** pop-down window, select Alignment-(1)(1). It should already be in the window. Accept the default layer name of **C-Road-Prof**. In the left center of the dialog box is a pop-down window called **Profile view style**. Choose the style called **Major Grids**. This style can be edited by the user if changes to the display of the Profile View are needed. Accept the defaults.

6. For the **Station Range** and the **Profile View Height**, accept the defaults. Click **OK**.

7. The software responds by asking you to: Select profile view origin. Type **18000,21000** as the coordinate for the location of the Profile View. This is up and to the right of the Plan View. Zoom into the grid representing the Profile View. The evenly spaced horizontal and vertical lines are the grid lines with spacing inherited from the **Major Grids** settings. Vertical blue lines are drawn wherever keypoints for the alignment's horizontal geometry exist, such as P.C.'s and P.T.'s.

8. Save and close your drawing.

Draw the Existing Ground Profile in the Profile View

The Profile View displays surface profiles and also provides the grid on which you can develop a layout profile or finished grade profile. Each Profile View displays new and/or existing profiles and offset profiles for a single horizontal alignment. Profiles can be created without choosing to display them in the Profile View. When a Profile View is developed, a complete list of profiles for the alignment in question is available so that it can be selected for inclusion in the view.

We have succeeded in developing a Profile View and the Existing Ground profile for the Access Road (or Alignment-[1][1]). The next task is to create a finished ground profile in the Profile View. Begin with the ***vertical tangents***, which are primitive geometry elements for the profile. At the Points of Vertical Intersection (or ***PVI***s), you add vertical curves. These curves differ from the horizontal alignments in that they are not simple arcs; rather they are parabolic curves. The curve on a hill is known as a crest curve, whereas the curve in a depression is known as a sag curve. The mathematics behind ***vertical curves*** will be discussed in Chapter 6 on profiles.

Vertical Tangents—Vertical grades

PVI—Points of Vertical Intersection

Vertical Curves—Parabolic curves

Exercise 2-10: Add Vertical Tangents

1. Open the drawing called Chapter-2e.dwg.

2. From the **Profiles** pull-down menu, choose the command **CREATE BY LAYOUT** and the software responds by asking you to: Select profile view to create profile. Pick the text box above the profile and the **Create Profile** dialog box appears.

3. Notice that it knows you are dealing with Alignment-(1)(1). In the **Name** field, type **Access Road FG**. Type **Access Road FG** in the **Description** field as well.

4. At the bottom of the dialog box in the pop-down window for **Profile Label Set**, choose the style called **Complete Label Set**. Accept the default for the **Profile Layer** option. Accept the default **Design Style** for the **Profile Style** option. Hit **OK**.

5. The **Profile Layout Tools—Access Road FG** toolbar opens up and the software command prompts the user to: Select a command from the Layout tools. Choose the first icon button for **Draw Tangents without Curves**. The software says to: Select the Start Point: Pick an Endpoint snap on the leftmost end of the Existing Ground linework. Choose your second point, when prompted at about station 3+50 and elevation 320. Choose your third point at about station 7+00 and elevation 345. The

Student Files

last point for the Access Road will be an Endpoint snap to the right most part of the Existing Ground linework. Hit <**Enter**> to terminate the sequence.

TIP

> When estimating the location of the stations and elevations at which to place these points, use the tooltips that pop up when the cursor is paused. Do not allow the cursor to be snapping or it tells the snap being computed. Rather, just place the cursor near a line but not on it, and the tooltip will pop up to tell you the station and elevation it is at. Use this to estimate the station and elevation requested in the procedure.

Exercise 2-11: Add Vertical Curves

Now add vertical curves to the profile tangents. Although the software does support both symmetrical as well as asymmetrical vertical curves, you use symmetrical at this point.

NEW
to AutoCAD
2007

The software now introduces the concept of Fixed, Free and Floating Vertical curves. This concept is similar to the philosophy for horizontal curves and lines in 2006 and is discussed in the Geometry chapter. Basically the concept involves "Constraint-based" commands and the creation of Fixed, Floating and Free vertical tangents and vertical curves. These objects can be edited dynamically and retain their tangencies.

A Fixed entity is fixed in its position and is defined by criteria such as a radius or located points. It is not dependant on other entities for geometry development or tangency. A Floating entity is always tangent to one entity and is defined by the parameters provided or is dependent on one entity to define its geometry. A Free entity is always tangent to an entity before and after it and must have at least two other entities to attach to which define its geometry.

1. From the **Profile** pull-down menu, choose the command **EDIT PROFILE**.
2. When prompted to select the profile, click on the finished ground profile tangents just drawn and the **Profile Layout Tools** toolbar appears. Click on the fifth button from the left; the tooltip should say **Add a Parabolic Curve**.

NEW
to AutoCAD
2007

Click on the sixth button from the left; the tooltip should say "about drawing a Parabola." This prompt will change as do others in the software, depending on the choice last selected by the user. Choose the option for Free Vertical Curve by Parameter. The software's response is to: `Select an entity before (and then after) to attach the curve to:`. Click near the first incoming tangent to the PVI and then the outgoing tangent.

3. The software's response is to: `Pick a point near a PVI or curve to add a curve:`. Click near the first PVI with the mouse.
4. The prompt says to: `Specify Curve Length or [Passthrough/K]`. Type in **300<Enter>**.
5. The software then assumes you will continue with additional PVIs and responds with: `Select an entity before (and then after) to attach the curve to:` again. Click near the second PVI and the prompt again says to: `Enter Curve Length or [Passthrough/K]`. Type in **300<Enter>**. <**Enter**> terminates the routine.
6. You see two vertical curves now added where the PVIs are.
7. Save and close the file.

Exercise 2-12: Create a Corridor for the Access Road

Now develop the remaining part of the road, that is, the finished ground *cross section* conditions for the pavement. Civil 3D introduces a corridor design capability that completes the road design cycle. It builds on the horizontal and vertical alignments that you have worked on so far. A *corridor* is a Civil 3D object, and as such is linked to the horizontal and vertical alignments, surfaces, and assemblies that were used to create it, so if one of these components is altered the corridor is correspondingly modified. The next step is to modify subassembly properties for your design criteria, create an *assembly* from the *subassemblies*, and create the corridor model.

> **Note:**
> Subassemblies are the primitive components of assemblies, and they instruct the software in how to handle lanes, curbs, guardrails, or shoulders and related side sloping. Autodesk provides many subassemblies to begin with and the user accesses them from Tool Palettes. The assemblies, on the other hand, are stored in the drawing and can be accessed from the **Project Tool-space** under **Assemblies**. They contain the definition of the typical section being designed and are developed by piecing together subassemblies.

TIP   These subassemblies are some of the first examples of parametric technology being introduced to the civil engineering community. Once these objects are created, the parameters within them can be changed on the fly. The dynamic modeling abilities of the software then can cause a ripple-through effect as the changes are applied to all related objects in the corridor.

Cross Section—A transverse view of the surface

Corridor—A swath

Assembly—Proposed roadways section

For our Access Road we have the following design requirements: The road will be a crowned section with a pavement width of 12 feet. The cross slope of the lane is −2.08% from the crown to allow for drainage off the roadway when it rains. A shoulder for this road will be 4 feet wide with a cross slope of −4% from the edge of pavement. The side sloping from the shoulder will be 3:1 until it hits existing ground.

> **Note:**
> A 3:1 side slope is a condition describing an elevation rise of 1 foot for every 3 feet horizontally and is usually denoted by this ratio.

Exercise 2-13: Set Up Toolspace and Subassembly Settings

1. Open the drawing called Chapter-2f.dwg.
2. A **Tool** palette should be open in Civil 3D that says **Civil 3D Imperial**. If it is not displayed, then click on the **General** pull-down menu; there is a **TOOL PALETTE** window command that should be turned on. Then right click on its spine to select **Civil 3D Imperial**.

Student Files

TIP   Depending on your software installation, you may already have a **Toolspace** configured with the catalogs built into it. Often the software installs without the rehabilitation commands. Let us add these to the **Toolspace Palette**, which also provides an exercise of how to configure the palettes. If you already have Roadway Rehabilitation and Reconstruction tools in the **Tool** palette, skip these steps and go to the next exercise, called Modify Subassemblies for Our Access Road.

Subassembly—Fundamental building blocks for assemblies

3. Right click somewhere on the spine of the **Tool** palette and select **New Tool Palette** from the menu that appears.
4. A small field opens, allowing you to provide a name for the new **Tool** palette. Enter **Roadway Rehab** for the new **Tool** palette name. You now

see a new tab, albeit an empty one, in the palette with the name you provided. Now you can populate this tab with subassemblies for road rehab.

5. Go to the **General** pull-down menu and select **Catalog. . .** and a window containing a library of catalogs opens. Click the picture that says **Corridor Modeling Catalogs (Imperial)**. This opens a window that has five different catalogs that can be used for a variety of road conditions. Begin with the one called **C3D Imperial Rehab Subassembly** catalog.

There are now 6 Catalogs, with the new one being Channels and Retaining Walls. This can also be invoked under the Corridors pulldown.

6. Several items, two pages' worth actually, should be available now. Take them all for this session. Right click in the window and choose **Select All**. They all highlight.

7. Click the **i-drop** symbol on any of the highlighted subassemblies and drag the selection over to your empty **Tool** palette. Release the mouse button and drop the selected assemblies onto the new **Tool** palette.

8. Next, click the hyperlink at the top of the Catalog page named **Catalog Top**. Click on the selection called **C3D Imperial Generic Subassembly** catalog. This opens a two-page window of link possibilities. Click on 2 to choose page 2 of 2.

9. Again, take them all for our session. Right click in the window and choose **Select All**. They all highlight.

10. Click the **i-drop** symbol on any of the highlighted subassemblies and drag the selection over to your empty **Tool** palette. Release the mouse button and drop the selected assemblies onto the new **Tool** palette.

11. Using the i-drop, drag this item into the **Toolspace** tab you created. This adds these to the ones you brought across a moment ago.

12. Close the Catalog. Your **Tool** palette is now ready to assist in performing reconstruction and rehabilitation computations.

These acts have created a **Toolspace** palette that can be used for rehab design for roadway corridors, and it will remain in place for future projects when you open Civil 3D.

Exercise 2-14: Modify Subassemblies for Our Access Road

At this point you should have a Toolspace configured with the subassembly tools needed to proceed. Let us move along and modify the properties of the subassemblies needed for the road. They include the **BasicLane**, the **BasicShoulder**, the **BasicGuardrail**, the **BasicSideSlopeCutDitch**, and the **LinkOffsetAndSlope** subassemblies to develop our Access Road design criteria. The **LinkOffsetAndSlope** subassembly can be found in the **Imperial Generic** palette.

1. From the **Basic Roadway** palette, right click on the **BasicLane** subassembly and select **Properties**.
2. Look under the **Advanced** menu parameters at the bottom of the **Tool Properties** window.

Feel free to stretch this dialog big enough to see all of the elements so you can access them.

3. Set the **Side** to **Right** and notice the lane width is 12'. Alter the parameter **%Slope** to **−2.08**. Set the depth to **1.0**. Then set the **Side** to **Left** and notice the lane width is also 12'. Alter the parameter **%Slope** to **−2.08** and the depth to **1.0**. Click **OK** to close it.

4. Now right click on the **BasicSideSlopeCutDitch** subassembly and choose **Properties** from the pop-up menu.
5. At the bottom of the **Tool Properties** window, look under the **Advanced** menu parameters. Set the **Side** to **Right**.
6. Change the cut slope to **3.0** and the fill slope to **3.0**.
7. Now set the **Side** to **Left** and change the cut slope to **3.0** and the fill slope to **3.0**. Hit **OK** to close.
8. Now right click on the **BasicShoulder** subassembly and choose **Properties** from the pop-up menu.
9. At the bottom of the **Tool Properties** window, look under the **Advanced** menu parameters. Set the **Side** to **Right**.
10. Change the width to **4.0** and ensure that the **%Slope** is **−4.0**. Set the **Side** to **Left**. Change the width to **4.0** and ensure that the **%Slope** is **−4.0**. Hit **OK** to close.
11. These three subassemblies are now set up for your design. For the fourth subassembly, **LinkOffsetAndSlope**, change its properties as you place it to show that they can be set on the fly.

Now you can create the assembly from the subassemblies on the **Tool** palette. The assembly contains the design control for the proposed typical section of the road.

Exercise 2-15: Create the Assembly

1. Type **Z** for **Zoom** at the command prompt and hit **<Enter>**. Type **C** for **Center** and type **17000,20000**. When asked for a height, type **50**.
2. This has placed you to a location where the assembly can be created. There is nothing special about this location; it is simply out of the way and zoomed to a height of 50'.
3. From the **Corridors** pull-down menu, select **Create Assembly**. The **Create Assembly** dialog box opens.
4. Enter the name as **Access Road**. Leave the remaining settings as default and hit **OK**.
5. The command prompts you to: Specify assembly baseline location. Pick a point in the middle of the screen, and you see a small symbol with a red line show up. This indicates the assembly location to which you will attach subassemblies.

If you look in the **Prospector** and expand the tree for **Assemblies**, you see the assembly name. Now attach the subassemblies to both the left and the right sides of the assembly location.

Exercise 2-16: Attach Subassemblies to the Assembly Marker

1. From the **Access Road Sections** palette, click the **BasicLane** subassembly. The properties for the subassembly can be altered at that time. Look in the **Properties** palette and ensure that the parameter for **Side** is set to **Right**. Set the depth to **1.0**. Ensure that the width is set to **12.0** and the **%Slope** is **−2.08**. You set these earlier so they should be correct.
2. The command prompts to: Select marker point within assembly. Pick the assembly baseline location, which is the circular symbol in the middle of the assembly that you created. The lane should show up on the right side of the assembly marker.
3. Now look back to the **Properties** palette and change the **Side** from **Right** to **Left** and place the **BasicLane** subassembly at the circular symbol on the assembly marker again. The left side should now have a **BasicLane**.
4. From the **Access Road Sections** palette, click the **BasicShoulder** subassembly. Look in the **Properties** palette and set the parameter for **Side** to **Right**. Set the depth to **0.67**. Ensure that the width is set to **4.0** and

the **%Slope** is −**4.00**. The command prompts to: `Select marker point within assembly`.

5. Pick the perimeter edge of the circle symbol located at the top right corner of the **BasicLane** subassembly. The shoulder should extend to the right of the lane.

6. Now look back to the **Properties** palette and change the **Side** from **Right** to **Left** and place the **BasicShoulder** subassembly at the top left corner of the **BasicLane** subassembly. The left side should now have a **BasicShoulder**.

7. At this point, the assembly should have a left and right travel lane and shoulders on both sides.

8. From the **Tool** palette, click the **BasicGuardrail** subassembly. Look in the **Properties** palette and select **Right** side from the pop-down parameter for **Side**. The command prompts to: `Select marker point within assembly`.

9. Pick the perimeter edge of the circle symbol located at the top right corner of the **BasicShoulder** subassembly. The guardrail should extend to the right of the shoulder.

10. Look in the **Properties** palette and select **Left** from the pop-down parameter for **Side**. The command prompts to: `Select marker point within assembly`.

11. Pick the perimeter edge of the circle symbol located at the top left corner of the **BasicShoulder** subassembly. The guardrail should extend to the left of the shoulder.

12. From the **Tool** palette, click the **LinkOffsetAndSlope** subassembly. Again, the **Properties** dialog box opens each time a subassembly is placed. In this case, ensure that the **Slope Type** is set to **%Slope** and that the **%Slope** is set to −**2**. Also set the **Offset from Baseline** to **22.0** (which produces a 6' area past the guardrail before sidesloping kicks in). A positive number for the offset indicates that it heads off to the right, whereas a negative offset indicates to the left.

13. Then pick the perimeter edge of the circle symbol located at the right, center **BasicGuardrail** subassembly. The Link should extend to the right of the guardrail.

14. Look back into the **Properties** palette and set the **Offset from Baseline** parameter to −**22.0**, so it will be applied to the left.

15. Then pick the perimeter edge of the circle symbol located at the left, center **BasicGuardrail** subassembly. The Link should extend to the left of the guardrail.

16. From the **Tool** palette, click the **BasicSideSlopeCutDitch** subassembly. Review the settings for the parameters in the **Properties** palette and accept the defaults. Set the **Side** to **Right**.

17. Then pick the perimeter edge of the circle symbol located at the right, center **LinkOffsetAndSlope** subassembly. The Link should extend to the right of the Link.

18. Go back to the **Properties** palette and set the **Side** to **Left**.

19. Then pick the perimeter edge of the circle symbol located at the left, center **LinkOffsetAndSlope** subassembly. The Link should extend to the left of the Link.

20. Press <**Enter**> twice to finish the command.

You have now created the assembly. Figure 2-8 shows how it should appear. Keep in mind that the designer can build the right side independently of the left if he or she chooses. The assembly does not have to be symmetrical, and on actual projects, it often may not be symmetrical. If a developer is developing a project on the east side of the road, it may proffer to improve that side of the road while allowing a future developer to improve the west side of the road.

Figure 2-8   Access Road Assembly

The assembly for the Access Road is now complete. It should appear similar to Figure 2-8.

Save the File. Now move forward to creating the roadway. In this task, the pieces to the Corridor puzzle come together. You have all of the components for the Corridor. To sammarize, you developed the primitive geometry for the alignment and developed a completed alignment. You then computed and drew an Existing Ground profile inside a Profile View. From there you designed a finished ground profile for the Access Road. You then set the settings for some subassemblies and created an assembly. Now create the Corridor and the finished design surface.

Exercise 2-17: Create a Corridor

Student Files

1. Open the drawing called Chapter-2g.dwg.
2. We will restore a pre-saved view to place our view around the centerline.
3. Type **V<Enter>** for **View** at the command prompt to obtain the **View** dialog box. Double-click the **Access Road Plan View** and hit **OK**. You are at the centerline of the Access Road. There is also have a view setup for the profile should you need it.
4. From the **Corridor** pull-down menu, select **Create Simple Corridor**.
5. Enter **Access Road** for the name and hit **OK**.
6. The command prompt requests that you: Select the baseline alignment. Pick the green centerline in the view.
7. The command prompt then asks you to: Select a profile (Figure 2-9), or hit <Enter> to obtain a list of available profiles.

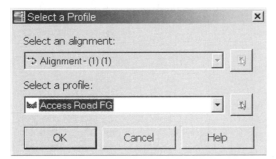

Figure 2-9   Select a Profile

8. Hit <Enter> and the **Selection** dialog box appears. Choose the **Access Road FG** and hit **OK**.
9. The command prompt then asks you to: Select an assembly (Figure 2-10), or hit <Enter> to select it from a list. Hit <Enter> and select **Access Road** from the list.

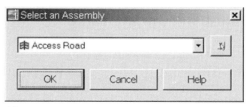

Figure 2-10   Select Assembly

10. A large window of data will display for **Logical Name Mapping**, as in Figure 2-11.

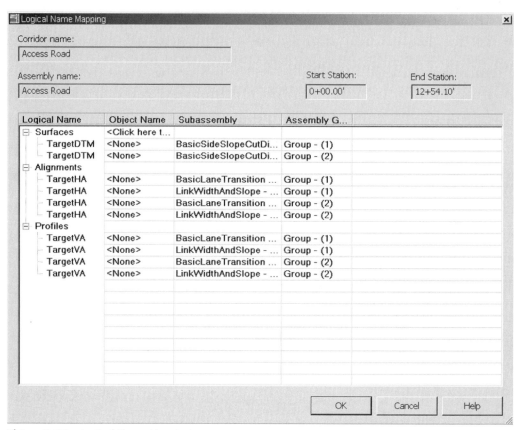

Figure 2-11   Logical Name Mapping

Exercise 2-18: Logical Name Mapping

Logical name mapping is a technical way of assigning the particular alignments, profiles, and surfaces required by the subassemblies to create the complete assembly. For instance, the **BasicSideSlopeCutDitch** subassembly needs to know which surface it will be computing to, and so the Logical Name Map for this would be the Existing Ground surface.

1. In the second column and in the row for **Surfaces** is a cell that says `<Click here to set all>`.
2. When you click on the cell, a dialog box called **Pick a Surface** displays. Choose the **Existing Ground** and click **OK**. You will see the Target DTM is now Existing Ground. Click **OK** and the corridor will generate, as shown in Figure 2-12.

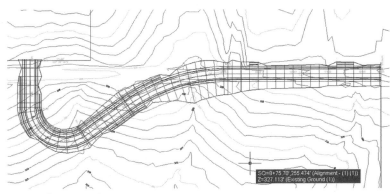

Figure 2-12   Corridor Created

3. Close the **Panorama** if it appears and inspect the results of the process-ing. A Corridor model will be evident in the drawing.
4. Restore a pre-saved view to place your model in a bird's-eye perspective with shading to help see the road better. Type **V<Enter>** for **View** at the command prompt to obtain the **View** dialog box. Double-click the **ISO View** and hit **OK**.
5. You see the road is 3D, complete with representation of the guardrails, as shown in Figure 2-13.

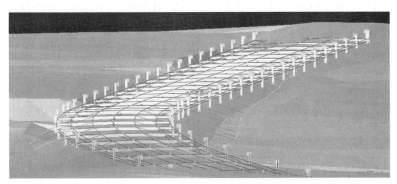

Figure 2-13   Corridor ISO

6. Type **ZP<Enter>** to return to the previous view when you are finished viewing.
7. Save and close the file.

Exercise 2-19: Viewing Additional Results of the Corridor Design

The Corridor model is now built, but it is necessary to discuss a few meth-ods that exist to review the model. One is to investigate what the finished sections look like. Follow this procedure to observe the section conditions.

1. Open the drawing called Chapter-2h.dwg. The previously created corri-dor is shown.
2. Under the **Corridors** pull-down menu, choose **View Corridor** section. The command prompt asks you to: Select the Corridor. Pick any part of the corridor that you can.
3. Once the corridor is selected, the first station of the section displays along with a **Control** toolbar that allows you to move up and down the alignment, viewing any section desired. This is displayed in Figure 2-14.

Student Files

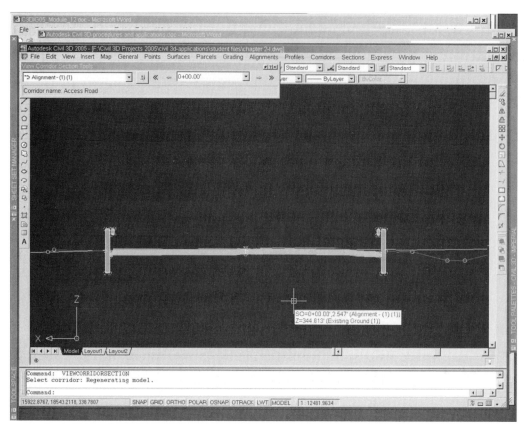

Figure 2-14   Corridor Section

4. Take a few moments to zoom in and out and review the section that appeared. Notice where the lanes, the ditch, and the tie-out slopes are.
5. Then move forward to the next station and observe the display.
6. Continue observing and when you are ready, close the toolbar; the normal view will return as it was before you entered this facility.

Feature Lines—Civil 3D objects running inside corridors or in gradings

Many aspects of the Corridor model can be manipulated and many of them will be in the Chapter 7 on corridors. However, here is a brief example of the powers. You follow a procedure that identifies some basic *feature lines* of the corridor, such as the centerline, the edge of the pavement (or EP), the bottom of the ditch, and the daylight cut and fill lines. These will be inspected for quality control purposes, but they serve a purpose later in the establishing construction information because they can also be exported as alignments or grading feature lines. Grading feature lines are discussed in Chapter 8 on site grading.

Exercise 2-20: Corridor Feature Lines

1. Select the Corridor model by picking it and it will highlight. Right click and select **Corridor Properties** . . . from the shortcut menu.
2. The **Corridor Properties** dialog box displays. Select the tab for **Feature Lines**, Figure 2-15.
3. By default, all of the feature lines are assigned to the **Standard** feature line style. In order to see some of the objects within the corridor better, assign Feature Line Styles to the Daylight_Cut, and the Daylight_Fill, and the Edge of Travel Way (ETW) feature lines. When

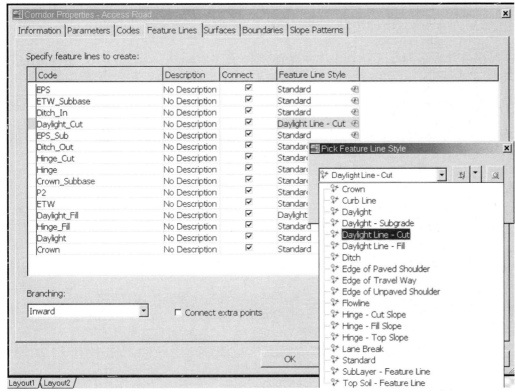

Figure 2-15   Feature Lines

you return to Civil 3D, they will appear differently as a result of this style modification.

4. In the **Code** column, pick the **Daylight_Cut** code and look in the **Feature Line Style** column. Pick the **blue** icon in this column and a dialog box called **Pick Feature Line Style** will display. In the pop-down window, select the style called **Daylight Line—Cut,** and hit **OK**.

5. In the **Code** column, pick the **Daylight_Fill** code and look in the **Feature Line Style** column. Pick the **blue** icon in this column and a dialog box called **Pick Feature Line Style** will display. In the pop-down window, select the style called **Daylight Line—Fill,** and hit **OK**.

6. In the **Code** column, pick the **ETW** code and look in the **Feature Line Style** column. Pick the **blue** icon in this column and a dialog box called **Pick Feature Line Style** will display. In the pop-down window, select the style called **Edge of Travel Way**.

7. Then click the **down arrow** button on the right side of the pop-down window. Choose **Edit Current Selection** from the options (Figure 2-16).

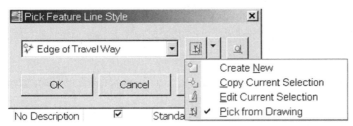

Figure 2-16   Edit Current Selection

8. The **Feature Line Style** dialog box will display. Choose the **Display** tab. Change the **Feature Line Color** to **Yellow** and change the **Lineweight** to **0.8mm**. Hit **OK**.

9. Hit **OK**. Hit **OK** one more time to exit the main **Corridor Properties** dialog box. The system will regenerate the Corridor. To see all of the alterations, click the **LWT** button at the bottom of the screen to **ON**.

10. The corridor should appear as shown in the graphic. The cut is red and fill is shown as green per the alterations just made to the display of the corridor. The edge of pavement is yellow and a heavy lineweight. If the **Panorama** displays, close it.

11. Save the file and close the drawing.

At this point in the project you have created the terrain model, a parcel for the design of our site, an alignment, a profile, an assembly, and a corridor that provides access to the commercial shopping center being designed. Refer to Figure 2-17. Some basic adjustments to the display of our data have been made along the way.

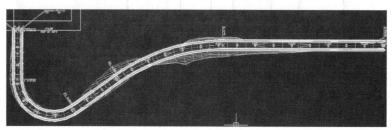

Figure 2-17   Cut and Fill Colored

Site Design

The next task is to create some building pads for the stores and a parking lot for the commercial site being designed. The design criteria for the site are as follows:

BRL—Building Restriction Line

- A Building Restriction Line (**BRL**) of 25 feet prevents any structures from being within 25 feet of the property line.
- The finished floor of the stores will be at 350 feet.
- The parking lot slope is indicated with callouts in the drawing or with points placed at various critical locations in the file. You will draw 3D Polylines to help guide the drainage over the parking lot.
- The parking lot ties into the access road edges of pavement at 344.58 feet. The crown of the Access Road is at 344.78 feet.
- A 6-foots wide sidewalk with a −2% grade exists around the stores except on the north and west sides.
- The majority of the Drainage flow will be directed toward the Access Road, where it will be conveyed until station 3+56, at which a drain inlet will be placed to divert it offsite.
- The site ties out to natural ground at a 3:1 side slope.
- A layer in the drawing called **Site Geometry** contains the buildings for the shopping center.
- A layer called **Site Drainage Criteria** indicates where the Feature Lines will go that will guide the drainage and become Feature Lines.

Exercise 2-21: Design the Shopping Center Building Finished Floor Slabs and the Parking Lot

1. Open the drawing called Chapter-2i.dwg. The previously created corridor is shown with the visual adjustments that made to it. There are

buildings in the file with their elevation set to 0.0 and an accompanying 6' wide sidewalk.

2. Change the properties of the buildings to place them on the **3D-Data** layer. To do this, double-click the building rectangle and the **Properties** dialog box will display. Change the elevation parameter to **350.0**. Hit **Cancel** to escape from the task.

3. Repeat this for the sidewalk, but make the elevation **349.88**, which represents a −2% drop over 6'.

4. Then offset the edge of sidewalk away from the buildings using the **OFFSET** command and a distance of **0.4'**. This represents the Top of Curb of the sidewalk.

5. Then offset the Top of Curb again, away from the building, a distance of **0.1'**. This now represents the Flowline of the Curb.

6. Double-click the Flowline of the Curb to open the **Properties** dialog box and change the elevation to **349.38**, which is a 0.5' drop from the Top of Curb.

7. You will see that this coincides with two points (Points 1 and 2) placed in the file at the same elevations, that is, 349.38. These two points represent the start of our flow divide and now you can see how they were arrived at.

8. Ensure that your current layer is set to **3D-Data**.

9. Type in **3P** at the command prompt to invoke the **3DPOLY** command. Use a Node snap and snap to Point 1 for the first point. Use a Node snap and pick Point 8. Continue by picking another Node snap and selecting Point 9. Hit <**Enter**> to terminate. You might notice that the 3D Polyline is now set to 3D due to snapping to the nodes that had elevations.

10. Type in **3P** at the command prompt to invoke the **3DPOLY** command again. Use a Node snap and snap to Point 2 for the first point. Use a Node snap and pick Point 8. Hit <**Enter**> to terminate.

11. Type in **3P** at the command prompt to invoke the **3DPOLY** command again. Use a Node snap each time to snap to Point 4, Point 5, Point 6, and Point 7. Hit <**Enter**> to terminate.

At the north end of the side, notice that Point 3 has been placed at an elevation of 349.6. This represents a −2% cross-slope over that pavement distance. This pavement area is for the loading and unloading of trucks and supply vehicles. The asphalt pavement flows away from the buildings at −2% and then at −1% to the east and west of Point 3. Now you will create grading for the perimeter of the site and parking lot.

12. In the **Grading** pull-down menu, select the **CREATE FEATURE LINES** command. Choose the two cyan-colored lines touching Point 3, representing the perimeter of the parking lot.

   When the dialog displays click the Style field On and select Edge of Travelway for the feature lines.

13. Each of these are grading feature lines. Click on the West feature line and right click. Select the **Elevation Editor**...

14. Refer to Figure 2-18 and set the first elevation to **349.6** if it is not already set. Then establish all of the grades for the segments at −0.5%, except for the last segment. Set the last segment to −1%, which should set the last elevation to **344.72**.

15. Click on the Eastern feature line and right click. Select the **Elevation Editor**...

16. Refer to Figure 2-19 and set the first elevation to **349.6** if it is not already set.

17. Then in the **Grade** column, set the first segment's grade to −**1.0%**.

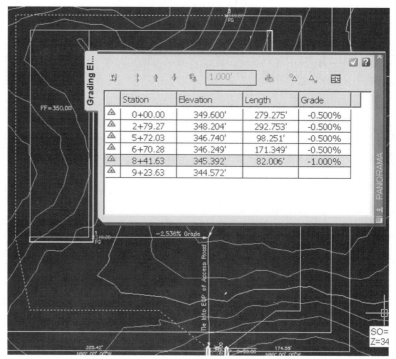

Figure 2-18   Elevation Editor

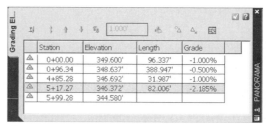

Figure 2-19   Elevation Editor Data

18. Set the next segment to **−0.5%**.
19. Set the next segment to **−1.0%**.
20. Set the last segment to **−2.185%**, which should set the last elevation to **344.58**.

The 3D linework should now all reside on the **3D-Data** and the **C-Topo-Grad-Flin** layers for easy selection later. Next set the grading for the perimeter of the site as it ties out to natural ground.

Exercise 2-22: Establish a 3D Grading Perimeter

1. From the **Grading** pull-down menu, select the **Grading Layout Tools**. These are now called Grading Creation Tools. A toolbar called **Grading Layout Tools** will display, shown in Figure 2-20. This is now the Grading Creation Toolbar.

Figure 2-20   Grading Layout Tools

A new button was added to the toolbar to allow for editing and managing the grading criteria. It was placed right after the pop-down window for criteria.

Use of this tool is detailed in our Chapter 8 on site grading; however, it is necessary to set some of the parameters now in order to move forward with this exercise. The first button ⬛ from the left sets the **Grading Group**. The second button ⬙ sets the **Target Surface**. The third button ≋ sets the **Grading Layers**.

2. Select the first icon button ⬛ from the left of the toolbar. This invokes the **Create Grading Group** dialog box to allow you to name the Grading Group. Clear the field at the top and type in **Parking Lot Grading Group**. Then hit **OK**. You may note that this creates Parking Lot Grading Group in the **Prospector** under Site, Site 1, Grading Groups (Figure 2-21).

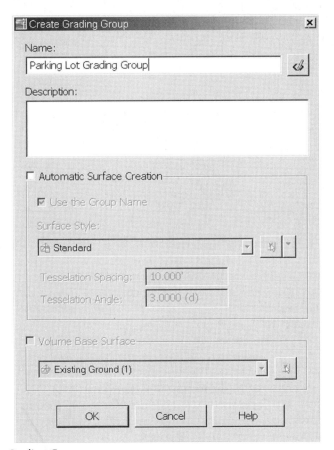

Figure 2-21   Create Grading Group

3. Select the second icon button ⬙ from the left of the toolbar. This invokes the **Target Surface** dialog box to allow you to select the surface to which you are going to tie out. Choose **Existing Ground** when the dialog box displays.
4. Select the third icon button ≋ from the left of the toolbar. This invokes the **Grading Layers** dialog box to allow you to develop layers for the grading feature lines. The default should be **C-TOPO-GRAD-FLIN** for the feature lines and **C-TOPO-GRAD** for the Grading. Hit **OK**.

5. Now select the fourth button item, the pop-down window 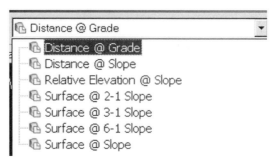 from the left to **Select a Criteria Set**. This opens a dialog box that might default to **Standard**. Open the pop-down window and choose the **Basic Set**. The pop-down window now has several options available. Choose the option called **Surface @ 3-1 Slope** (Figure 2-22).

Figure 2-22   Pop-down options

6. Now hit the button just to the right of the pop-down window in the **Grading Layout Tools** toolbar. It should have a tooltip called **Create Grading** if you pause your cursor over the button. If not, click the **down arrow** and select **Create Grading** (Figure 2-23).

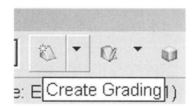

Figure 2-23   Create Grading

7. When this command is invoked, the command prompt will ask you to select the feature. Click on the Western perimeter feature line.
8. The command prompt will request you to: select which side the grading should occur on. Choose a point to the West of the site. The software will then ask: Apply to Entire Length? and the default is **Yes**. Hit <**Enter**> for Yes.
9. Without leaving the command, the software will ask to: select another Feature Line. Select the Western perimeter line and pick a point to the East of the site for the direction.
10. When complete, hit <**Enter**> and close the toolbar. You should see the perimeter graded out to the Existing Ground. The red indicates Cut activity and the Green indicates Fill. That perimeter is the Daylight where the proposed ties into the existing.
11. Save and close the file.

The next step is to create a finished ground surface representing the Access Road and the grading for the parking lot and the commercial shopping center.

Exercise 2-23: Create a Finished Ground Surface

1. Open the drawing called Chapter-2j.dwg. The previously created corridor and parking lot grading are shown in Figure 2-24.

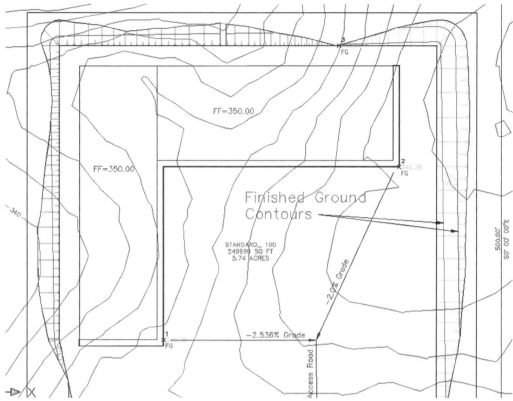

FF=350.00

FF=350.00

Finished Ground
Contours

STANDARD_ 100
249999 SQ FT
5.74 ACRES

-2.536% Grade

Figure 2-24   Corridor and parking lot grading

2. Go to the **Prospector** and expand the **Open Drawings** collection. Then expand Chapter-2j.dwg.
3. Then expand **Surfaces**, right click on **Existing Ground**, and choose **Properties**. Set the **Surface Style** to **Contours (Background)**. Then **Edit the Current Style** and **Set the Contour Interval** to **2'** and **10'**. Hit **OK** and **OK**.
4. Then expand Sites, Site 1, and Grading Groups. You see the Parking Lot Grading Group. Left click on the **Parking Lot Grading Group** and you will see a preview of the group in the right pane. Right click on the **Parking Lot Grading Group** and choose **Properties**. The **Properties** dialog box will open.
5. In the **Information** tab, turn **on** the toggle for **Automatic Surface Creation**. This will invoke the **Create Surface** dialog box.
6. The **Type** will be a **TIN Surface**, which is the default. The layer will default to **C-Topo**.
7. Ensure that the name **Parking Lot Grading Group** is in the **Name** field under **Information** in the **Properties** column.
8. Provide the **Description** of **Parking Lot Grading Group**.
9. Under the **Properties**, **Style** choose **Border and Contours**. Click **Edit Current Selection** and choose the **Contours** tab, shown in Figure 2-25. Change the Contour interval to **2'** and **10'**. Hit **OK**. Hit **OK** to execute.

Notice that contours now exist for the finished ground around the parking lot. Figure 2-24 shows some leaders pointing to the finished ground contours. The next step is to add in the feature lines for the drainage vectors, the sidewalks, and the buildings.

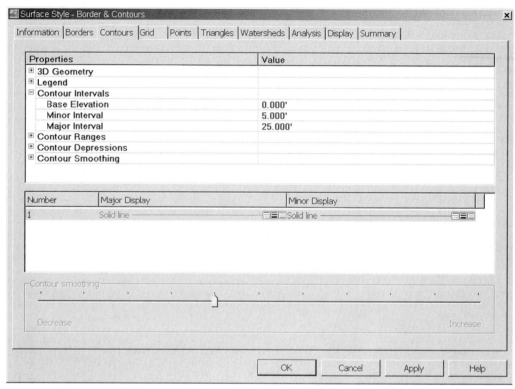

Figure 2-25   Surface Style

Exercise 2-24: Creating Feature Lines

1. In the **Prospector**, expand the **Surfaces** collection and then expand the **Parking Lot Grading Group** item. Expand the **Definition** item and you see Breaklines. Right click on **Breaklines** and hit **Add...** to invoke the **Add Breaklines** dialog box.
2. Type **Parking Lot Breaklines** into the **Description** field. These will be a Type of **Standard** and the **Mid-ordinate distance** will be **1'**, as shown in Figure 2-26. Hit **OK**.
3. The command prompt requests that you: `select the objects`. Select the two building pads, the sidewalk breaklines, the two drainage vectors in the middle of the parking lot, and the last vector on Points 4 through 7. Hit <**Enter**> when finished picking the breaklines.
4. Turn off the **3D-Data** and **C-Topo-Existing Ground** layers and you will see the Parking Lot contours.

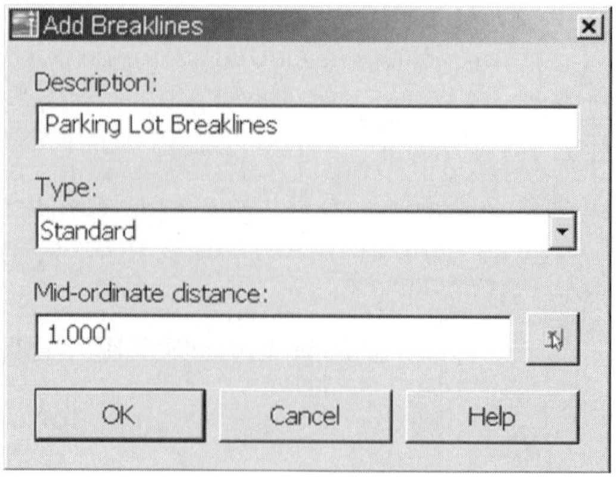

Figure 2-26   Add Breaklines

Figure 2-27 shows that if you place your cursor inside the parking lot, a tooltip pops up and automatically provides elevational information for the Existing Ground as well as the Parking Lot surfaces. The next step is to develop contours for a corridor from the Corridor model.

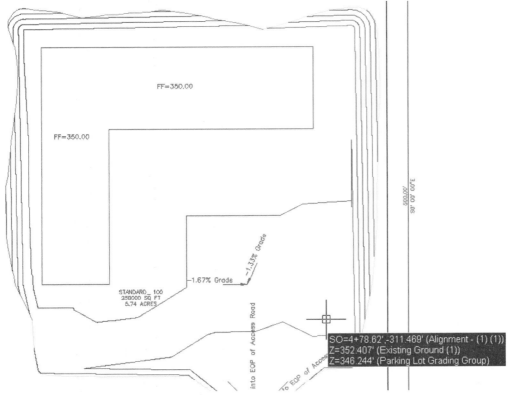

Figure 2-27   Site Contours

Exercise 2-25: Create Contours

1. Zoom into the corridor, or type **V** for **View** and select the **Plan View**, set it current, and hit **OK**.
2. Click anywhere on the corridor model and right click. Choose **Corridor Properties**. The **Properties** dialog box will display.
3. Choose the **Surfaces** tab, shown in Figure 2-28.
4. Click the first button on the top left that has a tooltip of **Create a Corridor Surface**. It uses the Data type of **Links** and the Specify code of **Top**.
5. The name **Access Road Surface** appears in the window below.

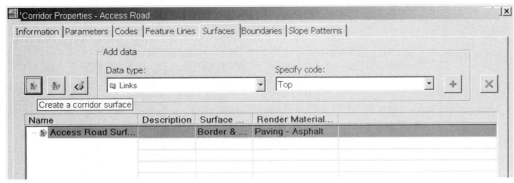

Figure 2-28   Corridor Properties, Surfaces

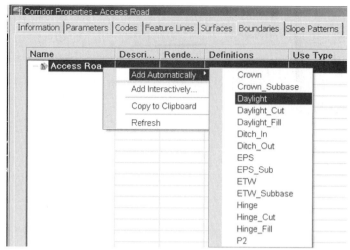

Figure 2-29   Corridor Properties Boundaries

6. Click the button with the plus sign to the right of the Specify code pop-down menu with a tooltip of **Add Surface Item**. It should be set to **Top**. The name **Access Road** can be expanded now and you should see **Top**. This is the surface data from which the surface will be created.
7. Now select the **Boundaries** tab, as shown in Figure 2-29. This option trims the contours for the roadway right at the daylight cut and fill lines to ensure clean tie-out of the contours.
8. Right click on the **Access Road Surface Name**.
9. Choose the **Add Automatically** option and then choose **Daylight** from the submenu. Hit **OK**.
10. The contours for the Corridor should be displayed in AutoCAD now.

Congratulations! You have developed a corridor and a shopping center, all in 3D. Now place some property corner points on the parcel.

11. Zoom back to the **Plan View** of the parking lot. From the **Points** pull-down menu, choose **Create Points**....
12. Select the first command in the **Points** toolbar called **Miscellaneous: Manual**. Using the Endpoint snap, set a point on each corner of the parcel. When prompted for the Description type: **IPS**. For the elevation, just hit **Enter**.
13. Save your file and close the drawing.

Now the last step shows preparation of the Plan & Profile sheets for plotting the project.

Exercise 2-26: Setting Up a P&P Sheet

1. Open the drawing called Chapter-2k.dwg. The previously created Corridor and Parking Lot are shown with their respective contours, as shown in Figure 2-30 and Figure 2-31.

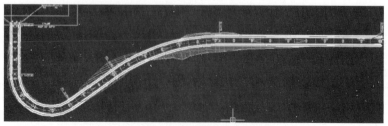

Figure 2-30   Corridor and Parking Lot are shown with Contours

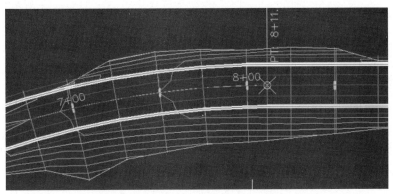

Figure 2-31 Close-up of Corridor and Parking Lot are shown with Contours

2. Type **V** for **View** and choose the **Profile** view, set it current, and hit **OK**. The profile is now displayed.
3. Click on a grid in the **Profile** and then right click. Choose **Edit Profile View Style**.
4. Select the **Graph** tab. On the right side of the dialog box is a parameter to set the **Vertical Exaggeration**. Set it to **10.00**. Hit **OK**.

TIP

Selecting an exaggeration of 10.0 means that the vertical scale for the profile will be 1/10th that of the horizontal scale. This often-used factor can be switched to 5.0 for a flatter profile or even sometimes to 1.0 for a profile that is not exaggerated.

5. Select the tab at the bottom of the screen called **Layout 1** in order to go to Layout 1.
6. Right click on the **Layout 1** tab and choose **Page Setup Manager**. When the dialog box appears, select the **Modify** button to get to the **Page Setup** dialog box shown in Figure 2-32.
7. Set the **Plotter name** to **DWF6 ePlot.pc3**.
8. Set the **Paper size** to **Arch D, 36 x 24**. Hit **OK** and **Close**.
9. From the **View** pull-down menu, select **Viewports**.
10. The command will prompt you with the following: Command: _-vports
11. Specify corner of viewport or [ON/ OFF /Fit/ Shadeplot/ Lock/Object/Polygonal/Restore/2/3/4] <Fit>:. Choose 2. The prompt continues with: Enter viewport arrangement [Horizontal/Vertical] <Vertical>: Type H. Then hit **<Enter>** when asked: Specify first corner or [Fit] <Fit>:.
12. Click on the viewports and then choose the **PROPERTIES** command in AutoCAD. Set the **Standard scale** for these viewports to **1:50**.
13. Pan the **Plan View** so you can see the corridor.
14. Pan the Profile into the Profile View so you can see it.
15. Choose the **View** pull-down menu and select **Regen All**.
16. Right click on **Layout 1**, and choose the **Rename** option. Rename it to **Plan & Profile**.
17. Now click on the **Layout 2** tab. After it is finished regenerating, right click on **Layout 2**. Choose the **Page Setup Manager**. In the **Current Page** window, click on **Layout 1** and hit the **Set Current** button, so it will inherit the properties of the previous viewport.
18. Stretch the viewport to take advantage of the whole sheet. Then click on the viewports and then choose the **PROPERTIES** command in AutoCAD. Set the **Standard scale** for these viewports to **1:20**. If this will not fit on the paper, use a **1:30** scale.

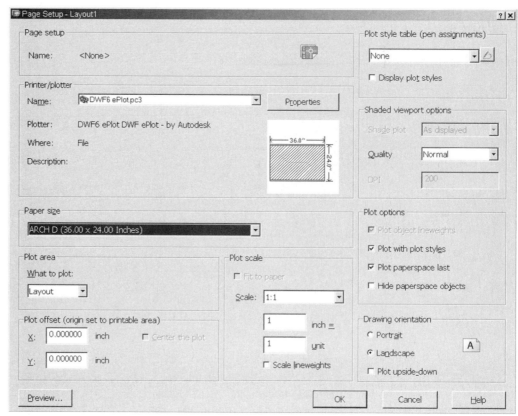

Figure 2-32   Page Setup

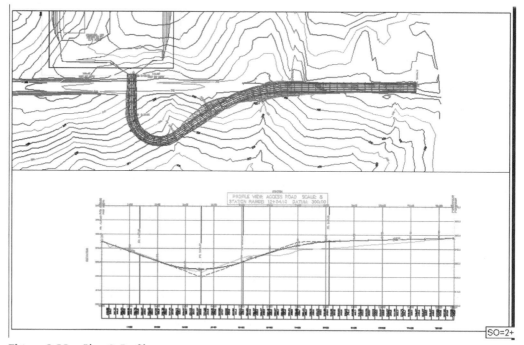

Figure 2-33   Plan & Profile

19. Right click on the **Layout 2** tab, and choose the **Rename** option. Rename it to **Plan Parking Lot.**
20. Choose the **View** pull-down menu and select **Regen All.**
21. These sheets or layouts are now ready for plotting. Refer to Figure 2-33.

Note that you can make changes to the various styles involved so that finished ground contours appear differently than the Existing Ground contours. Refer to Figure 2-34. More detailed concepts like this will be covered in the later chapters.

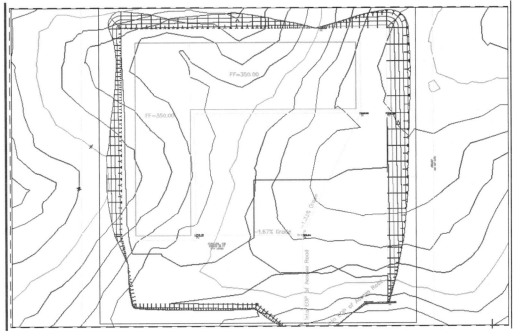

Figure 2-34   Completed site

Congratulations! You have completed the chapter's goal.

Job skills will benefit from this chapter because it covered an entire project from start to finish using Civil 3D. Although it was fairly simple in scope, it touched on developing multiple surfaces, parcels, property corner points, centerlines, profiles, assemblies, and corridors. The project was completed by creating construction plans in paper space. These functions could be called Minimum Proficiency Requirements.

Minimum Proficiency Requirements are those tasks that everyone in design should be capable of performing. A manager in the company should be able to approach any designer or technician and ask that a surface be assembled, a centerline be created, a profile be computed, or a plot be output. Of course, the real value in being able to do these tasks is to understand the mathematics behind them. This chapter assumes that the reader understands at least the rudimentary math behind the automated computations; however, the remaining chapters in the book do not. The advanced chapters that follow begin by going over the math behind Civil 3D, provide instruction in many of the formulas, and teach the algorithms that one must need to know to use the software properly.

CHAPTER TEST QUESTIONS

Multiple Choice

1. One of the things that this chapter prepares the user to do is:
   a. Develop an alignment for an access road to the shopping center
   b. Design a hydraulic grade line for a storm sewer
   c. Design the shopping center buildings, walls. and roof trusses
   d. None of the above

2. Which of the following are Civil 3D objects?
   a. Corridors
   b. Surfaces
   c. Earthworks
   d. A and B

3. Mass points are:
   a. Often from Photogrammetric sources
   b. In 3D with elevations
   c. Fairly evenly spaced across the site and are intended to densify the surface data
   d. All of the above

4. The following are types of surfaces:
   a. TIN and GRID
   b. Superelevation
   c. Building Restriction Lines
   d. Earthworks

5. Proximity breaklines are:
   a. 2D polylines
   b. That have elevations at each vertex in the polyline
   c. And are converted to 3D polylines for surface creation
   d. All of the above

6. A Wall breakline is:
   a. Vertical
   b. Nearly vertical
   c. Flat
   d. None of the above

7. In a Profile View which of the following are correct?
   a. The Horizontal Axis of the grid reflects the stationing.
   b. Vertical Axis represents elevational data.
   c. The grids represent specified spacing criteria.
   d. All of the above.

8. A vertical alignment may have the following components:
   a. A PI, PC, and PT
   b. A PRC and a PCC
   c. A broken back curve
   d. A PVI, PVC, and PVT

9. Vertical curves can be which of the following?
   a. Sag curves
   b. Crest curves
   c. Parabolic
   d. All of the above

10. A crest curve is:
    a. Usually at the bottom of a valley
    b. At the top of a hill
    c. A flat segment of roadway
    d. None of the above

True or False

1. True or False: Working with Civil 3D is easier and more effective when a template is selected with no styles preset within it.

2. True or False: The precision for contour labeling should always be set for two decimal places.

3. True or False: The Contour Label Control Line should be set to fall on the **Zero** layer.

4. True or False: The user does not have to set Surface Names in Civil 3D.

5. True or False: A template for Surface Naming can be established and it can be automatically incremented such that Surface 2 is created after Surface 1.

6. True or False: A Grid surface is for a surface of evenly spaced rectangular data whose elevations are interpolated from the data is desired.

7. True or False: Breaklines can be Proximity breaklines, Wall breaklines, and Breaklines from a file.

8. True or False: The Wall breakline option is used to represent perfectly vertical components of surfaces.

9. True or False: Parcels can have only four sides in Civil 3D.

10. True or False: The Profile View displays surface profiles but will not display a grid.

11. True or False: The Profile View displays new and/or existing profiles but not offset profiles for a single horizontal alignment.

12. True or False: Vertical tangents are segments that go only straight up in the air.

13. True or False: Vertical curves are often parabolic.

14. True or False: Vertical curves usually have a constant radius.

15. True or False: Subassemblies are made from assemblies.

16. True or False: Subassemblies allow for user definable parameters that can change the design of the corridor.

17. True or False: Assemblies govern the proposed corridor development.

18. True or False: A 3:1 side slope is a condition describing an elevation rise of 3 feet for every 1 foot horizontally.

19. True or False: Corridor feature lines have fixed properties and always look the same.

20. True or False: A Building Restriction Line (BRL) outlines the location of wetlands.

STUDENT EXERCISES, ACTIVITIES, AND REVIEW QUESTIONS

1. How are the Existing Ground and finished ground interrelated?

2. How much horizontal distance is needed to tie out a sideslope of 3:1 when the elevation difference between the tie-out point and the ground is 10 feet?

3. How much horizontal distance is needed to tie out a sideslope of 4:1 when the elevation difference between the tie-out point and the ground is 10 feet?

4. What is a breakline as a concept when looking at a roadway of a ditch compared with the purpose of a breakline as used by the software in preparing a TIN model?

5. Can a contour have different elevations on its vertices?

6. What is the difference between subassemblies and assemblies?

7. How does the software differentiate between a slope and a grade?

8. Provide some examples of where Styles come into play when designing and drafting a project in Civil 3D.

9. What types of data components can be added to a surface?

10. Are Parcels automatically annotated on creation, and if so, what controls this?

11. Why are vertical curves added to proposed roadway profiles?

12. Can a corridor be viewed in 3D?

The Simple, but Time Honored Point

3

Chapter Objectives

After reading this chapter, the student should know/be able to:
- Create, edit, report, and label Civil 3D Point objects.
- Set Point settings involved in the aesthetics of point display.
- Use Geodetics with a combination of MAP commands and Point commands.
- Use and create Point Styles and Point Groups and Description Key Sets.

INTRODUCTION

It is often proclaimed that the civil engineering business, with its adjunct partner of surveying, is a highly skilled and educated career and that individuals make a lot of money drawing invisible lines on the ground. These are items such as centerlines of roads and utilities and property, easement, and *Right of Way (Row)* lines. They can be drawn in a CADD file, but engineers use Points to identify them on the ground in the field, the idea being that these invisible lines begin and end at these points. The setting of these points has been referred as *COGO*, or coordinate geometry.

Points are locations in space defined by *X*, *Y*, and *Z* values in the *Cartesian* coordinate system. They have been used to identify objects for much of the history of civil engineering and surveying; hence they are simple and time honored.

In the earlier portion of this text the reader learned some of the background involving CADD related to civil engineering and the interface elements of the Civil 3D software. In Chapter 2 the reader explored the basic usage of the software, created a surface from Existing Ground data, and developed a shopping center on a piece of property with an access road and corridor providing ingress/egress to the shopping center. This provides an excellent background on which the remainder of the textbook will build. The detailed functionality of Civil 3D will be presented from this chapter forward.

Right of Way or ROW—A property identification

COGO—Coordinate geometry software

Cartesian—A word variation on *Descartes*

REVIEW OF FUNCTIONS AND FEATURES TO BE USED

This chapter uses engineering functions to create point data. There are fifty different ways to compute point data, almost all of which are crucial to developing project data in 2D as well as 3D. What sets this chapter apart from training manuals or software **Help** tools is that all of the exercises are paired with an AutoCAD file that has data set up to perform the commands instantly. No setup work is required to try out these functions. The reader who knows some of these routines can easily try new routines by simply opening the file and running through that specific procedure.

Bearing/Bearing Intersect—
Computation method
Distance/Distance
Intersect—Computation
method

Designers use several concepts in creating points that comprise a *bearing/bearing intersect*, a *distance/distance intersect*, and a variety of combinations of these operations. This chapter discusses the theory of what these mean before readers commence performing computations. A review of some of the distance and angular concepts is necessary before performing computations based on them.

Job skills will benefit from knowledge of point management because all phases of civil engineering use point data. Engineers begin using them when surveying the base information, whether collected via total station or GPS, followed by engineers setting points for proposed objects, and finished grade spot shots. Contractors also use them in construction by way of stakeout points, stringlines, and GPS data.

POINT BASICS

Points are very simple in concept. They occupy a specific location in 3D space and are defined by a numerical value for X, Y, and Z in the Cartesian coordinate system. These work well because AutoCAD's default system is also a Cartesian coordinate system. A simple method for keeping track of the X, Y, and Z axes is to use the "3-finger" rule. Make a fist with your left hand. Then extend your pointer finger straight out in front of you. Lift your thumb straight up. Extend your middle finger to the right so it is at a 90 degree angle to the pointer finger. Hold that position. The X axis is equivalent to the pointer, the Y axis is your thumb, and the Z axis is represented by your middle finger. Now if you maintain that relationship, you can rotate your hand around and the axes are all related to each other. There can be many X axes depending on what angle you point your pointer finger in; however, the other fingers always maintain that orthogonal relationship, just like in a CAD system.

The two-dimensional Cartesian coordinate system is a subset of the three-dimensional coordinate system and is used by those performing 2D drafting functions. Civil 3D is a three-dimensional system, but it is flexible enough to operate in either two or three dimensions. Both options will be investigated in the following discussions.

Real Numbers—Mathematical
expression

The coordinate system is such that the location of a point, which we can call P, is established by three **real numbers** indicating the positions of the perpendicular projections from the point to three fixed, perpendicular, measured lines, called axes. The horizontal coordinate is denoted as X, the vertical coordinate is denoted as Y, and the "height" coordinate is called Z.

Correspondingly, the axes are called the X axis, the Y axis, and the Z axis, and the point that fills that space is defined as P = (X, Y, Z). The Y axis is located 90° from the X axis. Together the X and Y axes produce a plane. The Z axis projects orthogonally from the X-Y plane and is positive in one direction and negative in the other.

A point that has a coordinate of (0, 0, 0) is located at the origin of the coordinate system. Points can also occupy locations along the negative portions of the axes. Therefore, Points defined as follows can all be legitimate:

$P = (10, 30, 55)$
$P = (12.3345, 111.3333, 0.0000)$
$P = (-45.33, -33.22, -10.00)$
$P = (10.55, 22.66, -66.67)$
$P = (2.02030405 \times 10^6, 3.01223344 \times 10^6, 1.43276566 \times 10^6)$
$P = (2.02030405 \times 10^{-6}, 3.01223344 \times 10^{-6}, 1.43276566 \times 10^{-6})$

Civil engineers usually express these values in terms of the Y axis being a Northing, the X axis being an Easting, and the Z axis as an Elevation and refer to them in that order. Note that the X and Y axes are reversed from the standard Cartesian order in civil engineering. Therefore, a civil engineering/surveying point is denoted as $P = $ (N: 5000.0, E: 7500.00, Elev: 120.0), whereas in the Cartesian system the same point would be represented as $P = $ (X: 7500.0, Y: 5000.0, Z: 120.0).

What can be accomplished with points and why are they used? Some uses for points in civil engineering and surveying follow:

- Property corners for parcels of ownership
- Roadway alignment curves and tangent locations
- Building corners and Column locations
- Stakeout locations for the construction of features such as Edges of Pavement

In fact, virtually everything drawn in a CAD system such as AutoCAD uses point-based information somewhere because AutoCAD is a vector-based drawing system. This means that each graphic element must have a starting point and some formulaic method for constructing the location of where that element exists. For instance, objects are defined in several ways and CAD systems do not always agree on the definitions, which can complicate translations between software applications. For instance, a horizontal arc can be located and defined using the radius point, a start angle, and an ending angle for the arc.

Example 3-1: In Figure 3-1 we have an arc defined by the center point of $X = $ 17.5219 $Y = $ 7.3619 $Z = $ 0.0000, the radius length of 10.0000, a start angle of 22.5000, and an ending angle of 45.0000.

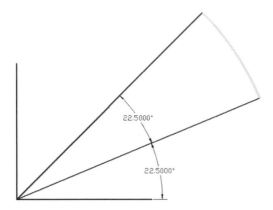

Figure 3-1   Angles

Another CAD program might use an entirely different algorithm to locate the same arc. For instance, the same arc could be described as having a center point of $X = $ 17.5219 $Y = $ 7.3619 $Z = $ 0.0000, and a start angle dictated by a matrix containing the sine of the angle from the horizontal axis (sine 22.5 degrees = 0.3826834) and congruent with the sine of the angle from the vertical axis (sine $-67.5 = -0.92387953$). The start angle is positive from the horizontal axis because it increases in the counterclockwise direction and negative from the vertical axis because it increases in a clockwise direction. This matrix places the arc in the correct quadrant. It then has a sweep angle of 22.5 degrees from the angle computed from the matrix.

Before using Civil 3D to help solve point-based problems, it is important to know the algorithms and formulas behind these computations. Let us discuss some of the mathematical formulas involving point data by reviewing how to calculate a vector between two points.

A line (or vector) is defined as the connection between the beginning point and its corresponding ending point. Therefore, a line would exist if it connected from a point called *P*1 to a point called *P*2. A line would also exist if it began at a point *P*1, had a known angle from the *X* or *Y* axis, and had a distance component. The basic formula for a line is $AX + BY + C = 0$. In solving for *Y*,

$$Y = -AX/B - C/B$$

Substituting $m = -A/B$ and $b = -C/B$ the equation reduces to $Y = mX + b$, where the *Y* value is the location along the *Y* axis, *m* is the slope of the line, *X* is the coordinate value along the *X* axis, and *b* is the *Y* intercept (or the position the line crosses the *Y* axis).

Example 3-2: If, for a line, the slope (*m*) = .5 and the *Y*-intercept (*b*) = 3, then any value of *Y* along the line is a function of where *X* is. So if you know *X*, you can compute where the Y coordinate falls on the line. Therefore, for a line where *X* = 10, you get:

$$Y = (.5)(10) + 3 = 8$$

The length of the line is computed using the principals of a triangle that has sides *A* and *B* with a hypotenuse of *C*. Sides *A* and *B* are the horizontal and vertical components of the right triangle, and *C* is the hypotenuse.

Pythagorean's theorem of $A^2 + B^2 = C^2$ reduces to $C = \sqrt{A^2 + B^2}$

Following the same thought process, then the length of a line bounded by two coordinates would be solved as shown here. These coordinates are in the *X*, *Y*, *Z* format.

$P1 = (10, 20, 0)$
$P2 = (5, 15, 0)$

Solve for ΔX or the change in *X* values = (10 − 5) = 5. This is the horizontal component.
Solve for ΔY or the change in *Y* values = (20 − 15) = 5. This is the vertical component.
Solve for ΔZ or the change in *Z* values = (0 − 0) = 0.

$$\therefore C = \sqrt{25 + 25} = \sqrt{50} = 7.0710678$$

If the line has a *Z* (or elevational) coordinate, then expand the formula as follows:

$P1 = (10, 20, 5)$
$P2 = (5, 15, 10)$

Solve for ΔX or the change in *X* values = (10 − 5) = 5. This is the horizontal component.
Solve for ΔY or the change in *Y* values = (20 − 15) = 5. This is the vertical component.
Solve for ΔZ or the change in *Z* values = (5 − 10) = −5. This is the elevational component. Note that the negative change in *Z* values is made into an absolute value by squaring it in the equation. Therefore, −5 squared = 25.

$$\therefore C = \sqrt{25 + 25 + 25} = \sqrt{75} = 8.66025$$

GEODETICS AND GLOBAL COORDINATE SYSTEMS

Another concept important to using points in Civil 3D is that of geodetic coordinate systems and how they relate to the Cartesian coordinate system. Civil 3D is built on the Autodesk MAP 3D program, which supports State plane coordinates. The State plane coordinate system is a method by which the curved spheroid that simulates the Earth is projected to a flat representation of that data (such as that on paper or computer screen, i.e., the Cartesian coordinate system). The idea is that if the shortest distance between two points is a line and that line is 10,000 feet long on a flat plane, if it were draped onto a curved surface it would then be longer than 10,000 feet due to its curvature. Many mathematical systems have been devised to account for the curvature of the Earth. To observe some of the available choices in Civil 3D, let us run through the following exercise.

TIP Remember that you performed an exercise in geodetic transformations in Chapter 1 when discussing MAP 3D. This example allows a look at a little of what is behind those settings.

Exercise 3-1: Geodetics Example 1

1. In Civil 3D select the **MAP** pull-down menu. Then choose **Tools >> Assign Global Coordinate System** as shown in Figure 3-2. An **Assign Global Coordinate System** dialog box displays,

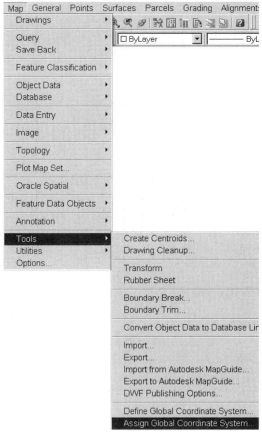

Figure 3-2   Assign Global Coordinate System

2. Choose the **Select Coordinate System** button for Code.

3. In the **Select Global Coordinate System** dialog box, under the **Category** pop-down window, choose **USA, Virginia** from the choices. Note that most of the world is supported. (Feel free to run this exercise again using your state or country.)

4. After choosing **USA, Virginia**, investigate the **Coordinate Systems in Category** window. There are solutions for NAD27, NAD83, and the HARN system. These include North and South zones for the state and the later ones include Metric and Imperial units.

5. The North American Datum of 1983, referred to as NAD83, uses the Geodetic Reference System of 1980, or GRS 80, as a reference *ellipsoid* simulating the Earth's "globe." Select **NAD27** from the window and hit the **Properties** button.

6. Look at the **Projection** tab and observe that the North American Datum of 1927, referred to as NAD27, uses the Lambert Conformal Conic, double standard parallel projection with an origin latitude of 37°-40′ and an origin longitude of −78°30′. A False Origin exists in that 2,000,000 feet are added to the Easting coordinate of data. In the Fairfax County vicinity, a legitimate NAD27 coordinate might be N: 400,000 E: 2,000,000.

7. A False Origin specifies the number of linear units to add to either the *Y* or *X* coordinates. If the user chooses an origin latitude that is south of the mapped area, then the Northing number is zero.

8. Cancel out of the dialog box and inspect the **Properties** for NAD83, North Zone, U.S. Foot. It uses the same projection but has a different False Origin. It adds 6,561,666.6667 to the Northings and 11,482,916.6667 to the Eastings. In Fairfax County, a typical coordinate might in the range of E: 7,000,000 N: 11,000,000. Hit **OK** and close out of the boxes.

9. Any time a new Civil 3D project is created, it may be a good idea to establish your *geodetic datum* at the same time. Check with your organization's surveyor for his or her recommendation on this.

10. Therefore, if you were working on a project in northern Virginia and were required to use a NAD83 coordinate system, you would select **USA, Virginia >> NAD83, North Zone, U.S. Foot.**

Civil engineering projects in many locales around the country are required to be in a specific coordinate system. For instance, in Fairfax County, Virginia, engineers and surveyors must produce data in Virginia, North Zone, NAD83 coordinates, and the reader now knows that Civil 3D would easily support this.

To understand the concept of geodetics, this chapter takes a look at the concept of a datum and how the coordinates in a datum are computed. The theory of a horizontal geodetic datum actually involves several concepts, beginning with a specification of a reference surface for the Earth, which is estimated as an ellipsoid. Refer to the National Geodetic Survey for detailed information on geodetics, because an in-depth discussion is not within the scope of this text.

Basically, imagine that a planar coordinate system is draped onto a spherical surface representing the Earth. First it is necessary to agree on the equation for the sphere that represents the Earth, because the Earth is not a perfect circle; in fact it is somewhat pear-shaped. The equation for this is an ellipsoid, a mathematical surface that best fits the shape of the *Geoid* and is the next step of approximation of the actual shape of the Earth. The Earth's physical surface includes mountains, valleys, and oceans. Without argument it is an irregular surface and is too intensive to be used as a computational surface. However, a smoothed approximation of the Earth, called a Geoid, can be easily used. Several definitions exist for a Geoid; however, a good rule of

Ellipsoid—A mathematical surface

Geodetic Datum—A mathematical representation of the Earth

Geoid—A surface equal to Sea Level

thumb might be as follows: A Geoid is a surface that is equal to Sea Level if it is not impeded or disturbed by any land masses. The computations for these representations of the size and shape of the Earth were determined using a variety of methods, including satellite measurements, and are accurate to about 2 meters.

Mathematicians have developed many spherical, mathematical models to represent the Earth. For the North American Datum 1983 (NAD 83) solution, you use the Geodetic Reference System of 1980 (GRS 80) and the WGS84 (World Geodetic System) as the mathematical ellipsoid for the Earth. Then a projection for the coordinate system must be defined. The most commonly used projections are Transverse Mercator for a region that is longer than it is wide and the Lambert Conformal Conic for a region that is wider than it is long.

Next, perform an example of establishing a geodetic zone for your project and determine some of the benefits that are achieved when doing so.

Exercise 3-2: Geodetics Example 2

You can be in any blank drawing when you perform this example.

1. In Civil 3D select the **MAP** pull-down menu. Then choose **Tools >> Assign Global Coordinate System**. An **Assign Global Coordinate System** dialog box displays.
2. Choose the **Select Coordinate System** button for Code.
3. In the **Select Global Coordinate System** dialog box, Figure 3-3, under the **Category** pop-down window, choose **USA, Virginia** from the choices.

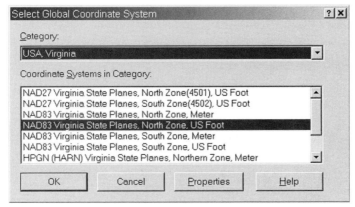

Figure 3-3   Select Global Coordinate System

4. Select **NAD83 Virginia State Planes, North Zone, U.S. Foot**. Hit **OK** and hit **OK** again.
5. Type in **Z** <Enter> for **Zoom** and the **C** <Enter> for **Center**. When requested, type in **7000000,11000000** <Enter>. Enter a height of **1000**. This coordinate is in the vicinity of a state plane coordinate for northern Virginia under the NAD 83 system.
6. From the **Points** pull-down menu, select the **Create Points...** command.
7. The **Create Points** toolbar displays. Select the arrow on the first drop-down icon. Ensure that the **Manual** option is checked. Refer to Figure 3-4.
8. Select the first icon button representing the placing of a manual point.
9. The software replies with the following prompt: `Please specify a location for the new point:`
10. Type in **11626006, 6865528**<Enter>.
11. `Enter a point description <.>:` Hit <Enter>.

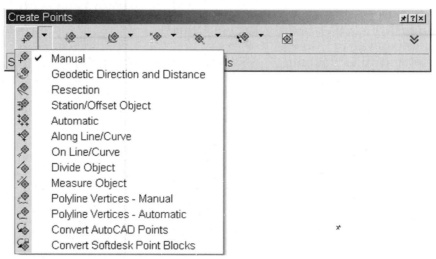

Figure 3-4   Create Points

12. Specify a point elevation <.>: Hit **<Enter>**.
13. It responds with N: 6865528', E: 11626006'
14. A point shows up in the center of the screen. Close the **Create Points** toolbar.
15. From the **Points** pull-down menu, choose **Utilities > Geodetic Calculator**.
16. A **Geodetic Calculator** dialog box displays that has a button in the top left corner called **Specify Point**. Click the button and using the Node snap, snap to the point you placed.
17. The dialog box reports that the point at a local Easting of 11626006.956119300000 and Northing of 6865528.191941710000 also has a latitude and longitude, as shown in Figure 3-5.

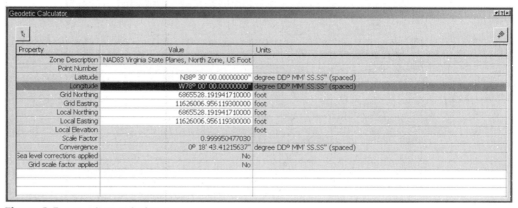

Figure 3-5   Geodetic Calculator

This example shows that the point has a specific and unique place on the face of the Earth. This point is located near Culpeper, Virginia. Civil 3D can deal with points by latitude and longitude or by coordinates values.

POINT AESTHETICS

Now that the chapter has covered various concepts of how to compute with points, discussion continues on issues regarding how points appear. A point conveys nothing else in its purest sense beside its position; however, civil engineers and surveyors do expect more from a point. For instance, in addition to the location of a point, points represent the locations of a variety of objects in the field, such as property corners, building corners, trees, manholes, or even locations of linear objects such as edges of pavement or sidewalks. Therefore, knowing what the point represents as well as its X, Y, and Z location are critical pieces of information.

In order to visualize what a point represents, you may want the ability to view the point differently depending on what the point's representation is. Sometimes there is text associated with the point that simply states what it is, such as "Ground Shot" or "Waterline." Another option is to vary the shape or symbol of the point. For example, the point shape might change depending on whether it is a property corner, perhaps displayed as a small circle O, or if it represents a ground shot it might appear as an **X** shaped tick mark.

Civil 3D has a myriad of ways to depict points and what they represent. Some short exercises that follow begin an introduction to this. They create a New Point Style, which affects how the point appears; a New Label Style, which affects how the labeling for the point appears; a New Point File Format, which allows points to be imported or exported with your choice of formats; and a Description Key Set, which controls the point codes indicating what the point actually represents. In the procedures that follow, you also create a customized table for reporting point data.

Exercise 3-3: Creating Point Styles

1. Create a new drawing in Civil 3D, use the template **_Autodesk Civil 3D Imperial By Layer.dwt.**
2. Choose the **Settings** tab on the **Toolspace** palette.
3. Expand **Point >> *Point Styles*** and notice a preset style called **Standard**, as shown in Figure 3-6. You could edit that or create a new one by right clicking on **Standard**. This yields **Edit, Copy, Delete, and Refresh**, which are style management functions. Instead, create a new style.

Point Styles—A definition of point appearance

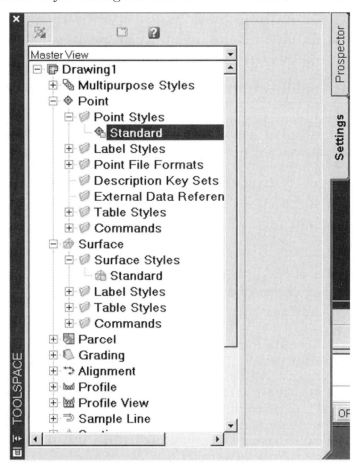

Figure 3-6   Point Styles

4. Right click on **Point Styles** and choose **New**....
5. A display occurs; note the tabs for establishing the Point Style. They are **Information, Marker, 3D Geometry, Display,** and **Summary,** as shown in Figure 3-7.

Information	Marker	3D Geometry	Display	Summary

Property	Value
⊟ **Information**	
Name	Prop-Corner
Description	
Created By	Harry
Date Created	11/5/2004 11:43:29 AM
Modified By	Harry
Date Modified	11/5/2004 12:43:24 PM
⊟ **Marker**	
Marker Rotation Angle	0.000
PDMODE	✕ 3
Point Marker Size	0.008
Point Marker Size Type	Use drawing scale
Point Marker Type	Use AutoCAD BLOCK symbol for marker
Point Marker Fixed Scale	(1.0000',1.0000',1.0000')
Orientation Reference	World Coordinate System
Marker Block	ip
⊟ **3D Geometry**	
Point Display Mode	Use Point Elevation

Figure 3-7   Point Style dialog box

6. The **Information** tab is used to change the style name from **New Point Styles** to the name you prefer; in this case call it **Prop-Corner**. Then select the **Marker** tab to identify the aesthetic appearance of a point symbol in AutoCAD. The marker can be an AutoCAD Point symbol or, more likely, a Civil 3D marker, which is selected from the panel of buttons. The buttons on the left indicate what the point should look like when placed. The two additional buttons on the right are superimposition buttons that can be laid over the style chosen on the left. Experiment with the options. Choose the "✕" and note the **Preview** display on the right. Choose the cross "+" and observe the **Preview**. Now choose the superimposed circle and note the **Preview**. The **Custom marker style** options are shown in Figure 3-8.

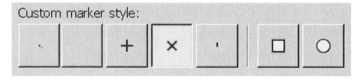

Figure 3-8   Custom marker styles

7. Now select the button for nothing on the left but leave the superimposed circle turned on. This option might be used by a surveyor in denoting a property corner. The size and inches option in the top right of the box indicates that the symbology will plot a 0.1" at the scale of the drawing. If you desire it to be larger or smaller, this value can be adjusted accordingly. The user can also choose an option of **Use Fixed Scale, Use size in absolute units,** or **Use size relative to screen,** whereby the size of the marker is a percentage of the drawing screen size, in addition to **Use drawing scale.**

8. Continuing this exploration, note the lower left quadrant of the dialog box. You see an option called **Marker rotation angle**. This value can rotate the selected marker so that it always comes in, in a rotated position. One last option is that the user can also select to use an AutoCAD block instead of a Civil 3D marker.

TIP   If blocks have been preset for the user they can simply be selected from the window. If your window is empty or if your organization wants you to use in-house-developed symbols, then right click inside the window and choose **Browse....** Then locate the drive, folder, and block you would prefer to use.

9. In this case, use **AutoCAD Block Symbol** for **Marker** and select the **Iron Pipe**. Check the **Preview** and notice a double circle symbol shows up. So any symbol the user desires can be established for the marker.

10. Select the **3D Geometry** tab. Expand the **3D Geometry** item. Poke into the pop-down window for **Point Display Mode**. Notice that a point can actually reside in three dimensions at the elevation of the point, it can be "flattened" to an elevation, or the elevation can be exaggerated by a scale factor. Selecting any of these options makes the **Point Elevation** or the **Scale Factor** option available as needed. Leave the option as it was at **Use Point Elevation**.

11. Now select the **Display** tab. This is where the layer, color, and so on are set for the Point Style in both 2D and 3D views. You can also preset if the marker and its associated label are to be turned on by default. To create a layer for the point, pick a point in the **Layer** column for the **Marker** row. A **Layer** dialog box appears. Hit the **New...** button and a **Create Layer** dialog box displays. Click under the **Values** column for **Layer Name**. Enter a layer name of **VF-Node-Ctrl-Ipf**. For the **Color** option, choose the button to select a color and choose **red**. Everything else will remain default. Hit **OK**. You are back in the **Layer Selection** dialog box. Select the new layer **VF-Node-Ctrl-Ipf**. Hit **OK**. You are now back in the **Point Styles—New Point Styles** dialog box, and the layer for the marker is set.

12. Now select the **Layer** cell for the **Label** row. Hit the **New...** button and a **Create Layer** dialog box displays. Click under the **Values** column for **Layer Name**. Enter a layer name of **VF-Ctrl-Text**. For the **Color** option, choose the button to select a color and choose **green**. Everything else will remain default. Hit **OK**. You are back in the **Layer Selection** dialog box. Select the new layer **VF-Ctrl-Text**. Hit **OK**. You are now back in the **Point Styles—New Point Styles** dialog box, and the layer for the marker is set.

13. Now right click on **Point Styles** and choose **New...** again.

14. The **Information** tab is used to change the style name from **New Point Styles** to the name you prefer; in this case call it **Planimetrics**. Choose the "." and note the **Preview** display on the right. Leave the other settings in this tab as default.

15. Now select the **Display** tab. This is where the layer, color, and so on are set for the Point Style in both 2D and 3D views. To create a layer for the point, pick a point in the **Layer** column for the **Marker** row. A **Layer** dialog box appears. Hit the **New...** button and a **Create Layer** dialog box displays. Click under the **Values** column for **Layer Name**. Enter a layer name of **VF-Node-Topo**. For the **Color** option, choose the button to select a color and choose **blue**. Everything else will remain default. Hit **OK**. You are back in the **Layer Selection** dialog box. Select the new layer **VF-Node-Topo**. Hit **OK**. You are now back in the **Point Styles—New Point Styles** dialog box, and the layer for the marker is set.

16. Now select the **Layer** cell for the **Label** row. Hit the **New...** button and a **Create Layer** dialog box displays. Click under the **Values** column for **Layer Name**. Enter a layer name of **VF-Topo-Text**. For the **Color** option, choose the button to select a color and choose **green**. Everything else will remain default. Hit **OK**. You are back in the **Layer Selection** dialog box. Select the new layer **VF-Topo-Text**. Hit **OK**. You are now back in the **Point Styles—New Point Styles** dialog box, and the layer for the marker is set to **Planimetrics**.

17. Hit the **Summary** tab to inspect your settings. When ready, hit **Apply** and **OK**. Notice that there is a new Point Style name called **Prop-Corner** in the **Settings palette**.

18. Save this drawing as **My Point Style-Initials.Dwg**, substituting your initials where it says "initials" in the filename.

Exercise 3-4: Creating Label Styles

In this exercise a new Label Style is created.

1. Open My Point Style.Dwg.
2. Choose the **Settings** tab on the **Toolspace** palette.
3. Expand **Point >> Label Styles** and notice a preset style called **Standard**. You could edit that or create a new one by right clicking on **Standard**. This yields **Edit, Copy, Delete**, and **Refresh**, which are style management functions. Instead, create a new style.
4. Right click on **Label Styles** and choose **New....**
5. A dialog box displays; note the tabs for establishing the Label Style. They are **Information, General, Layout, Dragged State**, and **Summary**. In the **Information** tab, change the name to **My Point Label Style**.
6. Select the **General** tab in Figure 3-9, and change the **Text Style** to **Romans** (which has been preset in the drawing). Notice that the text style changes in the **Preview** to the right.

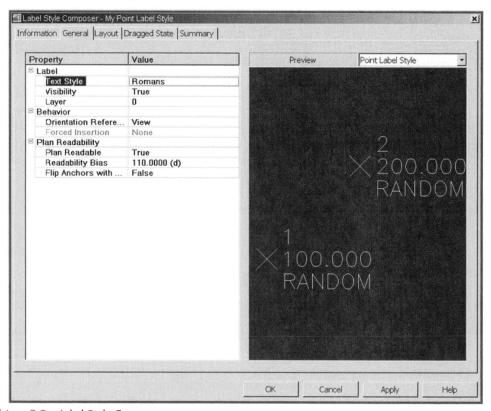

Figure 3-9   Label Style Composer

7. Leave the remaining defaults as they are. Hit **Apply** and **OK**.
8. In the **Toolspace**, go to the **Prospector** tab.
9. In the **Master View**, expand the item for your open drawing and expand the **Point Groups** item.
10. Right click on the **All Points** and select **Properties**. The **Point Group Properties** dialog box opens as shown in Figure 3-10.

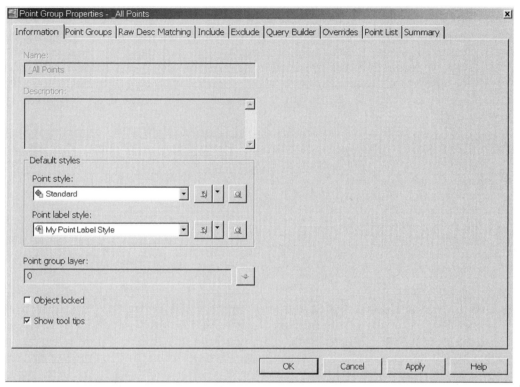

Figure 3-10   Point Groups

11. In the **Information** tab, come down to the **Point label style** option, pick in the pop-down window, and select **My Point Label Style**, which you just created. Hit **Apply** and **OK**.
12. In AutoCAD, type **Z** for **Zoom** and the **C** for **Center**. Type in a coordinate of **1000,1000** and a magnification or height of **25**.
13. From the **Points** pull-down menu, choose **Create Points**... to invoke the toolbar for point creation. Choose the first command, **Manual**.
14. Set a point at a coordinate of **1000,1000**. Give it a point description of **FG** and an elevation of **101.33**, and terminate the command. You see a point placed with the Romans text style you requested in Figure 3-11.

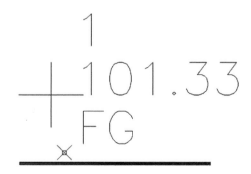

Figure 3-11   A Point

To see how your other settings change affects the display, Twist the drawing.

15. Type in **Dview** and select the point you just placed when asked by the prompt.

16. Choose **TW** for Twist and provide an angle of **215** (which would turn the point upside down). Hit **<Enter>** to complete the routine. Notice that the UCS icon and the point marker have rotated 215 degrees but the text orientation remained readable. The setting changed for **Orientation Reference to View** allows the point always to remain readable even when you Twist the view, usually for plotting purposes. This exercise shows how powerful the Civil 3D settings can be, and this is not necessarily a recommended setting for all points. Type **U** to Undo the creation of the point. Depending on how many commands you ran since you set the point, keep typing in *U* for Undo until the point is gone.

17. You can right click on the **My Point Label Style** setting and **Delete** it if you wish or right click on the **All Points** item and select **Properties**. When the **Properties** dialog box opens, in the **Information** tab, come down to the **Point Label Style** option, pick in the pop-down window, and select **Standard**. Hit **Apply** and **OK** and your points no longer use the style you created.

DESCRIPTION KEY SETS AND POINT GROUPS

Points are not actually Civil 3D objects in the same sense that surfaces and alignments are objects, although they share many characteristics with Civil 3D objects. For instance, points can be displayed in a drawing and modified graphically, and their manifestation is controlled using Styles and Label Styles. Yet they are technically not Civil 3D objects.

An excellent feature in Civil 3D is that Points can be defined such that they fall into Point Groups, which help immensely with organization. A Point Group is an object, and the point is a subset of the Point Group object. Whether a point is or is not an object does not make any real difference to the user in practice, although knowing this might assist in understanding better how points work.

There are two examples of how points will react differently than if they were objects. First, when a point is right clicked and the **Properties** option is selected, this invokes the AutoCAD **Properties** dialog box, which describes the Point Group that the point belongs to but not the point itself. Second, when a point is selected, the **Layer** toolbar lists the **Point Group** layer information but not the actual **Point** layer information.

Point groups are named collections of points that are used to organize and control points and their appearances in drawings. Point Groups have the following characteristics:

1. They have enduring properties that can be easily reviewed or modified, either beforehand or retroactively.

2. Displaying a points list shows the points included in a Point Group, and it can be updated automatically. This could occur when new points are input or modified that match the point group's properties or when the properties of a Point Group are modified.

3. A Point Group can be locked so as to avoid unintentional modifications. Some aspects of Point Groups can be observed, accessed, and managed at the project level through the **Toolspace**.

A strong feature of Point Groups is that it allows you to manage many points at one time by using their group name. Point Groups give the computer the ability to sort through only the points needed for accessing or review. They also allow you to exclude unwanted points from a task such as exporting a point dump. Point Groups organize points that have similar descriptions such as "EP" for edge of pavement shots or points that are related by function such as DTM spot shots. There are many methods for developing Point Groups including defining them from a selection set or providing a range of point numbers. Further, these methods can be compounded on one another. In other words, the points in a Point Group called Road Points can also be included in a Point Group called Proposed DTM Points. Once they are established, they can be saved into the template for automatic use on future projects, thereby becoming part of the organization's CADD standards.

Job skills would benefit from the imaginative use of point groups beyond the obvious of grouping points according to their description or function. In other words, a point group called Monday's Points can be established for a surveyor importing point data for a job shot and collected on, say, Monday. This can be repeated for those obtained on Tuesday with a point group called Tuesday's Points. Then the surveyor can refer back to those groups to identify which points were collected on Monday versus those collected on Tuesday. Another possible use of point groups is to separate 3D points that are satisfactory for inclusion into a surface's development versus those that are two dimensional and not appropriate for use.

DESCRIPTION KEY SETS

The key to success in defining point groups is to have a strong description library for the point codes and their intended functions. If this is accomplished, then one can assign point styles based on point descriptions, assign point symbology based on descriptions, and automatically populate point groups based on the descriptions. Obviously, this would benefit anyone setting points because a standardized mechanism would be used every time someone sets a storm manhole point. If all users used the preset description of, say, DMH for storm drain manholes, then they would achieve a consistent look to their storm manhole symbologies and annotations in AutoCAD. In other words, the same symbol would always appear for the storm manhole along with a standardized description next to the point.

The CADD file layers would always be correct and the point would always reside in the correct point group. The tool to accomplish this is called **Description Key Sets**. Table 3-1 shows an example of many of the point description codes that one might find on a survey project. The first column shows the Description Key Code, the second column the Description of the code, the third column the Suggested Layer for the point, the fourth column the Symbol for the point, and the last column the Related Point Symbol Layer.

Description Key Sets—
A library of codes

In the following two exercises Description Key Set (formerly called a Description Key Library for Land Desktop users) and then a Point Group are created.

Code	Desc	Layer	Symbol	Symbol Layer
AC	AC UNIT	VF-NODE-TOPO	CG78	VF-TOPO-MISC
BB	BB/TOE	VF-NODE-BRKL-TOE	CG32	VF-BRKL
BD	BUILDING	VF-NODE-BLDG	CG32	VF-BLDG
BM	BENCH MK	VF-NODE-CTRL-BMRK	CG85	VF-CTRL
BP	BOLLARD	VF-NODE-TOPO-BOLL	CG33	VF-TOPO-MISC
BR	SOIL BORING	VF-NODE-TOPO-BORE	CG85	VF-TOPO-MISC
BSL	BRUSH LINE	VF-NODE-BLIN	CG32	VF-SITE-VEGE-BRSH
CD	CL DITCH	VF-NODE-DTCH-CNTR	CG32	VF-DTCH-CNTR
CF	CG FACE	VF-NODE-CURB-ROAD	CG32	VF-ROAD-CURB-FACE
CM	CONC MON	VF-NODE-CTRL-CMON	CG74	VF-CTRL
CO	CLEANOUT	VF-NODE-MHOL-SSWR	CG01	VF-SSWR-MHOL
CP	CLR	VF-NODE-ROAD-CNTR	CG32	VF-ROAD-CNTR
CV	ELECT UGND VLT	VF-NODE-POWR-EVLT	GV	VF-POWR-UNDR
DF	DRAINFIELD	VF-NODE-SSWR	CG32	VF-SSWR-DF
DK	BLDG DECK	VF-NODE-BLDG-DECK	CG32	VF-BLDG
EC	DRWY CONC	VF-NODE-DRIV-CONC	CG32	VF-DRIV-CONC
EG	EDG SHLDER	VF-NODE-PVMT-GRVL	CG32	VF-PVMT-GRVL-SLDR
ELM	ELEC MTR	VF-NODE-INST-EMTR	CG88	VF-POWR-INST-EMTR
EM	EVERGRN MED	VF-NODE-TREE-EVMD	EM	VF-SITE-VEGE-TREE
EP	EOP	VF-NODE-PVMT	CG32	VF-PVMT-EDGE
ES	EVERGRN SMLL	VF-NODE-TREE-EVSM	ES	VF-SITE-VEGE-TREE
EW	EDGE WATER	VF-NODE-TOPO-EWAT	CG32	VF-TOPO-EWAT
FC	FILLER CAP	VF-NODE-POWR-FCAP	CG33	VF-POWR-UNDR
FF	FIN FLR	VF-NODE-BLDG-FFLR	CG04	VF-BLDG
FH	FIRE HYD	VF-NODE-WATR-STRC	FH	VF-WATR-FIRE-HYDT
FLM	FENCE METAL	VF-NODE-SITE-FENC-METL	CG32	VF-SITE-FENC
FLV	FENCE VINYL	VF-NODE-SITE-FENC-VINL	CG32	VF-SITE-FENC
FLW	FENCE WOOD	VF-NODE-SITE-FENC-WOOD	CG32	VF-SITE-FENC
FLY	FLY	VF-NODE-CTRL-FLY	FLY	VF-CTRL
FO	FIBER OPT	VF-NODE-COMM-FOPT	CG32	VF-COMM-FOPT
GD	GROUND SHOT	VF-NODE-TOPO-SPOT	CG36	VF-TOPO-SPOT
GR	GRD RAIL	VF-NODE-SITE-GDRS	CG32	VF-SITE-GRDS
GW	GUY WIRE	VF-NODE-POLE-GUYW	GW	VF-POWR-POLE
HC	HANDICAP MARKINGS	VF-NODE-MRKG-PRKG	HANDI	VF-PRKG-STRP
HR	HEDGEROW	VF-NODE-HEDG	CG32	VF-SITE-VEGE-HEDG
HS	HEAD STONE	VF-NODE-TOPO	CG78	VF-TOPO-MISC
IN	INLET	VF-NODE-STRM-INLT	CG04	VF-STRM-STRC
IPF	I PIPE FOUND	VF-NODE-CTRL-IPF	CG41	VF-CTRL
IPS	I PIPE SET	VF-NODE-CTRL-IPS	CG41	VF-CTRL
IRF	I ROD FOUND	VF-NODE-CTRL-IRF	CG41	VF-CTRL
IRS	I ROD SET	VF-NODE-CTRL-IRS	CG41	VF-CTRL
IV	IRRI VALV	VF-NODE-WATR-IRVA	CG94	VF-WATR-VALV
JW	WALL	VF-NODE-SITE-JWLL	CG32	VF-SITE-WALL
LA	LANDSCAPE	VF-NODE-VEGE-MISC		VF-SITE-VEGE-MISC
LE	EVERGRN LG	VF-NODE-TREE-EVLG	EL	VF-SITE-VEGE-TREE
LP	LIGHT POL	VF-NODE-POLE-LITE	CG92	VF-POWR-POLE

MB	MAIL BOX	VF-NODE-TOPO	MAILBOX	VF-TOPO-MISC
MS	SDMH	VF-NODE-MHOL-STRM	CG01	VF-STRM-MHOL
MW	WATER MH	VF-NODE-MHOL-WATR	CG49	VF-WATR-MHOL
NV	INVERT	VF-NODE-STRM-STRC	CG04	VF-STRM-STRC
OH	OH WIRES	VF-NODE-POLE-OHWR	CG32	VF-POWR-POLE
OV	OVERHANG	VF-NODE-BLDG-OHAN		VF-BLDG
PD	TELE PED	VF-NODE-COMM-PED	CG34	VF-COMM
PH	PHOTO CTRL	VF-NODE-CTRL-PHOT	PH	VF-CTRL
PM	LANE STRIPES	VF-NODE-MRKG-ROAD	CG32	VF-ROAD-MRKG
POR	PORCH	VF-NODE-BLDG-PORC	CG32	VF-BLDG
PP	POWER POLE	VF-NODE-POLE-POWR	UPOLE	VF-POWR-POLE
RI	RIP RAP	VF-NODE-RRAP	CG32	VF-RRAP
RW	RET WALL	VF-NODE-SITE-RTWL	CG32	VF-SITE-WALL
SATV	SAT TV	VF-NODE-COMM-SATV	CG03	VF-COMM-SATV
SB	SHRUB	VF-NODE-SHRUB	CG134	VF-SITE-VEGE-BRSH
SM	SMH	VF-NODE-MHOL-SSWR	CG01	VF-SSWR-MHOL
SN	SIGN	VF-NODE-SIGN	SN	VF-SITE-SIGN
SP	TREE	VF-NODE-TREE-SPEC	CG133	VF-SITE-VEGE-TREE
STL	SEPTIC LID	VF-NODE-SSWR		VF-SSWR-MHOL
STON	PLANTED STONE	VF-NODE-CTRL-STON	CG43	VF-CTRL
STP	STEPS	VF-NODE-SITE-STPS	CG32	VF-SITE-STPS
STW	STONE WALL	VF-NODE-SITE-STWL	CG32	VF-SITE-WALL
TB	TOP OF BANK	VF-NODE-BRKL	CG32	VF-BRKL
TC	CG TOP	VF-NODE-CURB-ROAD	CG32	VF-ROAD-CURB-TOP
TF	TRANSF	VF-NODE-POWR-TRAN	TF	VF-POWR-TOWR
TL	TREE LARGE	VF-NODE-TREE-LARG	TL	VF-SITE-VEGE-TREE
TLN	TREE LINE	VF-NODE-TRLN	CG32	VF-SITE-VEGE-TROW
TM	TREE MED	VF-NODE-TREE-MED	TM	VF-SITE-VEGE-TREE
TR	TRAVERSE	VF-NODE-CTRL-TRAV	CG48	VF-CTRL
TS	TREE SMALL	VF-NODE-TREE-SM	TS	VF-SITE-VEGE-TREE
TX	TRAF SIG BOX	VF-NODE-POWR-STRC	TRAFCNTL	VF-POWR-STRC
UE	ELEC UGND	VF-NODE-POWR-UNDR	CG32	VF-POWR-UNDR
UG	GAS UGND	VF-NODE-POWR-UGAS	CG32	VF-POWR-UNDR
UM	UNKNOWN MH	VF-NODE-TOPO	MH_UTIL	VF-TOPO-MISC
UT	TELEPHONE	VF-NODE-COMM-UTEL	CG32	VF-COMM
UTV	UG CABLE TV	VF-NODE-CATV-UNDR	CG32	VF-CATV-UNDR
UW	IWATER UNGD	VF-NODE-WATR-UNGD	CG32	VF-WATR-VALV
VDH	VDOT MON	VF-NODE-CTRL-VDOT	CG73	VF-CTRL
WA	WALL ALIGN	VF-NODE-SITE-WAAL	CG32	VF-SITE-WALL
WE	WELL	VF-NODE-WATR-STRC	CG01	VF-WATR-WELL
WK	WALK	VF-NODE-TOPO-WALK	CG32	VF-TOPO-WALK
WM	WATER METER	VF-NODE-WATR-METR	WM	VF-WATR-MAIN
WT	WETLANDS	VF-NODE-WLND	CG01	VF-SITE-VEGE-WLND
WV	WATER VALV	VF-NODE-WATR-VALV	WV	VF-WATR-VALV
YH	YARD HYDRANT	VF-NODE-WATR-YAHY		VF-WATR-VALV

Table 3-1   A Sample Listing of Description Keys

Exercise 3-5: Creating Description Key Sets

The Description Keys built here can be used in developing Point Groups in Exercise 3-6.

1. Launch Civil 3D and open My Point Style.Dwg. Make sure that the **Toolspace** is displayed.
2. Click the **Settings** tab of **Prospector** to view the **Settings** panel. Pick **Master View** in the pop-down window at the top. Expand the **My Point Style.Dwg** drawing name. Expand the **Point** item.
3. Right click on the **Description Key Sets** and select **New....**
4. In the **Description Key Sets Name** field, change the name from **New DescKey Set** to **MY-DESCRIPTIONS**. Click **OK** to create the new Description Key Set and observe that it now exists under the **Description Key Sets** item.
5. Now that the Description Key Set called **MY-DESCRIPTIONS** is created, the keys can be defined for point and label styles. Click on **MY-DESCRIPTIONS** and press the right mouse button. Select **Edit Keys....** This displays a **Panorama** dialog box listing all currently defined keys, shown in Figure 3-12. The new Description Key appears in capital letters.

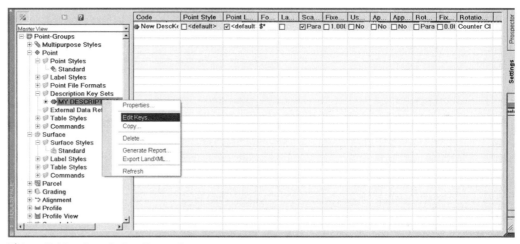

Figure 3-12   Description Key settings

6. In the first cell of the Code column, enter **AC***.
7. In the next column, click the **Point Style** toggle **On**.
8. Click on the word **<default>** in the Point Style column, and the **Select Point Style** dialog box displays.
9. Use the list arrow to observe the list of available styles; select **Planimetrics** from the list, and click **OK**.
10. In the next column, click the **Point Label Style** toggle **on**.
11. Click on the **<default>** in the Point Label Style column, and the **Select Point Label Style** dialog box displays. Choose **Standard**.
12. In the Format column, type **AC UNIT**. When these points are placed in Civil 3D, they will always be annotated as AC Units.
13. Then in the next column for Layer, click the **Layer** toggle **on**.
14. Click in the cell to the right of that toggle, and from the Layer list, select the **VF-Node-Topo** layer and click **OK**.
15. Now create another Description Key. Right click on the **AC** key and choose **Copy** from the shortcut menu. A **Copy of AC*** is created.
16. Click on the **Copy of AC*** and change the name to **IPF***.
17. Click on the **<default>** in the Point Style column, and the **Select Point Style** dialog box displays.

18. Use the list arrow to observe the list of available styles; select **Prop-Corner** from the list, and click **OK**.
19. In the next column, click the **Point Label Style** toggle **on**.
20. Click on the <default> in the Point Label Style column, and the **Select Point Label Style** dialog box displays. Choose **Standard**.
21. In the Format column, type **I Pipe Found**. When these points are placed in Civil 3D, they will always be annotated as I Pipe Found.
22. Then in the next column for Layer, click the **Layer** toggle **on**.
23. Click in the cell to the right of that toggle, and from the Layer list, select the **VF-Node-Ctrl-Ipf** layer and click **OK**.
24. Next create yet another Description Key. Right click on the **AC** key and choose **Copy** from the shortcut menu. A **Copy of AC*** is created.
25. Click on the **Copy of IPF*** and change the name to **MB***.
26. Click on the <default> in the Point Style column, and the **Select Point Style** dialog box displays.
27. Use the list arrow to observe the list of available styles; select **Planimetrics** from the list, and click **OK**.
28. In the next column, click the **Point Label Style** toggle **on**.
29. Click on the <default> in the Point Label Style column, and the **Select Point Label Style** dialog box displays. Choose **Standard**.
30. In the Format column, type **Mailbox**. When these points are placed in Civil 3D, they will always be annotated as Mailbox.
31. Then in the next column for Layer, click the **Layer** toggle **on**.
32. Click in the cell to the right of that toggle, and from the Layer list, select the **VF-Node-Topo** layer and click **OK**.
33. Click the check mark in the upper right of the **Panorama** dialog box to save and close the Desckey editor.
34. The **Settings** should appear as shown in Figure 3-13. Close the **Panorama**.

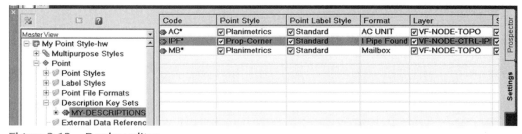

Figure 3-13   Desckey editor

Now there are Description Keys for three items to use in preparing Point Groups in the following exercise.

Exercise 3-6: Creating Point Groups

This task defines a Point Group to help organize your points. It helps in accessing points, reporting on them, or exporting them. The Point Group is made up of our Property corner control points. By default Civil 3D creates a point group called ALL.

1. While remaining in the same drawing, click the **Prospector** tab in the **Toolspace**.
2. Expand the drawing name item to view the Civil 3D object list.
3. Click on the **Expand tree** icon next to **Point Groups** to show the Point Group list.
4. Click on **Point Group** item, right click, and select **New...** from the shortcut menu.

5. In the **Information** tab, its name is Point Group-(1). Change the **Point Group** name to **Planimetrics**. Give it a description of **2D Planimetrics**. Refer to Figure 3-14.

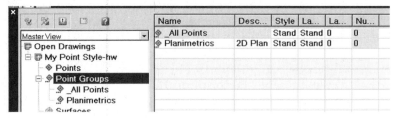

Figure 3-14   Point Group Editor

6. Click the **Raw Description** tab and toggle on **AC*** and **MB***.
7. Click the **Include** tab and notice that the same two Description Keys are selected here. Click **OK**.
8. The Point Group, Planimetrics, is now evident on the Point Group list.
9. Now create a Point Group for the Control points. Click on **Point Group** item, right click, and select **New...** from the shortcut menu.
10. In the **Information** tab, its name is Point Group-(1). Change the **Point Group** name to **Control**. Give it a description of **Control Points**.
11. Click the **Raw Description** tab and toggle on **IPF***.
12. This instructs Civil 3D to include only those points that match the raw descriptions of IPF*. As other Control keys are defined, they can be added here as well by revisiting this procedure. Click **OK**.
13. Under Point Groups you will now see All Points, Planimetrics, and Control.

> **Note:**
> This instructs Civil 3D to include only those points that match the raw descriptions of AC* and MB*. As other Planimetric keys are defined, they can be added here as well by revisiting this procedure.

Now that the data organization is created within Civil 3D, the next step is to create some points to sample how this should work.

Exercise 3-7: Create Points

1. Type **Zoom** and <**Enter**>. Then type **C** and <**Enter**>. When asked for the center point, type **1000,1000**. When asked for the Magnification or Height, type **500**.
2. From the pull-down menus, choose **Points >> Create Points...**
3. The **Create Points** toolbar displays. Select the first button, with the tooltip called **Miscellaneous: Manual**.
4. The command prompt says: `Please specify a location for the new point:` Type in **1000,1000**. The software then responds as follows.
5. `Enter a point description <.>:` **AC**
6. `Specify a point elevation <.>:` **112.33**
7. `Please specify a location for the new point:` **1100,1000**
8. `Enter a point description <AC>:` **IPF**
9. `Specify a point elevation <112.330'>:` **114.32**
10. `Please specify a location for the new point:` **1100,1100**
11. `Enter a point description <IPF>:` **MB**
12. `Specify a point elevation <114.320'>:` <**Enter**>

13. `Please specify a location for the new point:` **<Enter>** to terminate the command.
14. If, after the points are in place, there is an icon indicating that the object data are Out of Date, then right click on the item (**Point Groups, Planimetrics**) and select **Update**. That should eliminate the icon indicating Out of Date and place the points into the Point Group object. Repeat for any other Point Groups that are Out of Date.

The data shown in the **Master View** of the **Prospector** are updated automatically to show the current status of data in the drawing. The **Prospector** dynamically updates its contents as the user adds and removes data and civil objects. Icons adjacent to the data entry tell the user the status of the data, for example, if a surface is out of date or if a point group has changed. The **Prospector** should appear similar to Figure 3-15 after the Point Groups are created and the points are entered.

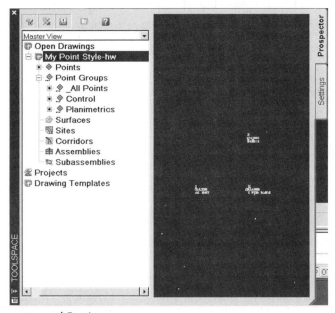

Figure 3-15   Prospector and Preview

Exercise 3-8: Creating Point File Formats

In this task a *Point File Format* is defined to help when importing or exporting point data to external locations. Formats indicate the relative positions of the point data in the string being read. For instance, what is the first item in the string: a point number or a Northing? What is the second item in the string: an elevation or an Easting? By default Civil 3D has a lot of predeveloped formats, but just in case you require one not already developed, it helps to know how to create a format.

Point File Formats—Definitions for importing or exporting points

1. While remaining in the same drawing, click the **Prospector** tab in the **Toolspace**.
2. Expand the drawing name item to view the Civil 3D object list.
3. Click on the **expand tree** icon next to **Point File Formats** to show the list.
4. Click on **Point File Formats** item, right click, and select **New...** from the shortcut menu.
5. When the **Select Format Type** dialog box opens, choose **User Point file** and hit **OK**.

6. The **Point File Format** dialog box opens. Type in a format name of **LatLong**, as shown in Figure 3-16.

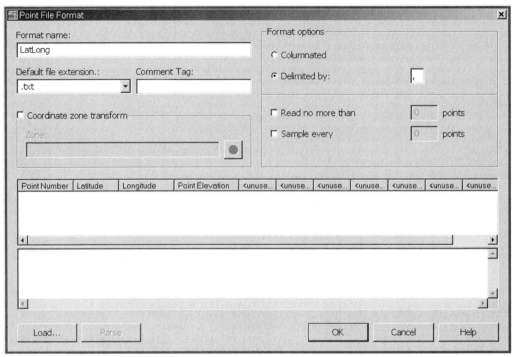

Figure 3-16   Point File Format

7. Choose the option for **Delimited by** and type a comma (,) in the open field.
8. Near the bottom is a wide window with column headers labeled **unused**. Click on the first header on the left. A **Selection** dialog box opens. Click in the pop-down window and choose **Point Number**. Hit **OK**.
9. Click on the second header. Click in the pop-down window and choose **Latitude**. Hit **OK**.
10. Click on the third header. Click in the pop-down window and choose **Longitude**. Hit **OK**.
11. Click on the fourth header. Click in the pop-down window and choose **Point Elevation**. Hit **OK**. Hit **OK** to close out.

You have created a format that allows for importing or exporting data using **LatLong** data.

EXTERNAL DATA REFERENCES

External Data References—
Additional data files

Civil 3D can create and maintain point database files that contain all the point information in the project. *External Data References* are adjunct databases that can amend the information in this database. The data stored include the point number, optional point name, Northing, Easting, elevation, and point description. External Data References can allow you to either substitute point elevations when points are accessed through a point group or substitute point raw description data.

This can be useful if your project is in Imperial units, but periodically you must report your elevations in Metric units.

External Data References allow you to link your custom point databases to Civil 3D. They are also called XDRefs. The XDRef is a pointer to a column of data in a custom Microsoft Access database. The key field is typically the point number. Then, when an XDRef is used to access a value for a point, the point number is looked up in the custom database, and the value from the specified column is used instead of the point's original value. These files are nondestructive in that they do not overwrite or alter the points in the drawing. Although this is not a highly used function, it can be very handy when the occasion arises.

Now that you have gained some experience in establishing Point Groups and related Description Key Sets, you learn how to set points using some of the many **Point Creation** tools provided within the software. These are obtained using the **Points** pull-down menu, and these commands are described next.

The **Points** pull-down menu (Figure 3-17), is very clear and succinct, characteristic of all the pull-down menus. Some software manufacturers load up the menus with many commands, and users are often lost in finding the right command at the right time. Civil 3D menus are compact and easy to understand. The Points commands are

Create Points... invokes the **Points Layout** toolbar for setting points. There are forty-eight commands in this toolbar for sophisticated point geometry creation.

Create Point Group... allows for the creation of Point Groups.

Edit... allows for modification of points.

Add Tables... allows for importing point information in tabular format.

Utilities... allows for Exporting points, Identifying where points are located, Transfer facilities, a Geodetic calculator, and the ability to create blocks from points.

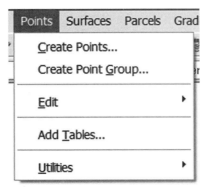

Figure 3-17   Points pull-down menu

POINT COMPUTATIONS

The text mentioned earlier that there are fifty different methods for computing point data. The **Create Points** toolbar (Figure 3-18) contains these commands.

Figure 3-18   Create Points toolbar

Each button has an accompanying pop-down arrow for making selection choices. The buttons in the toolbar include the following functions:

- **Miscellaneous**, which sets 2D points using basic techniques

- **Intersection**, which sets 2D points using intersectional criteria

- **Alignment**, which sets 2D points using alignment-based information

- **Surface**, which sets 3D points using Surface information and 3D criteria

- **Interpolation**, which sets 3D points typically densifying the data

- **Slope**, which sets 3D points using Grade and Slope criteria

- **Import Points**, which allows for bringing points into the drawing from other sources, drawings, projects, or programs

Each of these commands will be discussed independently, so the reader can use them on projects easily.

Exercise 3-9: Create Points Commands

1. Open the drawing called Chapter-3-Points.Dwg.
2. Type **Zoom** and **<Enter>**. Then type **C** and **<Enter>**. When asked for the center point, type **1000,1000**. When asked for the Magnification or Height, type **200**. You can also type in **V** for **View** and choose a saved view called **Points-A**. Hit **Set Current** and hit **OK**.
3. Select the **Points** pull-down menu and choose the **Create Points**... command. The toolbar displays. Investigate the first tools in the leftmost icon first.

Miscellaneous Point Creation Commands

Exercise 3-10: Miscellaneous—Manual Command

 Figure 3-19 shows **Manual**, which allows you to pick the origin of the point using a mouse or key-in value.

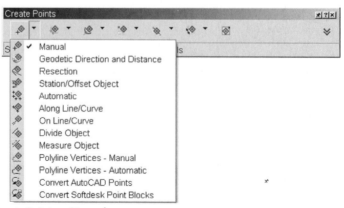

Figure 3-19   Create Points—Manual

TIP   This command can be used to set points manually wherever they are needed.

Job skills benefit because engineers always set points on objects representing property lines, easements, road centerlines, pipelines, and so on.

1. Select the first pop-down arrow and ensure that **Manual** is checked by clicking on it. This will also invoke the command.
2. Civil 3D responds in the command prompt with the following. Enter the responses as indicated.
3. Please specify a location for the new point:
4. Type in **1000,1000<Enter>**.
5. Enter a point description <.>: **IPF**
6. Specify a point elevation <.>: **122.45**
7. The software responds with: N: 1000.0000′ E: 1000.0000′
8. Please specify a location for the new point:
9. This time use your mouse and pick a point about 3″ (on the screen) to the right and 3″ (on the screen) downward from the first point. Again the software asks: Enter a point description <.>: **IPF**, Specify a point elevation <.>: **112.44**.
10. The software responds with: N: 941.6013′ E: 1077.4403′ (of course, your coordinates may differ because you are picking the point with your mouse).
11. Draw a line on the screen from the Node snap of Point 1 to the Node snap of Point 2.
12. Now select **Manual** again and this time when asked to: Please specify a location for the new point: Select the Midpoint snap and snap to the midpoint of the line you just drew.
13. When asked to: Enter a point description <.>: type **IPF** and when asked to: Specify a point elevation <.>: type **116.77.**

The result should appear similar to Figure 3-20.

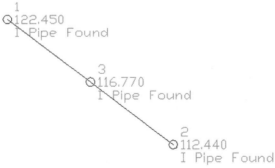

Figure 3-20   Result of Manual creation

Exercise 3-11: Miscellaneous—Geodetic Direction and Distance Command

 The next command, *Geodetic Direction and Distance*, allows for computing points based on the geodetic datum set in the drawing. Refer back to the earlier discussions on geodetics where a 1000′ line when draped along the face of the Earth is actually longer than 1000′.

Geodetic Direction and Distance—Direction and Distance using a **geodetic datum**

1. Type **Zoom** and **<Enter>**. Then type **C** and **<Enter>**. When asked for the center point, type **11626006, 6865528**. When asked for the Magnification or Height, type **300**. You can also type in **V** for **View** and choose a saved view called **Points-B**. Hit **Set Current** and hit **OK**.
2. From the **Create Points** toolbar, choose the pop-down arrow on the first button and select **Geodetic Direction and Distance**. Follow the procedure below.

When asked to `Please specify a location for the start point:` Type **11626006,6865528<Enter>**. The system responds with `Northing: 6865528.0000' Easting: 11626006.0000', Grid Northing: 6865528.0000' Grid Easting: 11626006.0000', Convergence: 0° 18' 43.40" Scale Factor: 1.000, Latitude: N38° 29' 591.00" Longitude: W78° 00' 00.01"`

3. `Geodetic azimuth <0° 00' 00.00">:` Hit **<Enter>**
4. `Geodesic distance <1.000>:` Type **100<Enter>**.
5. `Northing: 6865627.9936' Easting: 11626005.4554'`
   `Grid Northing: 6865627.9936' Grid Easting: 11626005.4554'`
   `Convergence: 0° 18' 43.40" Scale Factor: 1.000`
   `Latitude: N38° 30' 00.99" Longitude: W78° 00' 00.01"`
6. `Enter a point description <.>:` **IPF**
7. `Specify a point elevation <.>:` **0.0**
8. `Please specify a location for the start point:` **<Enter>** to terminate.

Azimuth—An angle measure

A Point 4 is created at a geodetic distance of 100′ along the curvature of the Earth in NAD 83 in northern Virginia at a geodetic *azimuth* of 0 degrees. In order to test how this handled the computation, draw a line from the initial coordinate to the point placed and list the line for its length and direction.

9. Type **L** for line, and specify the first point at **11626006,6865528**.
10. Specify the next point using a Node snap to the point created from the last command.
11. Then type **LI** for List and select the line.
12. Civil 3D responds with:

```
LINE Layer: "0"
Space: Model space
Handle = 629
from point, X=11626006.0000 Y=6865528.0000 Z= 0.0000
to point, X=11626005.4554 Y=6865627.9936 Z= 112.4500
In Current UCS, Length = 99.9951, Angle in XY Plane = 90
3D Length = 150.4793, Angle from XY Plane = 48
Delta X = −0.5446, Delta Y = 99.9936, Delta Z = 112.4500
```

Notice the length of 99.9951, when you clearly entered 100′. The 100′ was the length along the face of the Earth; hence, the horizontal length of the line is something shorter. Save your drawing and remain in the drawing with your initials for the next exercise.

Exercise 3-12: Miscellaneous—Resection Command

 The next command is a **Resection**. A resection creates a point at a location calculated from the measured angles between three known points.

TIP This function is often used to locate a coordinate when you can see three other known positions and can measure their angles.

Job skills benefit in that sailors use this same technique to locate their positions on the sea. So if you plan sailing during your retirement after your CADD days are over, this routine will be helpful.

In this exercise, you are located to the upper right of Points 1, 2, and 3. Measure their angles as shown, where the angle from Point 2 to 3 is **28.7538** degrees and the angle from Point 2 to 1 is **56.7010** degrees.

1. From the **Create Points** toolbar, choose the pop-down arrow on the first button and select **Resection**. Follow this procedure.
2. `Specify first (backsight or reference) point:` Snap to Point 2 using a Node snap.
3. `Specify second point:` Snap to Point 3 using a Node snap.
4. `Specify third point:` Snap to Point 1 using a Node snap.
5. `Specify the angle between the first and second point <0.0000 (d)>:` Type **28.7538**
6. `Specify the angle between the first and third point <0.0000 (d)>:` Type **56.7010**
7. `Enter a point description <.>:` **IPF**
8. `Specify a point elevation <.>:` **0**

Notice in Figure 3-21 that Point 5 shows up at the convergence of the linework indicating the angles shot by the surveyor. Now you have a coordinate for your location. Save your drawing and remain in the drawing with your initials for the next exercise.

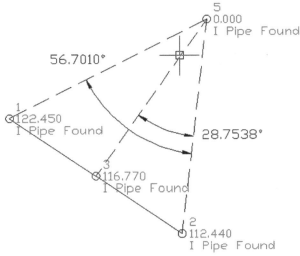

Figure 3-21   Result of Resection

Exercise 3-13: Miscellaneous—Station/Offset Object Command

 The next command, **Station/Offset Object**, creates a point based on the station and offset of a line or other object chosen for the computations. The offset is to the right if a positive value is indicated and to the left for a negative value.

 This command can be used to set a point such as a manhole at a specific location relative to a known linear or curved object nearby.

1. From the **Create Points** toolbar, choose the pop-down arrow on the first button and select **Station/Offset Object**. Follow this procedure.
2. `Select an arc, line, lot line, or feature line:` Pick the line connecting Points 2 to 1 near Point 2 but not on Point 2.
3. `Starting station <0.000>:` Hit **<Enter>**.
4. `Specify a desired station <0.000>:` Type **55<Enter>**
5. `Specify an offset <0.000>:` Type **-45<Enter>**
6. `Enter a point description <.>:` **IPF**
7. `Specify a point elevation <.>:` **0**
8. When asked to `Specify another desired station` hit **<ESC>** or **<*Cancel*>** several times to terminate the routine.
9. Notice Point 6 shows up 55′ from Point 2 in a direction toward Point 1 and 45′ to the left of the line. This works on several types of objects; however, only on independent objects. If you wish to set points using stations/offsets based on alignments, then use a command upcoming in the third button from the left called **Alignments**.

Save your drawing and remain in the drawing with your initials for the next exercise.

Exercise 3-14: Miscellaneous—Automatic Command

 The next command, **Automatic**, creates points based on the curvilinear objects selected by the operator.

 This method allows very rapid point placement on many objects at once and can be used to establish property corner points for a parcel.

1. In the same drawing, Chapter-3-Points.Dwg or Chapter-3-Points(with your initials).Dwg, turn on the Layer called **V-Misc-Objs**. Some lines and arcs will display. Type **Regen** if the arcs are not obvious.
2. From the **Create Points** toolbar, choose the pop-down arrow on the first button and select **Automatic**. Follow this procedure.
3. `Select arcs, lines, lot lines, or feature lines:` Select all of the linework that was on the Layer that you just turned on.
4. `Enter a point description <.>:` **IPF**
5. `Specify a point elevation <.>:` **<Enter>**
6. `Enter a point description <IPF>:` **<Enter>**
7. `Specify a point elevation <.>:` **<Enter>**
8. Repeat this for all other requests for point description and Elevation.
9. Hit **<Enter>** to terminate when complete.
10. Notice that points are automatically placed on all of the key geometric locations of the arcs and lines, including radius points of the arcs.

Student Files

Save your drawing and remain in the drawing with your initials for the next exercise.

Exercise 3-15: Miscellaneous—Along Line/Curve Command

 The next command, **Along Line/Curve**, creates points based on *curvilinear objects* at a distance along the object.

Curvilinear Objects—Arcs and spirals

1. In the same drawing, and from the **Create Points** toolbar, choose the pop-down arrow on the first button and select **Along Line/Curve**. Follow this procedure.
2. The software asks you to `Select an arc, line, lot line, or feature line:` **Pick the yellow arc on the Layer V-Misc-Objs.**
3. `Specify a distance:` Type **25**
4. `Enter a point description <.>:` **IPF**
5. `Specify a point elevation <.>:` **<Enter>**
6. It then asks you to again `Specify a distance:` **50**
7. `Enter a point description <IPF>:` **<Enter>**
8. `Specify a point elevation <.>:` **<Enter>**
9. `Specify a distance:` **<Enter>** to terminate.
10. You now have points (18 and 19) along the yellow arc at distances of 25 and 50 from the beginning of the arc. Refer to Figure 3-22.

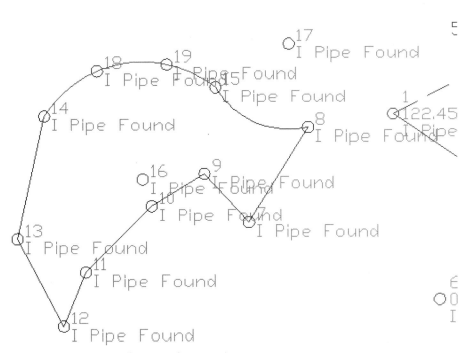

Figure 3-22   Result of Miscellaneous Along Line/Curve

Exercise 3-16: Miscellaneous—On Line/Curve Command

 The next command, **On Line/Curve**, creates points based on curvilinear objects at their key points. This command is similar to **Automatic** in its results.

Exercise 3-17: Miscellaneous—Divide Object Command

 The next command, **Divide Object**, creates points by the number of segments you request on an object. This command could be used to set stakeout points for landscape planting off to the side of a property line.

Greenspace—An area dedicated to the preservation of vegetation

> **TIP** In this case the landscape architect may have outlined that four equally spaced trees are to be planted in order to meet **Greenspace** requirements for the subdivision. You usually do not want to set stakeout points at the actual location of the prospective trees because as soon as the contractor begins digging, the stakes will be knocked out of the ground. So it usually is a good idea to set them safely off to the side of the construction area.

1. To begin this routine, **Draw** a line from **850,1050** to **930,1075**.
2. Then from the **Create Points** toolbar, choose the pop-down arrow on the first button and select **Divide Object**. Follow this procedure.
3. Civil 3D responds with `Select an arc, line, lot line, or feature line:` Select the line you just drew at the right end.
4. `Enter the number of segments <1>:` **3**
5. `Specify an offset <0.000>:` **25**
6. `Enter a point description <.>:` **IPF**
7. `Specify a point elevation <.>:` **<Enter>**
8. `Enter a point description <IPF>:` **<Enter>**
9. `Specify a point elevation <.>:` **<Enter>**
10. `Enter a point description <IPF>:` **<Enter>**
11. `Specify a point elevation <.>:` **<Enter>**
12. `Enter a point description <IPF>:` **<Enter>**
13. `Specify a point elevation <.>:` **<Enter>**
14. Save the drawing. You can close and open it as needed as you progress through these exercises.

The system places Points 20 through 23, in this case, where the line would have three segments including the endpoints, and at an offset of 25′. Of course you could set an offset of 0 and achieve the points right on the object. Save your drawing and remain in the drawing with your initials for the next exercise.

Exercise 3-18: Miscellaneous—Measure Object Command

 This exercise sets points using the **Create Points—Measure Object** command. Notice that the **Divide Object** and **Measure Object** commands seem like companion commands to the **DIVIDE** and **MEASURE** AutoCAD commands. They work essentially as they do in AutoCAD, except that they set point objects as shown in Figure 3-23.

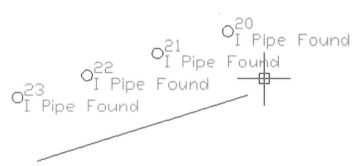

Figure 3-23   Result of Measure Object

> **TIP** This command could be used to set stakeout points to the side of a culvert. You usually do not want to set stakeout points right on the centerline of a culvert because as soon as the contractor begins digging, the stakes will be knocked out of the ground, rendering them useless. So it usually is a good idea to set them safely off to the side of the construction area.

Miscellaneous Polyline Vertices—Manual

 The next computations set points on polylines. The first example, using **Polyline Vertices—Manual**, sets points on a polyline using a manual approach. You might use this when you want to create points on a building pad that has a specific, flat elevation. Type in the desired elevation before setting the points, and the software will use it when creating the points.

Exercise 3-19: Miscellaneous—Polyline Vertices—Automatic Command

 The next computation, using **Polyline Vertices—Automatic**, sets points on polylines automatically. The previous command was pretty automatic, so how is this different? This one sets the elevations for the building pad automatically, assuming that the pad is set at the correct elevation to start with. You might use this when you want to create points on a building pad that has a preset, flat elevation.

1. Continuing on in the same drawing, erase the points from the previous command using the **ERASE** command in AutoCAD.
2. Now use the **MOVE** command in AutoCAD to move the building object up to an elevation of **100**. Type **M<Enter>**, select the pad and hit **<Enter>**. When the software says `Specify base point or displacement:` Type **0,0,100<Enter><Enter>**.
3. This has caused the building to move up to an elevation of 100′.
4. From the **Create Points** toolbar and using the pop-down arrow next to the first icon, choose **Polyline Vertices—Automatic**. The software responds with the following:
5. `Select a polyline object:` Select the polyline you just moved to elevation 100.
6. `Enter a point description <.>:` **IPF**
7. `Enter a point description <IPF>:` **<Enter>**
8. `Enter a point description <IPF>:` **<Enter>**
9. `Enter a point description <IPF>:` **<Enter>**
10. It will ask you to `Select a polyline object` when it has set points on your object. Hit **<Enter>** to terminate. Notice all the points are automatically set with an elevation of 100 because the software read it directly from the polyline.

Save your drawing and remain in the drawing with your initials for the next exercise.

Miscellaneous Convert AutoCAD Points

 The next tool uses **Convert AutoCAD Points** to create Civil 3D points from AutoCAD points. You might use this when you have a drawing that already has points; however, they are simple AutoCAD points, without any point numbers, elevation, or descriptions.

 This could occur when points have been created from a GIS system and you wish to use them in Civil 3D with intelligence.

Miscellaneous Convert Softdesk Point Blocks

 The last command in this area, **Convert Softdesk Point Blocks**, creates Civil 3D points from Softdesk points. This is a legacy item for users of Softdesk software. Refer back to the discussion in the introduction to this text for information about where Softdesk plays into the Civil 3D software.

You might use this command when you have a drawing that already has points in it that were generated by Softdesk software and you want them to be used in Civil 3D.

TIP

Intersection-Related Point Creation Commands

The commands that are investigated next are those related to intersections of a variety of objects. These types of commands have been used in surveying for a long time, but they were possibly referred to by other terms. For instance, the first command shown in Figure 3-24, called **Direction/Direction**, begins by asking for a point from a known position pointing and an angle toward a specific direction pointing toward infinity. It then asks for a second point and an angle also pointing toward infinity. Unless these angles are equal, producing a parallel set of vectors, they must intersect at some location. This is where a point will be created. Very often in surveying tasks, the angle would be estab-

Bearings—Angle measure   lished by using a *bearing*. This routine function then would typically be called a Bearing, Bearing intersect. Although the software does allow for the angle entry to be in bearings, it also allows for azimuths and other methods for establishing the angle; hence the name of the command is **Direction/Direction**, because it is not limited only to bearing entry. This next section of the text explores how these commands operate and identifies the reasons for using them.

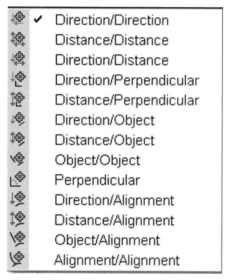

| | Direction/Direction |
| Distance/Distance |
| Direction/Distance |
| Direction/Perpendicular |
| Distance/Perpendicular |
| Direction/Object |
| Distance/Object |
| Object/Object |
| Perpendicular |
| Direction/Alignment |
| Distance/Alignment |
| Object/Alignment |
| Alignment/Alignment |

Figure 3-24   Direction/Direction

Exercise 3-20: Intersection—Direction/Direction Command

 This command is used to create a point that is the result of the intersection of two vectors.

This is also called a **Bearing/Bearing intersect** by many surveyors.

TIP

There are many reasons to use this routine, but one might be that a tangent is established for the centerline of roadway, let us call it Main Street. Another vector was established for First Avenue. While the surveyor was collecting the data, he or she determined that shooting a point at the intersection of First and Main was too dangerous due to traffic volume. However, the surveyor made a note that the vectors would continue as established, with no curves or additional bends until they met at the intersection. This would be a good example of using the **Direction/Direction** command to set the point at the intersection of First and Main.

1. Open the drawing called My Point Style-Intersections.Dwg.
2. Using the **V** for **View** command, highlight the view called **Points-A**, hit **Set Current**, and then hit **OK**.
3. Select the **Points** pull-down menu and choose the **Create Points…** command. The toolbar displays. First investigate the tools in the second button from the left. Notice the drop-down arrow to the right of the second icon from the left. It shows the information in the figure above starting with Direction/Direction. Select **Direction/Direction**.
4. The software requests that you `Specify a start point:` Type in an *X*, *Y* coordinate of **850, 950**.

A tick mark displays with a rubber-banding red vector emanating from the tick, indicating an angular data entry is required. The prompt allows for **B** for bearings entry, **Z** for Azimuth entry, or **A** for an AutoCAD-based angular entry (from the *X* axis).

5. The software responds with the following requests. `Specify direction at start point or [Bearing/aZimuth]:` Type **B** (for Bearing).
6. `Quadrants - NE = 1, SE = 2, SW = 3, NW = 4,` as shown in Figure 3-25.

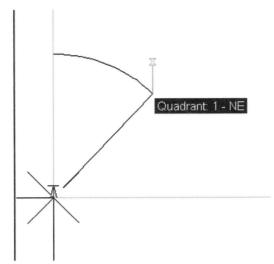

Quadrant 1 - NE

Figure 3-25   Quadrant angles

7. `Specify quadrant (1-4) or [aZimuth/Angle]:` Type **1**
8. `Specify bearing or [aZimuth/Angle]:` Enter **64.5631**
9. `Specify an offset <0.000>:` **<Enter>**

The software not only allows for a Bearing/Bearing intersect to be computed but also allows for a Bearing, offset at a specified distance to be computed to the intersection of another Bearing, offset at a specified distance. Hit **<Enter>** for the defaulted 0.000 for these offsets. This information sets the first vector and a tick mark with vector arrow displays. It now needs the second vector information to continue.

10. `Specify start point:` Enter **1000,950**
11. `Specify direction at start point or [Bearing/aZimuth]:` Type **B** (for Bearing).
12. `Quadrants - NE = 1, SE = 2, SW = 3, NW = 4`
13. `Specify quadrant (1-4) or [aZimuth/Angle]:` **4**
14. `Specify bearing or [aZimuth/Angle]:` **48.1752** (Figure 3-26).

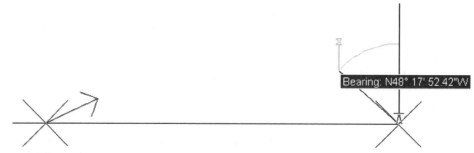

Figure 3-26   Entering Bearings

15. `Specify an offset <0.000>:` **<Enter>**
16. `Enter a point description <.>:` **IPF**
17. `Specify a point elevation <.>:` **<Enter>**
18. A point is now placed at the desired location (Figure 3-27).

Figure 3-27   Two angles established for Direction/Direction

Exercise 3-21: Intersection—Distance/Distance Command

 The next routine creates a point from the intersection of two circles. You establish a center point and a radius for the first circle; a center point and a radius for the second circle; and the system computes whether one or more solutions is found, tells you about it, and then asks you to select the one point that solves your problem or allows you to accept multiple points if they are found. This is called the **Distance/Distance** command.

 It may be used to compute an arc on a piece of property for which the radius and the starting and ending points of the arc are known, say from a curve table on a plat. Or perhaps little else is known and it needs to be drawn.

 In this on-the-job example, a parcel exists on a cul-de-sac. The side lot lines are known by distance and bearings on the plat, but the only information available to draw it is the arc radius. So the trick here is to locate the center of the arc because then AutoCAD can be used to draw the arc using the **Start, Center and End** option. This is a great use for the **Distance/Distance** command because the arc begins and ends at the ends of the property lines.

1. Continue in the same drawing, or Open the drawing called My Point Style-Intersections.Dwg.
2. Using the **V** for **View** command, highlight the view called **Points-B**, hit **Set Current**, and then hit **OK**. You see the property lines. An arc with a 50' radius will close the property.
3. Select the **Points** pull-down menu and choose the **Create Points…** command. The toolbar displays. First investigate the tools in the second button from the left. Notice the drop-down arrow to the right of the second icon from the left. Select **Distance/Distance**. The prompts follow.
4. `Please specify a location for the radial point:` Pick an Endpoint snap at the end of the open side of the left property line. The software then asks, `Enter radius <0.000>:` Type **50**
5. `Please specify a location for the radial point:` Pick an Endpoint snap at the end of the open side of the right property line. The software then asks, `Enter radius <0.000>:` Type **50**
6. The software computes two possible solutions and asks, `Point or [All] <All>:` Use the mouse to pick a point near the bottom green tick mark in Figure 3-28, because that is where the arc radius is intended to be.

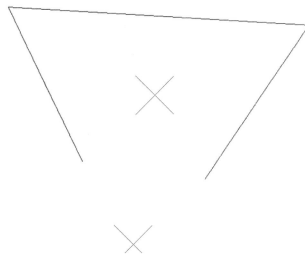

Figure 3-28   Distance/Distance command

7. It then asks, `Enter a point description <.>:` **IPF**
8. `Specify a point elevation <.>:` **<Enter>**
9. Hit **<ESC>** to terminate the command when asked to `Please specify a location for the radial point:`, because the software thinks you will continue with more problems.
10. Now type **A** for **Arc** in AutoCAD; using the Endpoint Osnap select the open end of the right parcel line. Then type **C** for **Center** of arc and use the Node Osnap to select the point you just located. Finish by using the Endpoint Osnap and select the open end of the left parcel line. An arc now exists representing the cul-de-sac.
11. Save the file.

Exercise 3-22: Intersection—Direction/Distance Command

Civil 3D performs many more related functions but varies the parameters in each solution. Continue with these functions so that you try each one at least once.

> **TIP** Perhaps you are trying to identify a conflict with a 200-year-old elm tree. The proposed roadway may be the direction vector and the tree needs to have a 55' clearance in order remain healthy. Does the proposed roadway conflict with this clear zone?

1. Continue in the same drawing, or open the drawing called My Point Style-Intersections.Dwg.
2. Using the **Layer** command, turn on the layer called **V-Misc-Objs**.
3. Using the **V** for **View** command, highlight the view called **Points-A**, hit **Set Current**, and then hit **OK**. You see a vector arrow and a circle that you need to compute points for. Compute one or more points for where the vector crosses the circle. It has a radius of 55'.
4. Select the **Points** pull-down menu if the **Create Points** toolbar is not displayed, and choose the **Create Points**... command. The **Create Points** toolbar displays.

You are still investigating the tools in the second button from the left. Using the drop-down arrow to the right of the second icon from the left, choose **Direction/Distance**. The prompts follow.

5. `Please specify a location for the radial point:` Pick a point at the center of the circle in the view using the Center Osnap.
6. `Enter radius <0.000>:` Type **55** for the radius.
7. `Specify start point:` Pick the left end of the vector arrow using the Endpoint Osnap.
8. `Specify direction at start point or [Bearing/aZimuth]:` Pick the end of the vector arrow at the arrowhead using the Endpoint Osnap.
9. `Specify an offset <0.000>:` **<Enter>** Accept 0.000 because you will not be offsetting the vector as it intersects with the circle.
10. The software solves the problem and offers two potential solutions, as shown in Figure 3-29. You can pick a point near the green tick mark to indicate your choice of solution or hit **<Enter>** for **All**. `Point or [All] <All>:` **<Enter>**

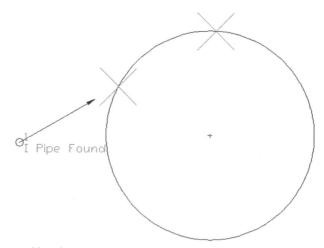

Figure 3-29   Two possible solutions

11. `Enter a point description <.>:` **IPF**
12. `Specify a point elevation <.>:` **<Enter>**
13. `Enter a point description <.>:` **IPF**
14. `Specify a point elevation <.>:` **<Enter>**
15. When asked to continue the routine with the prompt: `Please specify a location for the radial point:` Hit **<ESC>** to cancel out and terminate.

Two points exist on the perimeter of the circle where the vector would cross it, if it did. So in this example, there is a conflict between the proposed road and the tree's clear zone. The road must be moved a little farther away from the tree. Refer to Figure 3-30.

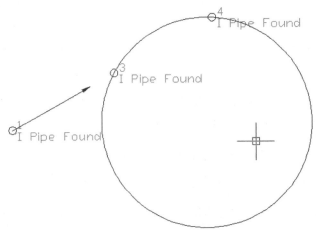

Figure 3-30   Two points are set

Exercise 3-23: Intersection—Direction/Perpendicular Command

 The next command, **Direction/Perpendicular**, can be used to compute the location of a point on a property line from the corner of a building.

TIP   We often must locate these to ensure that we have met minimum clearances to the building restriction lines or property lines as dictated by the local ordinances.

1. Continue in the same drawing, or open the drawing called My Point Style-Intersections.Dwg.
2. Using the **Layer** command, turn on the layer called **V-Misc-Objs**.
3. Using the **V** for **View** command, highlight the view called **Points-C**, hit **Set Current**, and then hit **OK**. You see a property parcel surrounding a house. Compute a point where the house corner is perpendicular to the right property line.
4. Using the drop-down arrow to the right of the second icon from the left, choose **Direction/Perpendicular**. The prompts follow.
5. `Specify start point:` Pick a point using the Endpoint Osnap on the property corner at the top of the parcel.
6. `Specify direction at start point or [Bearing/aZimuth]:` Pick a point using the Endpoint Osnap on the property corner at the bottom right of the parcel. A yellow vector appears on the right parcel line. Refer to Figure 3-31.
7. `Specify an offset <0.000>:` Type **10** indicating a 10' building restriction line.
8. `Please specify a location for the perpendicular point:` Using the Endpoint Osnap pick a point on the right corner of the structure as shown in the figure.
9. `Enter a point description <.>:` Hit **<Enter>**
10. `Specify a point elevation <.>:` Hit **<Enter>**
11. When asked to continue the routine with the prompt: `Specify start point:` Hit **<ESC>** to cancel out and terminate.

Student Files

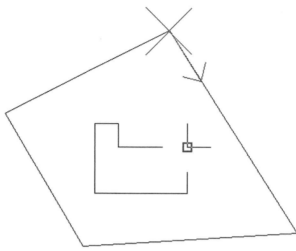

Figure 3-31   Direction/Perpendicular command

Exercise 3-24: Intersection—Distance/Perpendicular Command

 The next command, **Distance/Perpendicular,** can be used to compute the location of a point on a property line from the corner of a building to a circular set of parameters.

 This command can be used to set a point on the arc of a cul-de-sac radially from the corner of a building.

1. Continue in the same drawing, or open the drawing called My Point Style-Intersections.Dwg.
2. Using the **Layer** command, turn on the layer called **V-Misc-Objs**.
3. Using the **V** for **View** command, highlight the view called **Points-B**, hit **Set Current**, and then hit **OK**. You see a property parcel surrounding a house. Compute a point where the house corner is perpendicular (or radial) to the property line.
4. Using the drop-down arrow to the right of the second icon from the left, choose **Distance/Perpendicular**. The prompts follow.
5. `Please specify a location for the radial point:` Using the Center Osnap pick a point on the arc to locate the radius point of the arc.
6. `Enter radius <0.000>:` Using the Endpoint Osnap pick a point at the end of the arc so it can compute the radius from the two points.
7. `Please specify a location for the perpendicular point:` Using the Endpoint Osnap pick a point at the corner of the building nearest the arc.
8. `Enter a point description <.>:` **<Enter>**
9. `Specify a point elevation <.>:` **<Enter>**
10. When asked to continue the routine with the prompt: `Please specify a location for the radial point:` Hit **<Esc>** to cancel out and terminate.

You see a point placed on the arc, radially from the corner of the house in Figure 3-32.

Exercise 3-25: Intersection—Direction/Object Command

 The next command, **Direction/Object**, can be used to compute a point along a direction toward an object.

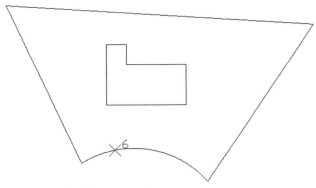

Figure 3-32   Distance/Perpendicular command

An example of use for this command is to set the location of a point on a property line from the corner of a building to a circular set of parameters. You want to compute a point where the culvert crosses the back property line.

Job skills benefit from being able to perform geometry commands such as these because they are required skills for every designer and surveyor.

Student Files

1. Continue in the same drawing, or open the drawing called My Point Style-Intersections.Dwg.
2. Using the **V** for **View** command, highlight the view called **Points-D**, hit **Set Current**, and then hit **OK**. You see a property parcel surrounding a house. Compute a point where the culvert crosses the back property line.
3. Using the drop-down arrow to the right of the second icon from the left, choose **Direction/Object**. The prompts follow.
4. `Select an arc, line, lot line, or feature line:` Pick the back property line.
5. `Specify an offset <0.000>:` **<Enter>**
6. `Specify start point:` Using the Endpoint Osnap pick a point at the top end of the culvert line.
7. `Specify direction at start point or [Bearing/aZimuth]:` Using the Endpoint Osnap pick a point at the bottom end of the culvert line.
8. `Specify an offset <0.000>:` **<Enter>**
9. `Enter a point description <.>:` **<Enter>**
10. `Specify a point elevation <.>:` **<Enter>**
11. `Select an arc, line, lot line, or feature line:` Hit **<ESC>** to cancel out and terminate.

A point is placed on the property line where the culvert crosses it. Refer to Figure 3-33.

Exercise 3-26: Intersection—Distance/Object Command

 The next command, **Distance/Object**, can be used to compute the location of point(s) at a distance from an object to another object.

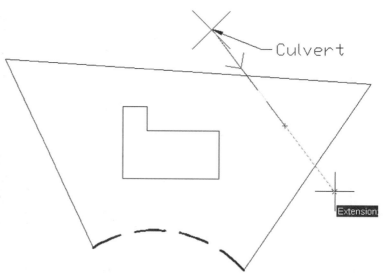

Figure 3-33   Direction/Object command

TIP

This would allow placement of point(s) on a property line from, say, the corner of a building to a minimum distance from the building corner.

Job Skills

A Job skill evident in a command such as this might include being able to compute point(s) for trees to be placed on the property line such that they are not closer to the house than 35'.

Student Files

1. Continue in the same drawing, or open the drawing called My Point Style-Intersections.Dwg.
2. Using the **V** for **View** command, highlight the view called **Points-D**, hit **Set Current**, and then hit **OK**. You see a property parcel surrounding a house. Compute point(s) on the right property line to plant trees no closer than 35' from the house.
3. Using the drop-down arrow to the right of the second icon from the left, choose **Distance/Object**. The prompts follow.
4. `Select an arc, line, lot line, or feature line:` **Select the right property line.**
5. `Specify an offset <0.000>:` **<Enter>**
6. `Please specify a location for the radial point:` **Using the Endpoint Osnap pick a point at the bottom right corner of the house.**
7. `Enter radius <0.000>:` **35**
8. `Point or [All] <All>:` **<Enter>**
9. `Enter a point description <.>:` **<Enter>**
10. `Specify a point elevation <.>:` **<Enter>**
11. `Enter a point description <.>:` **<Enter>**
12. `Specify a point elevation <.>:` **<Enter>**
13. `Select an arc, line, lot line, or feature line:` **Hit <ESC> to cancel out and terminate.**

Two points (Figure 3-34) are placed on the property line, each at a minimum distance of 35' from the house corner.

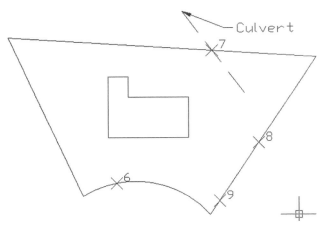

Figure 3-34   Distance/Object command

Exercise 3-27: Intersection—Object/Object Command

 The next command, **Object/Object**, can be used to compute the location of a point that results from the intersection of two objects.

 TIP
You may want to compute a point where the culvert would intersect the right (or east) property line if it were to be extended.

 Student Files

1. Continue in the same drawing, or open the drawing called My Point Style-Intersections.Dwg.
2. Using the **V** for **View** command, highlight the view called **Points-D**, hit **Set Current**, and then hit **OK**. You see a property parcel surrounding a house. Compute a point where the culvert crosses the right property line if it were to be extended.
3. Using the drop-down arrow to the right of the second icon from the left, choose **Object/Object**. The prompts follow.
4. `Select an arc, line, lot line, or feature line:` Select the right property line.
5. `Specify an offset <0.000>:` **<Enter>**
6. `Select an arc, line, lot line, or feature line:` Select the culvert line.
7. `Specify an offset <0.000>:` **<Enter>**
8. `Enter a point description <.>:` **<Enter>**
9. `Specify a point elevation <.>:` **<Enter>**
10. `Select an arc, line, lot line, or feature line:` Hit **<ESC>** to cancel out and terminate.

Notice a new point was placed on the right property line where the culvert would cross it if it did (Figure 3-35).

Exercise 3-28: Intersection—Perpendicular Command

 The next command, **Perpendicular**, can be used to compute the location of a point that results from the intersection of two objects. This command allows you to compute the perpendicular easily.

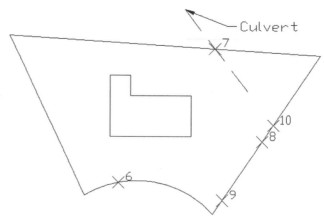

Figure 3-35   Object/Object command

Many objects in civil engineering are perpendicular to items and this routine simplifies these computations. For instance, side lot lines are usually perpendicular to front or back property lines, or side roads are typically perpendicular to main roads. Because computing a perpendicular is relatively easy, doing so can cut down on construction and surveying stakeout errors.

1. Continue in the same drawing, or open the drawing called My Point Style-Intersections.Dwg.
2. Using the **V** for **View** command, highlight the view called **Points-D**, hit **Set Current**, and then hit **OK**. You see a property parcel surrounding a house. Compute a point perpendicular to the back property line from a point outside the parcel.
3. Using the drop-down arrow to the right of the second icon from the left, choose **Perpendicular**. The prompts follow.
4. `Select an arc, line, lot line, or feature line:` **Select the back property line.**
5. `Please specify a location for the perpendicular point:` **890,1570**
6. `Enter a point description <.>:` **<Enter>**
7. `Specify a point elevation <.>:` **<Enter>**
8. `Select an arc, line, lot line, or feature line:` **Hit <ESC> to cancel out and terminate.**

A new point has been placed along the back property line in Figure 3-36, indicating the perpendicular intersection to the property line from the point beginning at 890,1570.

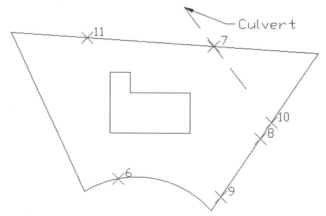

Figure 3-36   Perpendicular command

The next several commands all have to do with computing points from criteria that intersect with alignments.

Exercise 3-29: Intersection—Direction/Alignment Command

 The next command, **Direction/Alignment**, can be used to compute the location of a point that results from the intersection of an angle with an alignment.

> You might use this command to compute a point using the incoming bearing of a side road and where it intersects the main roadway, which has been defined as an alignment. The example might be similar to that used in the **Direction/Direction** command; however, this uses a predefined alignment.

1. Continue in the same drawing, or open the drawing called My Point Style-Intersections.Dwg.
2. Using the **V** for **View** command, highlight the view called **Points-E**, hit **Set Current**, and then hit **OK**. You see an alignment. Compute a point where a bearing crosses the alignment.
3. Using the drop-down arrow to the right of the second icon from the left, choose **Direction/Alignment**. The prompts follow.
4. `Select an alignment object:` Select someplace on the alignment.
5. `Specify an offset <0.000>:` **<Enter>**
6. `Specify start point:` **50,1500**
7. `Specify direction at start point or [Bearing/aZimuth]:` **B**
8. `The system reminds us of the quadrant numbers. Quadrants - NE = 1, SE = 2, SW = 3, NW = 4`
9. `Specify quadrant (1-4) or [aZimuth/Angle]:` **3**
10. `Specify bearing or [aZimuth/Angle]:` **60.4931**
11. `Specify an offset <0.000>:` **<Enter>**
12. `Enter a point description <.>:` **<Enter>**
13. `Specify a point elevation <.>:` **<Enter>**
14. `Select an alignment object:` Hit **<ESC>** to cancel out and terminate.

Notice that a new point has been placed on the alignment between stations 3+50 and 4+00.

Exercise 3-30: Intersection—Distance/Alignment Command

 The next command, **Distance/Alignment**, can be used to compute the location of a point that results from the intersection of a circular distance from a point with an alignment.

> This command could be used to compute a point 100′ from the end of an alignment.

1. Continue in the same drawing, or open the drawing called My Point Style-Intersections.Dwg.
2. Using the **V** for **View** command, highlight the view called **Points-E**, hit **Set Current**, and then hit **OK**. You see an alignment. Compute a point 100' from the end of the alignment.

3. Using the drop-down arrow to the right of the second icon from the left, choose **Distance/Alignment**. The prompts follow.
4. `Select an alignment object:` **Select someplace on the alignment.**
5. `Specify an offset <0.000>:`**<Enter>**
6. `Please specify a location for the radial point:` Using the Endpoint Osnap pick a point at the northern end of the alignment.
7. `Enter radius <0.000>:` **100**
8. `Enter a point description <.>:` **<Enter>**
9. `Specify a point elevation <.>:` **<Enter>**
10. `Select an alignment object:` Hit **<ESC>** to cancel out and terminate.

Notice that a point was placed at around station 6+00 on the alignment.

Exercise 3-31: Intersection—Object/Alignment Command

 The next command, **Object/Alignment**, can be used to compute the location of a point that results from the intersection of a circular distance from a point with an alignment.

 TIP You might use this command to compute a point on the alignment where a culvert crosses the alignment.

 Job Skills The job skill apparent in this routine is that designers and surveyors need these types of computations on a daily basis because utilities intersect with alignments routinely.

1. Continue in the same drawing, or open the drawing called My Point Style-Intersections.Dwg.
2. Using the **V** for **View** command, highlight the view called **Points-E**, hit **Set Current**, and then hit **OK**. You see an alignment. Compute a point on the alignment where a culvert crosses the alignment.
3. Using the drop-down arrow to the right of the second icon from the left, choose **Object/Alignment**. The prompts follow.
4. `Select an alignment object:` **Select someplace on the alignment.**
5. `Specify an offset <0.000>:`**<Enter>**
6. `Select an arc, line, lot line, or feature line:` Select someplace on the culvert that is around station 1+25.
7. `Specify an offset <0.000>:` **<Enter>**
8. `Enter a point description <.>:` **<Enter>**
9. `Specify a point elevation <.>:` **<Enter>**
10. `Select an alignment object:` Hit **<ESC>** to cancel out and terminate.

Notice that a point was placed at around station 1+25 on the alignment exactly where the culvert crossed the alignment.

Exercise 3-32: Intersection—Alignment/Alignment Command

 The next command, **Alignment/Alignment**, can be used to compute the location of a point that results from the intersection of a circular distance from a point with an alignment.

Student Files

TIP

> You might use this command to compute a point where a 12' edge of pavement on the main road meets the 12' edge of pavement for the side street.

1. Continue in the same drawing, or open the drawing called My Point Style-Intersections.Dwg.
2. Using the **Layer** command, turn on the layer called **V-Misc-Objs**.
3. Using the **V** for **View** command, highlight the view called **Points-F**, hit **Set Current**, and then hit **OK**. You see an alignment. Compute a point where a 12' edge of pavement on the main road meets the 12' edge of pavement for the side street.
4. Using the drop-down arrow to the right of the second icon from the left, choose **Alignment/Alignment**. The prompts follow.
5. `Select an alignment object:` Select someplace on the longer alignment (it has a curve in it).
6. `Specify an offset <0.000>:` **12**
7. `Select an alignment object:` Select someplace on the shorter alignment (it has no curve in it).
8. `Specify an offset <12.000>:` **<Enter>**
9. `Enter a point description <.>:` **<Enter>**
10. `Specify a point elevation <.>:` **<Enter>**
11. `Select an alignment object:` Hit **<ESC>** to cancel out and terminate.

Notice that a point was placed at around station 3+00 on the main road alignment at a 12' offset. If an offset of **-12** were provided, the point would be on the left side of the roadway.

Alignment Related Point Commands

These commands and exercises conclude the Intersection-based commands in Civil 3D. The next set of Point Creation commands have to do with Alignments. An alignment must exist for these to do their job.

Exercise 3-33: Alignment—Station/Offset Command

 The next command, **Station/Offset**, can be used to compute the location of a point(s) that have station and offset values relative to an alignment.

TIP

> This command might be used to locate Planimetric features such as fire hydrants or mailboxes at a variety of stations when they are a standard distance off the face of curb.

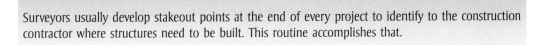 Surveyors usually develop stakeout points at the end of every project to identify to the construction contractor where structures need to be built. This routine accomplishes that.

Job Skills

1. Continue in the same drawing, or open the drawing called My Point Style-Intersections.Dwg.

2. Using the **V** for **View** command, highlight the view called **Points-G**, hit **Set Current**, and then hit **OK**. You see an alignment. Compute points representing mailboxes at the stations provided at offsets that are 16' to the right and left of the roadway centerline.

3. Using the drop-down arrow to the right of the third icon from the left, choose **Station/Offset**. The prompts follow.

4. `Select an alignment:` **Select someplace on the alignment.**

5. `Specify station:` **75**

6. `Specify an offset <0.000>:` **16** (which indicates 16' to the right of the alignment).

7. `Enter a point description <.>:` **MB**

8. `Specify a point elevation <.>:` **<Enter>**

9. `Specify station:` **75**

10. `Specify an offset <16.000>:` **-16**

11. `Enter a point description <MB>:` **<Enter>**

12. `Specify a point elevation <.>:` **<Enter>**

13. `Specify station:` **150**

14. `Specify an offset <-16.000>:` **<Enter>**

15. `Enter a point description <MB>:` **<Enter>**

16. `Specify a point elevation <.>:` **<Enter>**

17. `Specify station:` **150**

18. `Specify an offset <-16.000>:` **16**

19. `Enter a point description <MB>:` **<Enter>**

20. `Specify a point elevation <.>:` **<Enter>**

21. `Specify station:` **225**

22. `Specify an offset <16.000>:` **<Enter>**

23. `Enter a point description <MB>:` **<Enter>**

24. `Specify a point elevation <.>:` **<Enter>**

25. `Specify station:` **225**

26. `Specify an offset <16.000>:` **-16**

27. `Enter a point description <MB>:` **<Enter>**

28. `Specify a point elevation <.>:` **<Enter>**

29. `Specify station:` **300**

30. `Specify an offset <-16.000>:` **<Enter>**

31. `Enter a point description <MB>:` **<Enter>**

32. `Specify a point elevation <.>:` **<Enter>**

33. `Specify station:` **300**

34. `Specify an offset <-16.000>:` **16**

35. `Enter a point description <MB>:` **<Enter>**

36. `Specify a point elevation <.>:` **<Enter>**

37. `Specify station:` **<Enter>** to terminate.

Figure 3-37 shows eight points representing mailboxes that will be placed along the alignment at the specified stations and 16' to the right and left of the roadway up to station 3+00.

Exercise 3-34: Alignment—Divide Alignment Command

 The next command, **Divide Alignment**, can be used to compute the location of a point(s) that have station and offset values relative to an alignment.

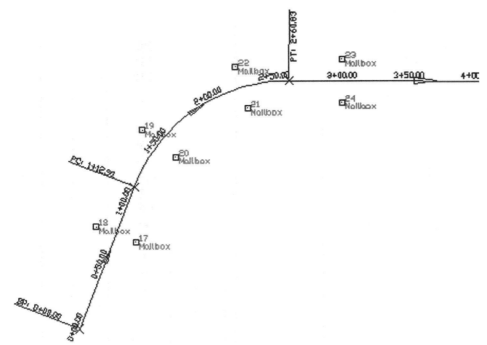

Figure 3-37   Station/Offset command

TIP

This might be used to compute points for the excavation for tree bulbs such that there are twenty-two trees planted at an offset of 50' to the right side of the roadway following construction.

1. Continue in the same drawing, or open the drawing called My Point Style-Intersections.Dwg.
2. Using the **V** for **View** command, highlight the view called **Points-G**, hit **Set Current**, and then hit **OK**. You see an alignment. Compute points for the excavation for tree bulbs such that there are twenty-two trees planted on the right side of the roadway following construction.
3. Using the drop-down arrow to the right of the third icon from the left, choose **Divide Alignment**. The prompts follow.
4. `Select an alignment:` Select someplace on the alignment.
5. `Enter the number of segments <1>:` **21**
6. `Specify an offset <0.000>:` **50**, which indicates that they will be placed to the right side.
7. `Enter a point description <.>:` **<Enter>**
8. `Specify a point elevation <.>:` **<Enter>**

Continue hitting **<Enter>** as the software prompts for point descriptions and point elevations until it is finished putting twenty-two points on the right side of the road. When it asks you to `Select alignment:` hit **<Enter>** to terminate. You see the points on the right side of the road.

Exercise 3-35: Alignment—Measure Alignment Command

The next command, **Measure Alignment**, can be used to compute the location of a point(s) that have station and offset values relative to an alignment.

TIP This might be used to compute points for the stakeout of the road. You may want points placed at 25' intervals at an offset of 50' to the left side of the roadway.

A large part of a surveyor's job is to produce construction stakeout points for roadways and this is exactly how they would be developed.

1. Continue in the same drawing, or open the drawing called My Point Style-Intersections.Dwg.
2. Using the **V** for **View** command, highlight the view called **Points-G**, hit **Set Current**, and then hit **OK**. You see an alignment. Place points at 25' intervals at an offset of 50' to the left side of the roadway
3. Using the drop-down arrow to the right of the third icon from the left, choose **Measure Alignment**. The prompts follow.
4. `Select alignment:` **Select someplace on the alignment.**
5. `Starting station <0+00.00>:` **<Enter>**
6. `Ending station <6+58.53>:` **<Enter>**
7. `Specify an offset <0.000>:` **-50**, which indicates that the points will be on the left side for the road.
8. `Enter an interval <10.000>:` **25**
9. `Enter a point description <.>:` **<Enter>**
10. `Specify a point elevation <.>:` **<Enter>**

Continue hitting **<Enter>** as the software prompts for point descriptions and point elevations until it is finished putting points on the left side of the road. When it asks you to `Select alignment:` hit **<Enter>** again to terminate. You see the points on the left side of the road in Figure 3-38.

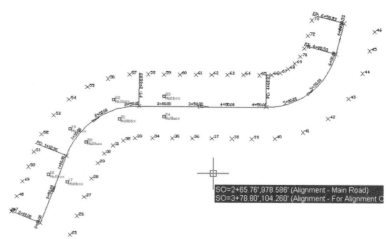

Figure 3-38   Measure Alignment command

Exercise 3-36: Alignment—At PC, PT, SC, Etc. Command

The next command, **At PC, PT, SC, Etc.**, can be used to compute the location of a point(s) that have station and offset values relative to an alignment.

You want points placed at the geometrically important locations of the alignment, namely the Points of Curvature (PCs), Points of Tangency (PTs), Spiral to Curve, Radius Points, and Points of Intersections (or PIs). If spirals are included, their corresponding points will be placed as well. These are defined and detailed in Chapter 4 on alignments. With these points identified, a surveyor has the criteria needed to lay out the centerline of the road on the construction site.

1. Continue in the same drawing, or open the drawing called My Point Style-Intersections.Dwg.
2. Using the **V** for **View** command, highlight the view called **Points-G**, hit **Set Current**, and then hit **OK**. You see an alignment. Place points at the geometrically important locations of the alignment.
3. Using the drop-down arrow to the right of the third icon from the left, choose **At PC, PT, SC, Etc**. The prompts follow.
4. `Select alignment:` Select someplace on the alignment.
5. `Starting station <0+00.00>:` **<Enter>**
6. `Ending station <6+58.53>:` **<Enter>**
7. `Specify a point elevation <.>:` **<Enter>**
8. `Specify a point elevation <.>:` **<Enter>**

Continue hitting **<Enter>** as the software prompts for point elevations until it is finished putting points on the alignment. When it asks you to `Select alignment:` hit **<Enter>** again to terminate. You see the points located at all of the critical geometric points of the road. The point descriptions are automatic because they relate to the point being located such as the BOA (Begin of Alignment), EOA (End of Alignment), PC, PT, PI, or RP.

Exercise 3-37: Alignment—Radial or Perpendicular Command

The next command, **Radial or Perpendicular**, can be used to compute the location of a point(s) that have station and offset values relative to an alignment.

You often need to compute points at radial or perpendicular locations from an alignment.

This command can be used to locate where a side street may intersect the roadway.

1. Continue in the same drawing, or open the drawing called My Point Style-Intersections.Dwg.
2. Using the **V** for **View** command, highlight the view called **Points-G**, hit **Set Current**, and then hit **OK**. You see an alignment. Place points at radial or perpendicular locations on the alignment.
3. Using the drop-down arrow to the right of the third icon from the left, choose **Radial or Perpendicular**. The prompts follow.

4. Select alignment: **Select someplace on the alignment.**
5. Specify a point that is radial or perpendicular to the current alignment: **Pick a point using the Endpoint Osnap at the end of the arc at station 5+77.23. The software will confirm this as Station: 5+77.23 with Offset: 74.096' in Figure 3-39.**

Figure 3-39   Radial or Perpendicular

6. **It will then ask:** Enter a point description <.>: **<Enter>**
7. Specify a point elevation <.>: **<Enter>**
8. Select alignment: **<Enter> to terminate.**

Exercise 3-38: Alignment—Import from File Command

 The next command, **Import from File**, can be used to import points that were generated based on an alignment and use station, offset values relative to that alignment. The software uses the next available point numbers and prompts for the information needed as it goes.

1. Continue in the same drawing, or open the drawing called My Point Style-Intersections.Dwg.
2. Using the **V** for **View** command, highlight the view called **Points-G**, hit **Set Current**, and then hit **OK**. You see an alignment. Import alignment based points.
3. Using the drop-down arrow to the right of the third icon from the left, choose **Import from File**.
4. A dialog box opens called **Import Alignment Station and Offset File**. Select the file called **Points-import file-sta-off.txt** in the Support files folder for this textbook and hit the **Open** button.
5. The software then prompts you to choose the format for this file. The options are as follows: 1. Station, Offset, 2. Station, Offset, Elevation, 3. Station, Offset, Rod, hi, 4. Station, Offset, Description, 5. Station, Offset, Elevation, Description, 6. Station, Offset, Rod, hi Description.
6. Enter file format (1/2/3/4/5/6): <0>: **1** (This file uses format 1, station, offset.)
7. It then needs to know what the delimiter is between data items. Our file uses commas.
8. 1. Space, 2. Comma
9. Enter a delimiter (1/2): <0>: **2**
10. Enter an invalid station/offset indicator <-99999>: **<Enter>**
11. Select alignment: **<Enter> to terminate.**

Notice that several points enter the file to the left and right of the alignment in Figure 3-40.

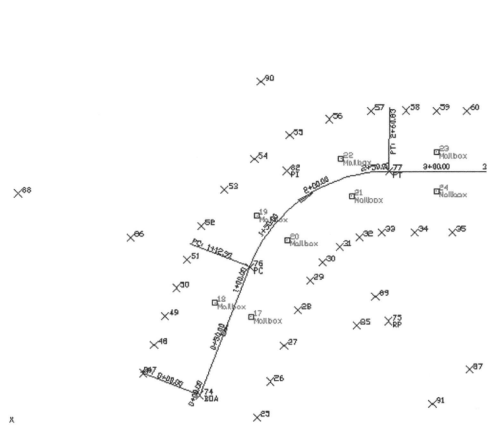

Figure 3-40   Import from File

This concludes the segment on placing Alignment-based points into your project. The file My Point Style-Intersections-Complete.Dwg shows the results of these examples.

Surface-Related Point Commands

This next segment delves into procedures for placing Surface-related points into a project drawing.

The next command in this toolbar item is "create new points" based on an alignment's profile information. This can be used for creating 3D stakeout data of the roadway's crown, for example. The command is called Profile Geometry Points. The user is requested to select the alignment and then to select the profile in question. The software then sets the points on the alignment with 3D elevations.

The ability to adjust point elevations based on a surface has also been added. This is a very beneficial routine if points have been set to a surface using the Random Points command, and the surface is modified and updated, then the points that were dependent on that surface can be updated automatically. Change the elevation for a point or a group of points by selecting a location on a surface. This command can be found under the **Points** pulldown >> **Edit... >> Elevations from surface....**

Exercise 3-39: Surface—Random Points Command

 The next command, **Random Points** (Figure 3-41), can be used to set points based on surface data.

 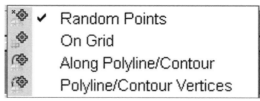

Figure 3-41   Random Points command

 The elevation values for these points are extracted directly from the surface information. This avoids having to hand calculate the elevations and place them in manually. Hand calculations are prone to typographical errors and miscalculation.

The software uses the next available point numbers and prompts for other information as it goes.

1. Open the drawing called Chapter 3-a.Dwg.
2. Using the **V** for **View** command, highlight the view called **Points-H**, hit **Set Current**, and then hit **OK**. You see a surface. Set some points based on this surface.
3. In the **Create Points** toolbar, and using the drop-down arrow to the right of the fourth icon from the left, choose **Random Points**.

You see some closed contours in the view (Figure 3-42), and you will be placing a couple of points within the closed contours to give others an idea of the elevations in this area. The prompts follow.

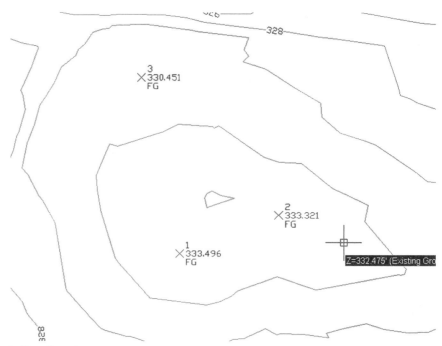

Figure 3-42   Closed contours

4. `Select a surface object:`
5. `Please specify a location for the new point:` Select a point within one of the open contours, as shown in Figure 3-42, for Point 1.
6. `Enter a point description <.>:` **FG**
7. `Please specify a location for the new point:` Select a point within one of the open contours, as shown in Figure 3-42, for Point 2.
8. `Enter a point description <FG>:` **<Enter>**
9. `Please specify a location for the new point:` Select a point within one of the open contours, as shown in Figure 3-42, for Point 3.
10. `Enter a point description <FG>:` **<Enter>**
11. `Please specify a location for the new point:` **<Enter>**
12. `Select a surface object:` **<Enter>** to terminate.

Exercise 3-40: Surface—On a Grid Command

The next command, **ON A GRID**, can be used to set points in a grid based on surface data. The elevation values are extracted directly from the surface information. This avoids having to hand calculate the elevations and place them in manually. The software uses the next available point numbers and prompts for other information as it goes.

You often need to place points on the corners of a grid overlaid onto a surface. It provides a representative sampling of the elevations occurring on that surface.

If this command is used on a Grid Volume type of surface, you can achieve cut and fill values of the proposed work as compared with those of the existing surface.

1. Continue in the same drawing, or open the drawing called Chapter 3-a.Dwg.
2. Using the **V** for **View** command, highlight the view called **Points-H**, hit **Set Current**, and then hit **OK**. You see a surface. Set some points based on this surface.
3. Using the drop-down arrow to the right of the fourth icon from the left, choose **On a Grid**.
4. `Select a surface object:` Pick any contour for the surface.
5. `Specify a grid basepoint:` **14900,18275**
6. `Grid rotation <0.0000 (d)>:` **<Enter>**
7. `Grid X spacing <5.0000>:` **50**
8. `Grid Y spacing <50.0000>:` **<Enter>**
9. `Specify the upper right location for the grid:` **15150,18475**
10. `Change the size or rotation of the grid/grid squares [Yes/No] <No>:` **n**
11. `Enter a point description <.>:` **FG**
12. Continue hitting **<Enter>** as the software prompts for point descriptions until it is finished putting points on the grid. When it asks you to `Select a surface` again, hit **<Enter>** to terminate. See Figure 3-43.

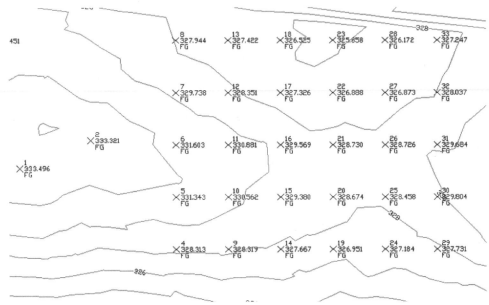

Figure 3-43   Points set On a Grid

Exercise 3-41: Surface—Along Polyline/Contour Command

 The next command, **Along Polyline/Contour**, can be used to set points on a polyline or contour at a user-defined interval. The elevation values are extracted directly from the surface information. This can be used to set points on an alignment directly from the surface in 3D. The software uses the next available point numbers and prompts for other information as it goes.

TIP This command can be used to set the rim elevations of manholes with a distance of 300' between manholes on the proposed road surface.

1. Continue in the same drawing, or open the drawing called Chapter 3-a.Dwg.
2. Using the **V** for **View** command, highlight the view called **Points-H**, hit **Set Current**, and then Hit **OK**. You see a surface and a polyline along the road running east to west. Set some points based on this surface right on the polyline.
3. Using the drop-down arrow to the right of the fourth icon from the left, choose **Along Polyline/Contour**.
4. `Select a surface object:` Pick any contour for the surface.
5. `Distance between points <10.0000>:` **50**
6. `Select a polyline or contour:` Select the Polyline with the heavy lineweight.
7. `Enter a point description <.>:` **FG**

Notice in Figure 3-44 that the polyline now has points every 50' along it.

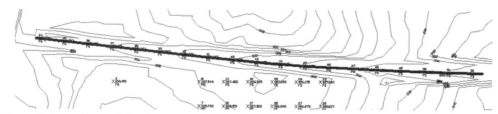

Figure 3-44   Along Polyline/Contour

Surface—Polyline/Contour Vertices Command

 The next command, **Polyline/Contour Vertices**, can be used to set points on a polyline or contour at the vertices of the polyline, and elevation values are extracted directly from the surface information. This can be used to set points on a building pad when the surface already has the pad elevations built into it. The software uses the next available point numbers and prompts for other information as it goes.

> **TIP** This routine can be used for many tasks such as establishing a daylight buffer for grading a site out. A polyline can be drawn at an offset from the property boundary and points can be established on this polyline from the existing ground surface. When those points are included in the Proposed ground surface, they ensure a tie-out to the existing ground. This example sets on the centerline based on the existing ground surface.

Notice that points now exist on the building corners.

This concludes our segment on placing Surface-based points into your project.

Exercise 3-42: Import Points

 Our next command is **Import Points** and can be used to import point data from an external file. A good option within this routine is the ability to create a point group as it executes.

1. Continue in the same drawing, or open the drawing called Chapter 3-a.Dwg.
2. Using the **V** for View command, highlight the view called **Points-L**, hit **Set Current** and then Hit **OK**. You will see existing ground contours. We will import point data from a file provided with your data set for this text.
3. Use the seventh icon from the left to choose **Import Points**. A dialog will display. Use the **PNEZD** format option. A display of other formats is shown in Figure 3-45. PNEZD stands for Point, Northing, Easting, Elevation, Description. Our file is comma delimited.
4. Select the source file in the Support files folder called **POINTS-IMPORT FILE-PNEZD.TXT** that is in your textbook data folder.
5. Toggle **on** the **Add Points** to Point Group option in Figure 3-46.
6. Hit the button to the right of the popdown window and provide a name for the new point group called **Import-EG Points**.
7. Turn **off** the toggles for the **Advanced Options** and hit **OK**.

Autodesk Uploadable File
ENZ (comma delimited)
ENZ (space delimited)
External Project Point Database
NEZ (comma delimited)
NEZ (space delimited)
PENZ (comma delimited)
PENZ (space delimited)
PENZD (comma delimited)
PENZD (space delimited)
PNE (comma delimited)
PNE (space delimited)
PNEZ (comma delimited)
PNEZ (space delimited)
PNEZD (comma delimited)
PNEZD (space delimited)

Figure 3-45   Formats

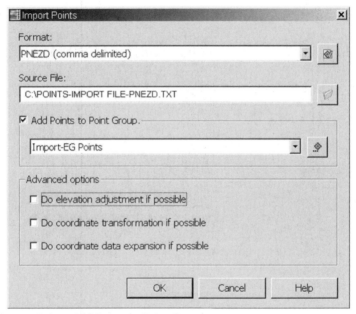

Figure 3-46   Import Points, Add Points to Point Group

8. The points in this file are imported into the project shown in Figure 3-47 and are in the in the preset view called **Points-L**.

9. Save this drawing as we will use it again later.

You will also notice that the **Prospector** in the **Toolspace** (Figure 3-48) shows the new Point Group called Import-EG Points and when selected it displays all of the points within the group.

Again, note that a unique feature of this text is that examples for practical use are provided for and shown as Tips that describe when these commands might be used on actual projects.

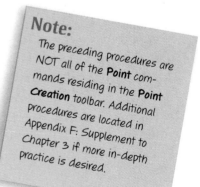

Note:
The preceding procedures are NOT all of the **Point** commands residing in the **Point Creation** toolbar. Additional procedures are located in Appendix F: Supplement to Chapter 3 if more in-depth practice is desired.

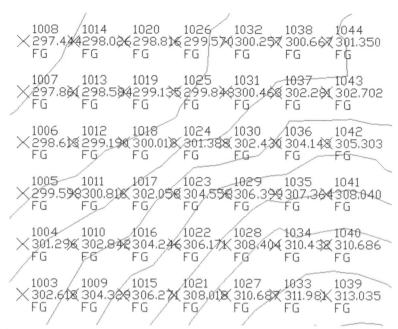

Figure 3-47   Imported Points

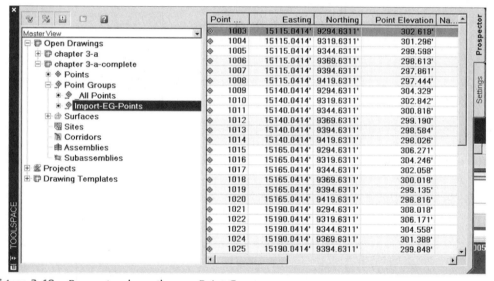

Figure 3-48   Prospector shows the new Point Group

TIP

One last note on the toolbar is that there is one additional icon on the far right of the toolbar. This allows for the toolbar to be expanded or contracted. If it is expanded some fields show up that allow for the creation of Point Styles, Point Label Styles, and Point Layers in the event the user would like to create or switch these on the fly. Also notice that a pushpin is included in the banner of the toolbar. This pins the toolbar, placing it in a fixed location, and also shrinks the toolbar down to a small placeholder when the cursor moves outside the toolbar. To display the expanded toolbar, move the cursor over the placeholder.

Continue with the Points Pull-Down

Next is a discussion of some other capabilities within the **Points** pull-down menu, Figure 3-49, in Civil 3D.

The second command in the **Points** pull-down menu is the **Create Point Group**... command. This allows the creation of point groups such as the one created in Exercise 3-42 on **Import Points**.

Points
Surfaces
Parcels
Grad
Create Points...
Create Point Group...
Edit
Add Tables...
Utilities

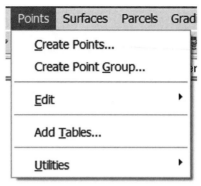

Figure 3-49 The **Points** pull-down menu

The **Edit**... commands modify points, as shown in Figure 3-50. The choices follow.

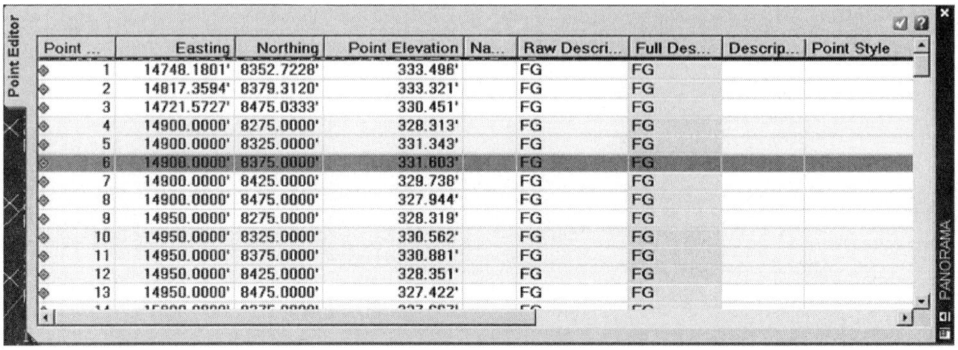

Point Editor
Point ... | Easting | Northing | Point Elevation | Na... | Raw Descri... | Full Des... | Descrip... | Point Style
1 14748.1801' 8352.7228' 333.496' FG FG
2 14817.3594' 8379.3120' 333.321' FG FG
3 14721.5727' 8475.0333' 330.451' FG FG
4 14900.0000' 8275.0000' 328.313' FG FG
5 14900.0000' 8325.0000' 331.343' FG FG
6 14900.0000' 8375.0000' 331.603' FG FG
7 14900.0000' 8425.0000' 329.738' FG FG
8 14900.0000' 8475.0000' 327.944' FG FG
9 14950.0000' 8275.0000' 328.319' FG FG
10 14950.0000' 8325.0000' 330.562' FG FG
11 14950.0000' 8375.0000' 330.881' FG FG
12 14950.0000' 8425.0000' 328.351' FG FG
13 14950.0000' 8475.0000' 327.422' FG FG
PANORAMA

Figure 3-50 Points—Edit...

1. **Points** allowing for the tabular editing of point data where the **Panorama** displays with cells of all pertinent point data. By clicking in a cell, you can edit the information on the fly. By checking the green check mark on the upper right of the box, you can effect the change immediately.
2. **Point Groups** (Figure 3-51) allows you to change the display order for point groups in a drawing and allows for updating out-of-date point groups. This dialog box can also be used to select a point group when you need it for an application such as adding a point group to a surface data set. The **Show Differences** button in the top left of the dialog box, allows for reviewing and updating out-of-date point groups. The **Update Point Groups** button, the second button in the top left of the dialog box, allows updating all out-of-date point groups. A point group is out of date when an exclamation mark within a yellow shield is displayed next to a point group Name.

Point Groups
Name
Description
Import-EG-Points
_All Points
OK
Cancel
Apply
Help

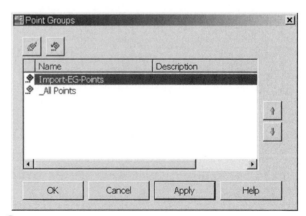

Figure 3-51 Point Groups

3. **Datum** updates points that were set, perhaps to an assumed elevation, to a corrected benchmark elevation. Points with no elevation values are not adjusted. Points that need to be changed individually can be altered via the **Edit Points** command. This command does not really warrant further procedure because it asks only one question, `Change in elevation or [Reference]:` to which you type in the amount of elevation change you want to accomplish.

4. **Renumber** allows points to be renumbered by using an additive factor. In other words, say that you have points already set from Point 1 through Point 115. Now a field crew is bringing in a new set of points that happen to overlap your point numbers. You cannot have multiple points with the same point number, so something has to give and someone must renumber his or her points. In Civil 3D, use the **Renumber** command to add a factor of, say, 1000 to the points already set and you then have points ranging from 1001 to 1115, thereby clearing the way for new points in the 1 to 115 range. This command does not really warrant further procedure because it asks only one question, `Enter an additive factor for point numbers:` to which you type in the amount change you want to add to the existing numbering.

Exercise 3-43: Creating Automated Tables—Add Tables

The next command in the **Points** pull-down menu is the **Add Tables...** command. Refer to Figure 3-52. This builds point tables based on the points or point groups in the drawing.

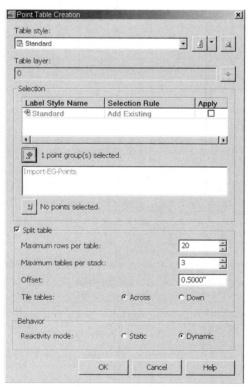

Figure 3-52   Creating Automated Tables

1. Continue in the same drawing, or open the drawing called Chapter 3-a.Dwg. Type **ZOOM, EXTENTS**.
2. Under the **Points** pull-down menu, choose the **Add Tables...** command. A **Point Table Creation** dialog box displays.

3. If you want a customized table for the points, select a predeveloped one under the **Table style** pop-down or use the arrow in the drop-down arrow next to the icon just to the right of the **Table style** window. Select **Create New** or **Edit Current Selection** to customize a table. In this example, click the **Edit Current Selection** and the **Table Style—Standard** dialog box opens, shown in Figure 3-53. Under the **Information** tab the name can be changed, which would make sense if you selected **New**. Otherwise, choose the **Data Properties** tab, where you can elect to wrap text or to have the table maintain view orientation (in the event that your view is twisted to accommodate a plotting orientation). Also you can effect how your titling and column headers display when the table splits, if it is set up to split. Title style, **Header styles**, and **Data styles** are selected here as well. In the **Structure** window, the table can be customized to show any related columns that you wish.

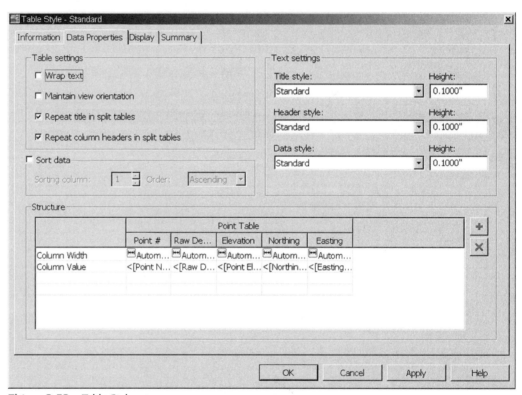

Figure 3-53   Table Style

4. You can customize this table to show latitude and longitudinal values along with the coordinates, point numbers, and elevations for the points in our drawing. The buttons to the right of the window allow you to **add** columns or **delete** them. Hit the **Plus** button to add a column. A new column becomes available next to the Easting column, albeit with no header yet.

5. Double-click on the blank header and a **Text Component Editor—Column Contents** dialog box appears. In the black window, type **Latitude** and hit **OK**.

6. Now double-click in the cell in the Latitude column and in the Column Value row. A **Text Component Editor—Column Contents** dialog box opens. Make sure you are in the **Properties** tab. In the **Properties** pop-down, select **Latitude** from the alternatives.

7. Then click the **blue arrow** button just to the right of the window, to send the option into the black window on the right.

8. Notice that you can choose the **Format** tab and change text properties if you wish. Hit **OK**. You are in the **Table Style—Standard** dialog box again.

9. Repeat this sequence to add in the Longitudinal information.

10. Hit the **Plus** button to add a column. A new column becomes available next to the Latitude column, again with no header yet.

11. Double-click on the blank header and a **Text Component Editor—Column Contents** dialog box appears. As shown in Figure 3-54, in the black window, type **Longitude** and hit **OK**.

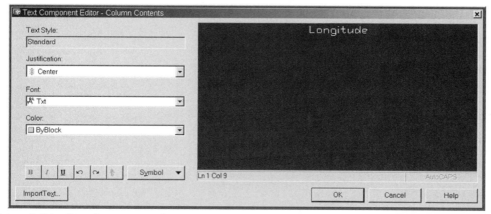

Figure 3-54   Text Component Editor

12. Now double-click in the cell in the Longitude column and in the Column Value row. A **Text Component Editor—Column Contents** dialog box opens, shown in Figure 3-55. Make sure you are in the **Properties** tab. In the **Properties** pop-down, select **Longitude** from the alternatives.

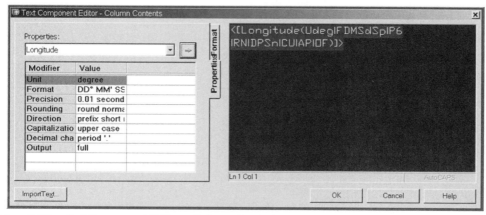

Figure 3-55   Text Component Editor—Column Contents

13. Then click the **blue arrow** button just to the right of the window, to send the option into the black window on the right.

14. Notice that you can choose the **Format** tab and change text properties if you wish. Hit **OK**. You are in the **Table Style—Standard** dialog box again.

15. Click on the **Display** tab (Figure 3-56). Here you can set colors, component layers, and such for appearance's sake. Hit **OK**.

16. You are in the **Point Table Creation** dialog box.

17. In the center of the dialog box, hit the button where it says **No Point Groups** selected.

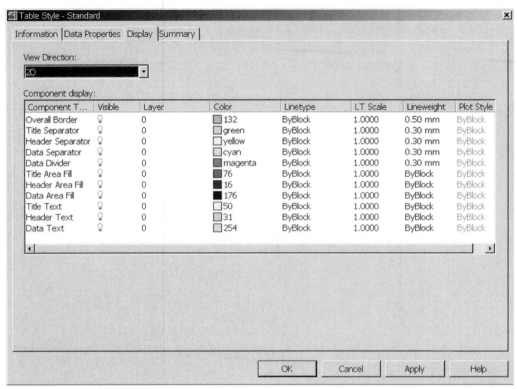

Figure 3-56   Table Style Display

18. Choose the Point Group created in **Import Points** exercise called **Import EG-Points**, and hit **OK**. The dialog box should now appear as in Figure 3-52.

19. Leave the **Split table** option checked **on** and retain the default values.

20. Hit **OK**. The software prompts for: `Select upper left corner:` Pick a point well outside the site and off to the right.

21. The table appears with the data it had initially as well as with the latitude and longitude because the file has a geodetic datum established for northern Virginia, NAD 83, U.S. foot.

Utilities

The next command in the **Points** pull-down menu is the **Utilities** commands, which include the following:

Export allows for exporting points from your drawing to other formats. This creates files like the one used to import the points in the **Import Points** exercise.

Quick View Project displays a tick mark where points exist in the external project point information, assuming one exists.

Draw Project Extents draws a rectangle around the extents of the project's points from external project point information, assuming one exists.

Zoom to Project Extents places your view as if it did a **ZOOM EXTENTS** of the project's points from external project point information, assuming one exists.

Transfer allows you to read an external file in one format and transfer it to another format.

Geodetic Calculator allows for computing a coordinate from a lat/long, or a lat/long from a coordinate, once a geodetic zone is set in the file. Exercise 3-2 was provided in the beginning of this chapter on this function.

CHAPTER TEST QUESTIONS

Multiple Choice

1. Some uses for Points in civil engineering and surveying comprise:

   a. Property corners for parcels of ownership
   b. Roadway earthworks computations
   c. Developing renderings and animations
   d. None of the above

2. Point groups have the following characteristics:

   a. They have enduring properties that can be easily reviewed or modified, either beforehand or retroactively.
   b. Displaying a points list shows the points included in a point group and it can be updated automatically.
   c. A point group can be locked to so as to avoid unintentional modifications.
   d. All of the above

3. Which of the following codes can be used in the Description Key Set?

   a. AC
   b. YARDINLET
   c. 1033
   d. All of the above

4. The **Point Creation** toolbar allows for the following commands:

   a. Automatically checking that all points on a project are correct
   b. Setting points by snapping to objects
   c. Setting all points to the correct elevational value required by the project
   d. None of the above

5. Points can be created for which of the following circumstances?

   a. Locating points related to alignments
   b. Locating points related to surfaces
   c. Importing points from external files
   d. All of the above

6. Points can be established when the user has which of the following criteria?

   a. Elevations
   b. A starting point and a distance and a grade
   c. A starting point and a slope
   d. All of the above

7. Points can be:

   a. Defined or set
   b. Edited or modified
   c. Reported on
   d. All of the above

8. Editing of points can be accomplished using which of the following?

   a. **Panorama**
   b. AutoCAD **MOVE** command
   c. **Prospector**
   d. All of the above

9. A point can be located if the user has which of the following criteria?

   a. An Easting, an X coordinate, and a Z coordinate
   b. Two Northings and a Time value
   c. A line depicted by a beginning and an ending location
   d. None of the above

10. Some typical needs for a designer to set points would be for:

    a. Stakeout purposes
    b. Identifying the locations of existing condition planimetrics
    c. Spot shots
    d. All of the above

True or False

1. True or False: COGO is an acronym for Consolidated, Orthogonal Geometry Organization.

2. True or False: Points are defined using the Cartesian coordinate system and consist of X, Y, and Z values.

3. True or False: The Y axis is located 180° from the X axis.

4. True or False: A line (or vector) is defined as the connection between the beginning point and its corresponding ending point.

5. True or False: A line exists if it begins at a point P1, has a known a slope distance, and has a distance component.

6. True or False: Geodetic transformations can be accomplished using the AutoCAD **MOVE** command.

7. True or False: A Geoid is a surface that is equal to Sea Level if it is not impeded or disturbed by any land masses.

8. True or False: Civil 3D has a myriad of ways to depict points and what they represent and Point Styles is the vehicle used to accomplish this.

9. True or False: Point Styles is limited to Civil 3D symbols and user-defined blocks may not be used.

10. True or False: Point Groups are named collections of points and are used to associate points that have relationships, such DTM points, Property Corner points, and so on.

11. True or False: Points in Point Groups cannot be overlapped with other Point Groups.

12. True or False: Description Key Sets are used to authorize the use of the Civil 3D software.

13. True or False: Point File Formats can be customized to include formats not delivered with the Civil 3D software.

14. True or False: External Data References, XDRefs, are AutoCAD files that can be cross referenced into the AutoCAD drawing.

15. True or False: The **Points** pull-down menu is a very sophisticated menu with over fifty **Point Creation** commands.

STUDENT EXERCISES, ACTIVITIES, AND REVIEW QUESTIONS

1. What is the difference between coordinate values of X, Y and N, E?

2. What is the value of Y for a line represented by a formula of $Y = (.25)(10) + 5$?

3. Find the length of the vector between points P1 = 10,10,0 and P2 = 15, 12,0.

4. Plot that same point in AutoCAD. What angle exists between this vector and the X axis?

5. What is the slope of this vector?

6. If a line starts at P3 = 5,5,0 and has a slope of .333 and a length of 10', what is the ending coordinate?

7. Set up a geodetic zone for your community.

8. Using the **Geodetic Calculator**, find the latitude/longitude for a State plane coordinate in your zone. This will vary according to zone so ask a surveyor or engineer for a value if you do not know one.

9. Using the **Points >> Create Points** toolbar, select **Miscellaneous Convert AutoCAD Points** to convert AutoCAD points to Civil 3D points.

   Type **Point** at the command prompt to create an AutoCAD point (which is not a Civil 3D point). The software responds with: `Specify a point:` Type a coordinate of **1850,1975,230.54**. Notice a dot shows up inside the parcel polyline you were working with in the last couple of exercises. It actually resides at an elevation of 230.54. Use **Miscellaneous Convert AutoCAD Points** to create a Civil 3D point at this precise location in 3D.

   Now use **Convert** to create a Civil 3D point at this precise location in 3D.

10. The next example is to set points using the **Create Points—Measure Object** command.

   To begin this exercise, Draw a line from 850,1025 to 930,1050.

   Then from the **Create Points** toolbar, choose the pop-down arrow on the first button and select **Measure Object**.

   Use a Starting station of **0.0000** and an Ending station of **83.815**. Set points using a 20' offset right, from the line, with an interval of **10.00**, a point description of IPF, and just hit <**Enter**> for the point elevation. Set seven points each at a 20' offset to the right of the line every 10'.

11. This exercise has you set points on polylines using the **Polyline Vertices—Manual command.**

   Using the **Zoom Center** command, choose a zoom center point of **2000,2000** with a magnification or height of **500**.

   Draw a polyline starting at an X, Y coordinate of **1750,1900** to **1850,2100** to **2100,2000** to **2100,1875**, and close it by typing C.

   From the **Create Points** toolbar and using the pop-down arrow next to the first icon, choose **Polyline vertices—Manual** and set a point on each vertex using a default elevation of **100.0** for the proposed building pad, and use the IPF description for each point as it is set.

Civil 3D, the Modern Curvilinead

4

Chapter Objectives

After reading this chapter, the student should know/be able to:
- Understand the mathematics behind the Civil 3D geometry tools.
- Understand distances; bearings; information on traverses used for establishing control on projects; and formulas for circles, arcs, triangles, and lines.
- Use Civil 3D for creating and editing these geometry items.
- Use the many **Transparent** commands that are in Civil 3D.
- Understand how to develop, label, and edit alignments and parcels.

INTRODUCTION

Civil 3D is a modern curvilinead. What is a curvilinead? It is an instrument that draws curvilinear and rectilinear objects. The 3D in Civil 3D could stand for Design, Delineation, Depiction because of its strengths in developing curvilinear and rectilinear geometry objects. This chapter begins with some of the formulas needed to design geometry and discusses curvilinear linework generation, roadway alignment, and parcel creation, editing, and drafting. The ability to modify these data and the development of curvilinear 2D geometry (Lines, Arcs, Spirals), parcels, alignments, and labels are discussed in detail.

REVIEW OF FUNCTIONS AND FEATURES TO BE USED

This chapter uses engineering functions to compute angles, slopes, and distances. Civil 3D has many tools for creating lines and curves, and this chapter examines how these are then used to create disciplinary objects such as alignments and parcels.

COMPUTATIONS OF LINES AND THEIR ANGLES

In a vector-based program, points create lines. Points occupy a position consisting of only an X coordinate, a Y coordinate, and a Z coordinate. A linear connection of the dots occurs with a clear start, a distance, and direction. Refer to Figure 4-1 for a line starting at 0, 0 and continuing through a point of 7.5, 2.5.

The points can be Civil 3D points or simply just AutoCAD points or picks on the screen, but they are points nevertheless. When drawing lines in AutoCAD, you can easily find out information about them by using the **LIST** command.

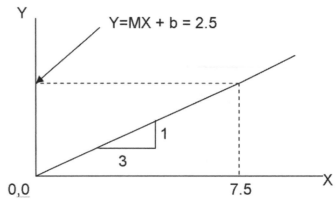

Figure 4-1 A vector

But where does the **LIST** command compute its results from? Can these results be accomplished by hand, using manual techniques? You certainly should be able to reproduce these results using manual techniques just in case you want to check the computer. One of the worst things that someone using a computer can do when asked, "How do you know that the value is correct?" is to say, "Because the computer said so." A better response would be: "I know this is a correct estimate of the value based on my knowledge of the software's algorithms and the data I developed." It is necessary to know the engineering and mathematical formulas behind these algorithms, and each chapter in this text explores them. This chapter discusses linear and curvilinear computations.

Example 4-1: Find the Angle of a Two-Dimensional Line from the X Axis

To find the angle from the X axis of the two-dimensional line with the same coordinate values as shown in Figure 4-1, follow this process.

Slope $= \Delta Y/\Delta X = (20 - 15)/(10 - 5) = 5/5 = 1$

The angle then is computed as: $(ATAN((ABS)slope)) = ATAN(1) = 45°$

Example 4-2: Find the Angle of a Three-Dimensional Line from the X Axis

To find the angle from the X axis of the three-dimensional line with the same coordinate values as shown in Figure 4-1, follow this process.

Slope $= \Delta Z/\sqrt{A^2 + B^2} = (5 - 10)/(7.0710678) = 0.70710678$

The angle then is computed as: $(ATAN((ABS)slope)) = ATAN(0.70710678) = 35.254°$
The formula for a line $- Y = mX + b$ was discussed in Chapter 3 under Point Basics.
The Y coordinate $=$ the slope of the line times the X coordinate plus the y intercept. Therefore, in Figure 4-1 $m = .333$, $X = 7.5$, and $b = 0$; $Y = 2.5$.

DISTANCES AND BEARINGS

Distances and bearings are used to show course and lengths for many items such as parcel lines, boundaries, alignments, pipelines, and so forth. The bearing has four quadrants, each no larger than 90 degrees. It always starts at either north or south and is followed by the angle and then the direction to which it turns, which is either east or west. See Figure 4-2.

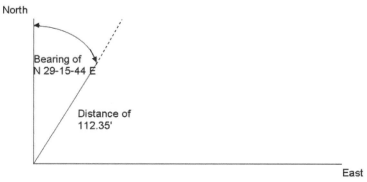

Figure 4-2   Bearing quadrants

3D LINEWORK USES SLOPE DISTANCES FOR DISTANCE MEASUREMENT

Distances measured can include horizontal lengths, vertical lengths, *slope distances*, or vertical angles. Figure 4-3 shows what these are in establishing data connecting Point *A* to Point *B*.

Slope Distances—The distance measured along an incline

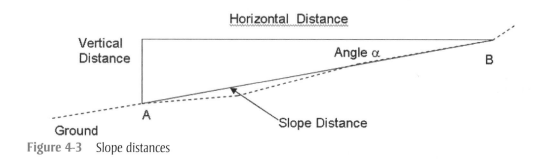

Figure 4-3   Slope distances

Example 4-3: Computing the Difference in Height Between Rods

As shown in Figure 4-4,

Rod reading = 7.21′
Rod reading = 2.63′
Difference in Elevation = 7.21 − 2.63 = **4.58′**

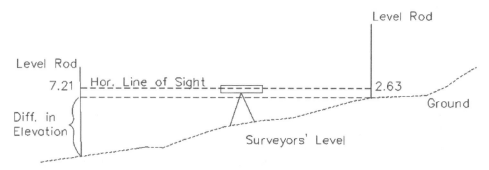

Figure 4-4   Computing the difference in height between rods

CONTROL SURVEYS

Control Surveys establish reference points and reference lines for preliminary and construction surveys. Vertical reference points called benchmarks are also established. These are then tied into state coordinate systems, property lines, road centerlines, or arbitrary grid systems. The various types of tie-in methods are shown in Figure 4-5.

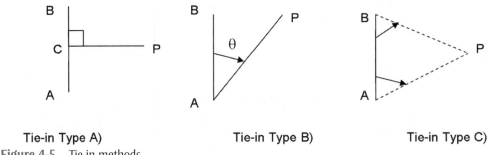

Tie-in Type A)   Tie-in Type B)   Tie-in Type C)

Figure 4-5   Tie-in methods

- Type A, rectangular tie-in, is also known as a right angle offset. Remember in Chapter 3 how many commands assisted us in developing these Perpendiculars?
- Type B, polar tie-in, is also known as the angle/distance technique.
- Type C, intersection tie-in, uses two angles from a baseline to locate a position.

Example 4-4: Stationing

Distances along baselines are called stations, as in Figure 4-6. Right angles can be achieved from the baseline. The distance along this right angle is called an offset.

Station 2 + 36.517 is 236.517′ from the intersection of Elm and Pine. The left corner of the building is 236.517′ from Elm and 40.01′ at a 90 degree angle from Pine.

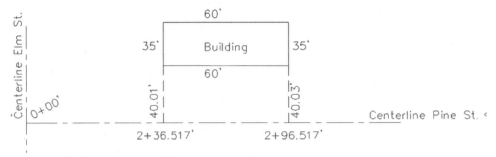

Figure 4-6   Stationing

Example 4-5: Doubling the Angle

A surveying technique to enhance precision when shooting data in the field is called Doubling the Angle. The angles are effectively shot and computed twice for each interior angle. Due to human error, ambient condition error, or equipment error, these measures are often slightly different. So the

surveyor often creates a mean of the two angles, anticipating that the mean
angular error is less than any one of the angles independently.

Station	Direct Angle	Double Angle	Mean
A	101-24-00	202-48-00	101-24-00
B	149-13-00	289-26-00	149-13-00
C	80-58-00	161-57-00	80-58-30
D	116-20-00	232-38-00	116-19-00
E	92-04-00	184-09-00	92-04-30
		Total Degrees	538-119-00 = 539-59-00
			(Is this correct?)

The total number of degrees within a closed polygon depends on the number
of sides and is computed using the following formula. ANGULAR CLOSURE =
$(N - 2) \times 180$; therefore, our polygon = 3×180 = 540-00-00. So subtracting
these yields 540-00 minus 539-59-00 = 01'-00" ERROR. See Figure 4-7.

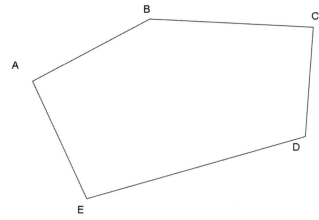

Figure 4-7   *Polygons*

VECTOR DATA AND FORMULAS

As a refresher in order to fully understand the computations that the Civil 3D
software performs, the following information is provided along with examples.
Formulas for computing the following data manually follow:

Circles
- Radius, Perimeter, Area

Right triangles
- Lengths of sides, Interior angles

Curvilinear Objects Including Arcs, Circular Curves, and Spirals
- Arc definition, Chord definition
- Define the components of an arc
- Formulas for computing Arc Length (L), Radius (R), Chord length (C),
  Mid-ordinate chord distance (M), External *secant* (E), and Tangent
  length (T)
- Define the components of a spiral
- Types of horizontal curves

Secant—Geometry term

Circle Formulas

> Radius = 1/2 × Diameter
> Pi radians = 180°
> 1 radian = 180/π
> 2 Pi radians = 360
> Perimeter = 2 × π × r or, π × Diameter
> Area = π × r^2

Right Triangle Definitions and Formulas

> $a^2 + b^2 = c^2$
> $a^2 = (c^2 - b^2)$
> $b^2 = (c^2 - a^2)$
> $c^2 = (a^2 + b^2)$

Sides

> Tan = Opposite/Adjacent
> Sin = Opposite/Hypotenuse
> Cos = Adjacent/Hypotenuse
> $A + B + C = 180°$

Note

> $A° = 180° - 90° - B°$ or, $90° - B° \rightarrow B° = 90° - A°$
> $A° = Tan^{-1}(b/a)$ or $Sin^{-1}(b/c)$ or $Cos^{-1}(a/c)$
> $B° = Tan^{-1}(a/b)$ or $Sin^{-1}(a/c)$ or $Cos^{-1}(b/c)$

Example 4-6: Triangles

If the right triangle in Figure 4-8 has a 27.5′ base and a height of 38.7′, how would you find:

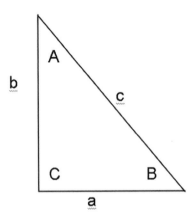

Figure 4-8 Triangles

> 1. The angle *B*.
>
> $B = Tan^{-1}(opp/adj) = Tan^{-1}(38.7/27.5) = 54.60261522$
> $0.60261522 × 60 = 36.1569$
> $0.1569 × 60 = 9.414 = 54 - 36 - 9$

2. The remaining angle of the triangle:

   $90 - 54.6026 = 35.40$

3. The length of the hypotenuse:

   $c = \sqrt{((27.5)^2 + (38.7)^2)} = 47.48'$

4. The area of the triangle:

   $(27.5 \times 38.7)/2 = 532.13 \text{ ft}^2$

5. What is the closest distance from C to any point along the hypotenuse?

   We know angle $B = 54.6026°$

   We know that Sin (54.6026) = opp/hyp
   If $X = Y/Z$, then $Y = X \times Z$ and $Z = Y/X$, then 27.5
   $\times$ Sin (54.6026) = 22.42'

6. Where does the shortest distance line intersect the hypotenuse in relation to point A?

   We know two sides of the new triangle so,
   $a^2 + b^2 = c^2$ and that $c = 22.42$ and $b = 27.5$
   $a = \sqrt{(c^2 - b^2)} = \sqrt{((27.5)^2 - (22.42)^2)} = 15.92'$

Because a right triangle is being divided into smaller right triangles, the interior always remains the same as with the original triangle.

Example 4-7: A Practical Problem

See Figure 4-9. A homeowner wants to remove the Ex. Tree and put a well there. County regulations say the well must be 50' from the drain field and 25' from the house. Does it work?

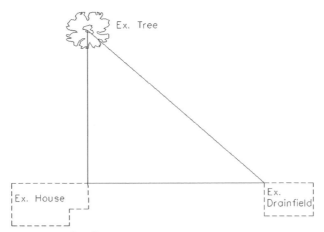

Figure 4-9   Locating a proposed well—1

The measured distance from the tree to the house = 26'; OK. The distance from the drain field to the house is 32.5'. Distance from the tree to the drain field = $\sqrt{(32.5^2 + 26^2)} = 41.60$; Not OK. See Figure 4-10.
Suggested new location for well could be:

An alternative #1 might be: $a = \sqrt{(c^2 - b^2)} = \sqrt{(50^2 - 32.50^2)} = 38'$
An alternative #2 might be: $b = \sqrt{(50^2 - 26^2)} = 42.71$, say, 43'

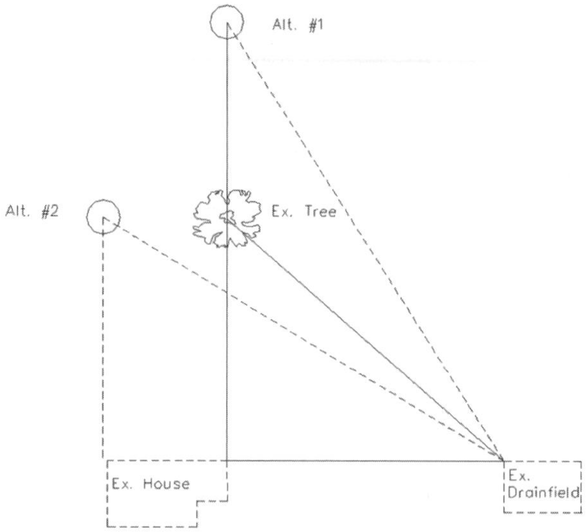

Figure 4-10   Locating a proposed well–2

From the corner of the drain field looking at the corner of the house, what is the angle to locate the proposed well?
For alternative #2,

$$A = \text{Tan}^{-1}\ (40/32.5) = 50.9°$$
$$= \text{Sin}^{-1}\ (40/50) = 53.13°$$
$$= \text{Cos}^{-1}\ (32.5/50) = 49.45°$$

CURVILINEAR OBJECTS—ARCS, CIRCULAR CURVES, AND SPIRALS—DEFINITIONS AND FORMULAS

Another aspect of drawing is the ability to compute curves and arcs. Arcs are defined as those curvilinear objects that have a fixed radius, and curves are those objects that may have a fixed radius but may also be dynamic. Spirals and parabolas fall into the curve category.

Some specific formulas for computing arc data and spiral data need to be discussed in detail before moving toward automated solutions.

For arcs there are two definitions, the Arc definition (Figure 4-11) and the Chord definition (Figure 4-12). The Arc definition is the most widely used and typically used by highway designers and other land development staff. It is also called the "Roadway definition." The Chord method is called the "Railway definition."

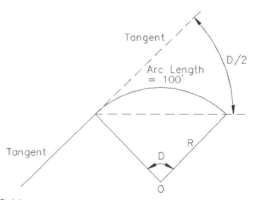

Figure 4-11   The Arc definition

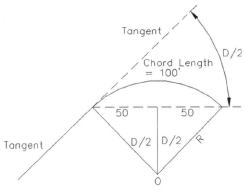

Figure 4-12 The Chord definition

The *Arc method* of curve calculation occurs where the degree of curve is measured over an arc length of 100′. The formulas are:

$D_a/100 = 360°/(2 \times PI \times R)$,
$D_a = 36000/(2 \times \pi \times R)$, or
$D_a = 5729.578/R$.
$R = 5729.578/D_a$. If $D_a = 1°$, then $R = 5729.578$ units and $\pi = 3.14159\ldots$
And $R = 5729.578/D_a$. If $D_a = 1°$, then $R = 5729.578′$ and $\pi = 3.14159\ldots$
And $R = 5729.578/D_a$. If $D_a = 1°$, then $R = 5729.578$ m and $\pi = 3.14159\ldots$

The *Chord method* of curve calculation occurs where the degree of curve is measured over a chord length of 100'. The formulas are:

Sin $(D_c/2) = 50/R$
$R = 50/Sin (1/2)D_c$, where D_c is in degrees.
If $D_c = 1°$, then $R = 5729.648$ units.

Circular Curves

According to Figure 4-13,

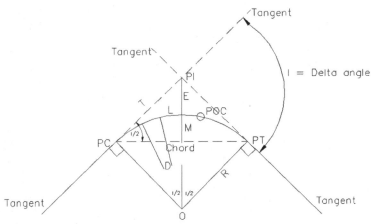

Figure 4-13 Circular Curves

PI = Point of Intersection
PT = Point of Tangency
T = Tangent length
L = Length of curve

E = External distance
D = D_a degree of curve, arc definition
D = D_c degree of curve, chord definition
POC = Point on Curve
PC = Point of Curvature
I = Deflection angle (Delta)
R = Radius
C = Chord length
M = Middle ordinate distance
PRC = Point of Reverse Curvature (not shown)
PCC = Point of Compound Curvature (not shown)

Arc Formulas

Note in the formulas, the difference between *the letter I* and *the number 1.*

$L = 100 \times I°/D_a°$ $L = RI$ (where I is in radians), *or* $RI\pi/180$
$R = 5729.578/D_a°$ $T = R \times Tan\ (I/2)$
$C = 2R \times Sin\ (I/2)$ $R/(R + E) = Cos\ (I/2)$
$E = R((1/Cos\ (I/2)) - 1)$ $(R - M)/R = Cos\ (I/2)$
$M = R(1 - Cos\ (I/2))$ $E = T \times Tan\ (I/4)$
$M = E \times Cos\ (I/2)$ $R = 50/(Sin\ (D/2))$

Example 4-8: Basic Curve Formulas

If $I = 45$, $L = 75$, then $R = 95.49$, $T = 39.55$, $C = 73.09$, $E = 7.87$,
   $M = 7.27$, $D_a = 60.0000$
75 = $(100 \times 45)/D_a$, therefore $D_a = (100 \times 45)/75 = 60.0000$
$R = 5729.578/60 = 95.49$; $E = 95.49((1/Cos(I/2)) - 1) = 95.49$
   $\times\ 0.08239 = 7.867$
$M = 95.49 \times (1 - (Cos\ (I/2))) = 95.49 \times (1 - .92387) = 7.268$
$C = 2(95.49) \times Sin\ (45/2) = 73.08$

Example 4-9: Curve Application

Two highway tangents intersect with a right intersection angle I = 12-30-00
at station 0 + 152.204 m. If a radius of 300 m is to be used for the circular curve, prepare the field book notes to the nearest minute in order to lay
out the curve with 20 m staking.

Solution:

$T = R \times Tan\ (I/2) = 300 \times Tan\ (6°15')$
   = 32.855 m
$L = RI\pi/180 = (300)(12.50)(\pi)/180$
   = 65.450 m
Station of PI = 0 + 152.204
$-T = -(0 + 32.855)$
Station of PC = 0 + 119.349
$+L = +(0 + 65.450)$
Station of PT = 0 + 184.799 m

Note:
The deflection angle for a chord length of 20 m from the PC is (20/L)(I/2). The field notes are set up in the following table.

Station	Chord Dist from PC	Point	Deflection Angle
0 + 184.799	65.450	PT	6-15-00
0 + 180	60.651		5-47-30
0 + 160	40.651		3-52-55
0 + 140	20.651		1-58-19
0 + 120	0.651		00-03-44
0 + 119.349	0	PC	00-00-00
Procedure:	$180 - 119.349 = 60.651$		
	$(60.651/65.45) \times 6.25 = 5.7917° = 5\text{-}47\text{-}30°$		

Example 4-10: Centerline Curve

Find the arc distance for the roadway centerline shown in Figure 4-14.
If this were a complete circle, the perimeter would be:

$2\pi r = 2\pi 200 = 1256.64'$
Therefore, $27°/360° \times 1256.64 = 94.25'$.
Shortcut $= (\text{angle} \times R)/57.30 = 94.25'$
Partial area $= \text{angle}/360 \times \pi r^2$
Shortcut $= (\text{angle} \times R^2)/114.59 = 9424.91 \text{ ft}^2$

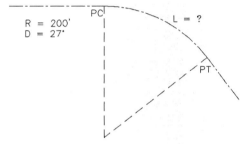

Figure 4-14   Arc distance

Example 4-11: Miscellaneous Applications for Curve Solutions

These concepts can be used in widely varying problems. Here is an example of a pipe problem in which you need to compute flow through a pipe. A 48″ diameter RCP is flowing 14″ deep (Figure 4-15). What are the area of flow and wetted perimeter?

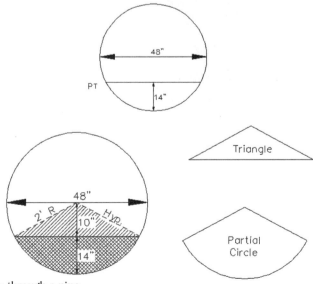

Figure 4-15   Flow through a pipe

Figure the angles of the triangle and base width (across water surface).

Cos angle = Adjacent/Hyponteuse
Angle = Cos − 1 (0.83'/2.0') = 65.4807°

Then, the total angle across the water surface = 2 × 65.4807 = 130.9614°
The base width = $\sqrt{(c^2 - a^2)}$ = $\sqrt{(2^2 - 0.83^2)}$ = 1.82'
Then, the total width of the water surface = 2 × 1.82 = 3.64'
Figure the flow area. Area (triangle) = ((base × height)/2) × 2; the 2s cancel out.

3.64 × 0.83 = 3.02 ft^2
Area of water = angle × R^2/114.59 = 130.9614 × 2^2/114.59 = 4.57 ft^2

Flow area = 4.57 − 3.02 = 1.55 sq. ft^2
The wetted perimeter = angle × R/57.30 = 130.95614 × 2/57.30 = 4.57'

This information could be used in Mannings formula as follows.
The RCP is sloped at a grade of 5%. We know that the $n = 0.013$ for Mannings friction coefficient. Then, $Q = A \times \left(\dfrac{A}{Wp}\right)^{2/3} \times \sqrt{S} \times \left(\dfrac{1.486}{n}\right)$

$$Q = 1.55 \times \left(\frac{1.55}{4.57}\right)^{2/3} \times \sqrt{0.05} \times \frac{1.486}{0.013} = 19.26 \text{ cfs.}$$

Spiral Definitions and Formulas

Spirals are used to create a smooth transition as one travels from a tangent to a curve and then back out to a tangent again. Figure 4-16 shows the parameters for coming into a spiral. The parameters are the same for leaving a spiral.

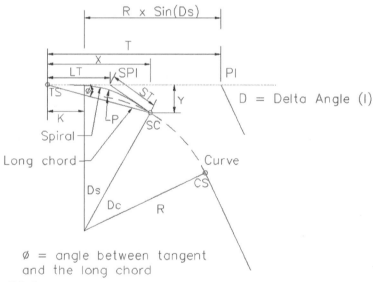

Figure 4-16   Spirals

TS = Tangent to spiral point
SC = Spiral to Curve
CS = Curve to spiral
PI = Point of Intersection
D = Total deflection angle
D_s = Spiral delta angle
D_c = Circular curve delta
T = Total tangent length
R = Radius of circular curve

L = Length of spiral
LT = Long tangent
ST = Short tangent
X = Total "Horizontal" Spiral length
Y = Total "Vertical" Spiral Length
P = Offset of the tangent line to the PC of the shifted curve
K = Abscissa of the PC of the shifted circular curve referred to the straight
  end of the spiral
SPI = Spiral PI

Formulas for Approximate Solutions to Spiral Problems

$$Y/L = \text{Sin } (\phi) \qquad\qquad Y = L \times \text{Sin } (\phi)$$
$$X_2 = L_2 - Y_2 \qquad\qquad X = \sqrt{(L_2 - Y_2)}$$
$$K = X/2 \qquad\qquad P = Y/4$$
$$LT = (\text{Sin } (2/3 \times \phi)) \times L/\text{Sin } (\phi) \qquad ST = (\text{Sin } (1/3 \times \phi)) \times L/\text{Sin } (\phi)$$

Types of Horizontal Curves

The curves shown in Figure 4-17 illustrate the various types of curves that are in the civil engineering business. The ***broken back curve*** is usually not allowed by review agencies because the tangent is not tangent to the curve.

Broken Back Curve—Non-tangent data

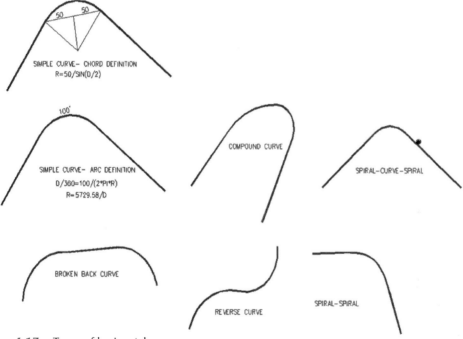

Figure 4-17   Types of horizontal curves

CIVIL 3D's DEFAULT ANGULAR SYSTEM

In Civil 3D, the default angle definitions occur using the Cartesian system, in which the positive X direction is to the right on the screen or monitor and negative X is to the left. Positive Y is up the screen and negative Y is down. Positive Z is basically in the user's face with negative Z going into the monitor. Angles increase counterclockwise from 0° along the X axis to 360°, back to the X axis. Figure 4-18 shows a 90° angle from the X axis.

Figure 4-18   Angles in AutoCAD

Another default in AutoCAD and Civil 3D is that the units for the angles tend to be set to decimal degrees and 0 places of precision. These should also be set up prior to usage so that the desired angles and criteria are to the company's liking. Some people prefer to use degrees, minutes, and seconds as the angle units type, and the precision is usually set to, at least, the nearest second. If the user prefers to use decimal degrees then the precision should be to at least four places. The software converts any output listings to the units the user sets. So if the units are set to decimal degrees and a **LIST** command is issued, the CAD system responds with decimal degrees as part of the listing. If the units are set to degrees, minutes, and seconds, then the system responds accordingly. These can be flipped at any point in the working session to achieve the desired units reporting.

CONVERTING DECIMAL DEGREES TO D-M-S

If the user must perform these computations manually, here are some examples of converting angular units from decimal degrees to degrees, minutes, and seconds.

Example 4-12: Angular Conversions Decimal Degrees to D-M-S

49.5566° breaks down to 49° and 0.5566 of a degree. Then, multiply 0.5566 degree × 60 (minutes per degree) to get the number of minutes, which = 33.396 minutes. Now you have 33 minutes and 0.396 of a minute. Then multiply 0.396 minute × 60 seconds per minute to get the number of seconds, which is = 23.76 seconds. Therefore, 49.5566° = 49°−33′−23.76″.

Example 4-13: Angular Conversions D-M-S to Decimal Degrees

To convert DMS 49°−33′−23.76″ to decimal degrees perform the following: 23.76″/60 (seconds per minute) = 0.396 minute. Then add the 33′ to the 0.396′ to get 33.396′. Then divide 33.396′ by 60 (minutes per degree) and get 0.5566 degree. Then add the 49° to the 0.5566 to get 49.5566°.

ANGLES IN CIVIL ENGINEERING

Now that you can compute the angles of linework manually and based on the Cartesian coordinate system, it is time to discuss in more depth how to talk about angles in terms that civil engineers and surveyors use. This industry uses angles to describe many things, such as property line directions, road centerline directions, and construction instructions for stakeout. Some of the types of important angles to understand are interior angles, azimuths, bearings, cardinal directions, deflection angles, and turned angles.

Interior angles (Figure 4-19) measure the angle between two linear objects. The angle can be acute or obtuse.

Figure 4-19   Interior angles

An azimuth can be measured from either the North or South. The North azimuth measures angles from the North axis, in a clockwise direction to 360°. The South azimuth measures angles from the South axis, clockwise to 360°.

Bearings (Figure 4-20) measure a heading and are broken into four quadrants: North, South, East, and West. Each quadrant has 90°. The quadrant usually referred to as quadrant 1 is the North-East quadrant and measures from North and turns toward the East. Therefore, a bearing of N 45°−30′30″ E is located by facing North and turning toward the East an angle of 45°−30′−30″. Quadrant 2 is South-East and begins by facing South and turning the angle toward the East. Quadrant 3 is South-West and also begins by facing South but turns toward the West. Quadrant 4 is North-West, begins by facing North and turning the angle toward the West.

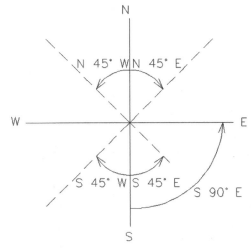

Figure 4-20   Bearings

There are also angles known as cardinal directions, which are North, South, East, and West, routinely seen in describing the heading of a hurricane. "Hurricane Ivan is heading North-North-West," which is halfway between North and North-West. In degrees that would be N 22°−30′00″ W.

Figure 4-21 illustrates a person standing at Point *B*, looking forward at the angle defined from Point *A* to Point *B* and then turning 28°−16′−51″ to the right to get to Point *C*. This shows deflection angles.

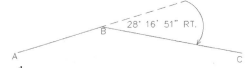

Figure 4-21   Deflection angles

Figure 4-22 illustrates a person standing at Point *B*, looking back at Point *A*, and then turning 208°−16′−51″ to the right to get to Point *C*. This shows turned angles.

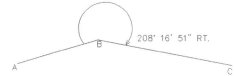

Figure 4-22   Turned angles

On the subject of angles, there are two other types of angle measurement, one called Grads and another called Radians. A Grad is 1/400th of a circle, where North is what would be zero, East is 100 g, South is 200 g, West is 300 g, and back to North is 400 g. This angular system is used in some international countries and by the United States Department of State. A Radian is 180°/π and is often used in computer programming and in the calculation of some arc formulas.

Exercise 4-1: Units

1. Begin the Civil 3D software.
2. Draw a line from **11,7** to **38, 18** and terminate the command.
3. Then type **LIST** at the command prompt. The software responds as follows.

```
Select objects:
```

Select the object, hit <**Enter**> and this listing will display.

```
LINE Layer: "0"
Space: Model space
Handle = 3F2
from point, X = 11.0000 Y = 7.0000 Z = 0.0000
to point, X = 38.0000 Y = 18.0000 Z = 0.0000
Length = 29.1548, Angle in XY Plane = 22
Delta X = 27.0000, Delta Y = 11.0000, Delta Z = 0.0000
```

Note the angle of 22 degrees. Is it an even 22°?

4. Next, at the command prompt, type **UN** for **Units** to bring up the **Units** dialog box. See Figure 4-23.
5. Note under the **Angle Type** pop-down window are the supported angle types. **Decimal Degrees** is the default. The default **Precision** is set to 0;

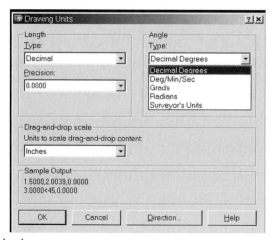

Figure 4-23   Units dialog box

hence there is no precision for display beyond the nearest degree. Set the **Precision** to **4** and hit **OK**.

6. Type **LIST** again, select the line, and hit **<Enter>**. Notice that the angle shown now is 22.1663 degrees.

7. Next, at the command prompt, type **UN** again to bring up the **Units** dialog box.

8. Under the **Angle Type** pop-down window choose **Deg/Min/Sec** and hit **OK**. Type **LIST** again, select the line, and hit **<Enter>**. Notice that the Angle in *XY* Plane = 22d9′59″.

9. Next at the command prompt type **UN** again to bring up the **Units** dialog box.

10. Under **Angle Type** pop-down window, choose **Surveyor's Units** and hit **OK**. Type **LIST** again, select the line, and hit **<Enter>**. Notice that the Angle in *XY* Plane is a bearing of N 67d50′1″ E.

Many Geometry commands can be performed using the **Parcels** and **Alignments** menus. Discussion of other tools as well will follow but here is a brief explanation of the menu pull-downs for Parcels and Alignments.

The **Parcels** pull-down menu (Figure 4-24) commands include:

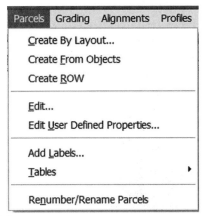

Figure 4-24   The Parcels pull-down menu

Create By Layout... invokes the **Parcel Creation Layout** toolbar.

Create From Objects allows for creating new parcels from entities.

Create ROW allows for creating automatic offsets for Right of Way creation.

Edit Parcel Segments... invokes the Layout toolbar for editing parcels that already exists.

Edit... invokes the **Layout** toolbar for editing parcels that already exist.

Edit User Defined Properties... allows editing of user defined properties that are attached to parcels.

Add Labels... adds specific labels to parcels.

Tables allow for creating property tables of lines, arcs, and so on.

Renumber/Rename Parcels allows the same.

The **Alignments** pull-down menu (Figure 4-25) commands are:

Create By Layout... invokes the **Alignment Creation Layout** toolbar.

Create From Polyline allows for creating new alignments from polylines.

Edit Alignment Geometry... invokes the Layout toolbar for editing alignments that already exist.

Edit... invokes the **Layout** toolbar for editing alignments that already exist.

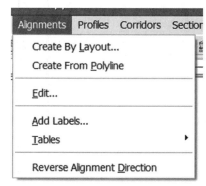

Figure 4-25   The Alignment pull-down menu

Add Labels... adds specific labels to alignments.

Tables allow for creating alignment tables of lines, arcs, and so on.

Reverse Alignment Direction allows the same.

GEOMETRY COMPUTATIONS

Now that the chapter has covered the basic and fundamental mathematics behind curvilinear and rectilinear geometry, discussion proceeds with the capabilities of Civil 3D in these areas. Again, enormous functionality exists within AutoCAD alone, and because it is a prerequisite to know AutoCAD, coverage of those routines within Civil 3D follows.

In addition to investigating some of the primitive geometry tools, the tools for creating parcels and alignments are also explored.

THE TRANSPARENT COMMANDS TOOLBAR

To begin, type in **Toolbar** at the command prompt. In the **Toolbar** dialog box, choose the **Toolbars** tab. In the menu group **Window**, select **Civil**. In the **Toolbars** window, turn **on** the check for **Transparent** commands. If it is already on, leave it on. Hit **Close**. The **Transparent Commands** toolbar should be displayed on the screen, as in Figure 4-26.

Figure 4-26   The Transparent toolbar

You can use **Transparent** commands within other commands that need data entry consisting of multiple points such as the AutoCAD **LINE** command. These can be used just like the AutoCAD **Transparent** commands by preceding them with an apostrophe. You can also pick them as needed from the toolbar. A user can set a series of points using that data entry format because it remains in that data entry mode until you change it or exit.

The Civil **Transparent** commands are used to enter values based on information that can be supplied when you are prompted for a radius, a point, or a distance. Most of these commands are used to specify point locations within another operation, such as the creation of property lines for parcel data. These commands provide great flexibility of obtaining information from data already existing within the drawing without forcing you to accumulate the information ahead of time. This allows you to calculate the location for a point from a variety of angles and distances.

Job skills will benefit because these routines save hours of manual computation for designers.

They are fun to use and provide excellent visuals as you work with them. The command definitions from left to right across the toolbar are described as follows with the accompanying abbreviation that can be typed in as the command is being used.

- **'AD**—Angle and Distance

- **'BD**—Bearing and Distance

- **'ZD**—Azimuth and Distance

- **'DD**—Deflection Angle and Distance

- **'NE**—Northing and Easting

- **'GN**—Grid Northing and Grid Easting

- **'LL**—Latitude and Longitude

- **'PN**—Point Number

- **'PA**—Point Name (alias)

- **'PO**—A Point in a Drawing

- Three new commands were added to the toolbar. They include: Profile Station from Plan, Profile Station and Elevation from Plan and Profile Station and Elevation from COGO Point. These commands prompt for and allow for the user to select information for the profile by referring to the plan view of the alignment. These can be beneficial in that an important location can be identified using plan view objects that can not be seen when viewing the profile. This could be the location of a horizontal utility manhole perhaps.

NEW
to AutoCAD
2007

- **'SO**—Station and Offset

- **'SS**—Side Shot From a Point

- **Transparent command filters** exist at this location in the toolbar. These allow for selecting point information by Point Number, by selecting the Point anywhere the point data are visible, or via a Northing/ Easting entry.

- 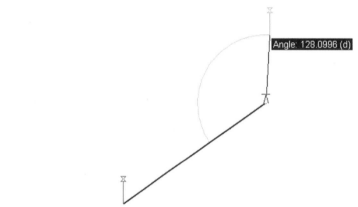 'PSE—Station and Elevation in a Profile View

- 'PGS—Grade and Station in a Profile View

- 'PGL—Grade and Length in a Profile View

- 'MR—An Object's Length

- 'ML—An Object's Radius

Exercise 4-2: The 'AD Command

You now explore how these commands work. Let us begin with the 'AD command. This command allows you to build linework using a turned angle and distance (Figure 4-27).

Angle: 128.0996 (d)

Figure 4-27   Turned angle and distance

TIP   This command can be used to describe a traverse based on an occupied point, a backsight, turned angles, and distances.

1. Open the drawing called Chapter-4-Geometry.Dwg.
2. Using the **V** for **View** command, highlight the view called **Geom-A**, hit **Restore**, and then hit **OK**. You draw linework using turned angles and distances from the line visible on the screen.
3. Type **L** for **Line**. Follow the prompts:
4. LINE Specify first point: Using an Endpoint Osnap pick the northern end of the line in the view.
5. Specify next point or [Undo]: Type **'AD** or select the first icon on the **Transparent Commands** toolbar.

6. >>Specify ending point or [.P/.N/.G]: Using an Endpoint Osnap pick the southern end of the line in the view.

7. Current angle unit: degree, Input: DD.DDDDDD (decimal)

TIP

If the prompt shows DD.MMSS then you can type 128d5'58.56" or change the drawing settings so it asks for DD.DDDDDD. This is done by going to the **Toolspace**, clicking on the **Settings** tab, right clicking on the drawing name, and choosing **Edit Drawing Settings**. Choose the **Ambient Settings** tab and expand the **Angle**. Change the **Value for the Format** to the desired setting.

8. >>Specify angle or [Counter-clockwise]: Type **128.0996**

9. >>Specify distance: **75**

10. Resuming LINE command.

11. Specify next point or [Undo]: (922.563 1060.79 0.0)

12. Specify next point or [Undo]:

13. Current angle unit: degree, Input: DD.DDDDDD (decimal)

14. >>Specify angle or [Counter-clockwise]: **279.8302** (Figure 4-28).

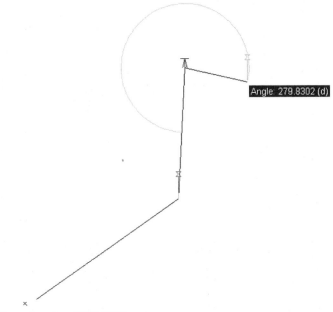

Angle: 279.8302 (d)

Figure 4-28   Turned angle of 279.8302

15. >>Specify distance: **100**

    Resuming LINE command.

16. Specify next point or [Undo]: (1020.06 1038.57 0.0)

17. Specify next point or [Close/Undo]:

    Current angle unit: degree, Input: DD.DDDDDD (decimal)

18. >>Specify angle or [Counter-clockwise]: **233.3313** (Figure 4-29).

19. >>Specify distance: **75**

20. Resuming LINE command.

21. Specify next point or [Close/Undo]: (1050.37 969.965 0.0)

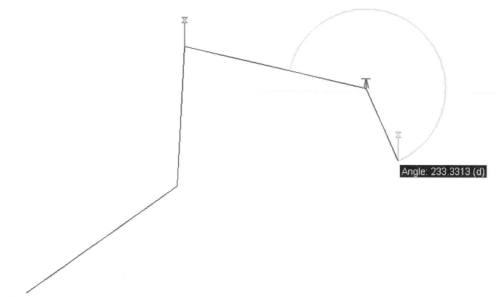

Figure 4-29 Turned angle of 233.3313

22. `Specify next point or [Close/Undo]:`
23. `Current angle unit: degree, Input: DD.DDDDDD (decimal)`
24. `>>Specify angle or [Counter-clockwise]:` ***Cancel***
    `Resuming LINE command.`
25. `Specify next point or [Close/Undo]:` **<ESC>**
26. Save your file with the same name and a suffix of your initials.

Notice the almost video game–style graphics that appear to help you through the computations. The yellow marker simulates a station setup while the green "rodman" is placed at the backsight and the red "rodman" is placed at the foresight location. The red arc indicates that the computations are using a turned angle from the backsight to the foresight.

Exercise 4-3: The 'BD Command

This command allows you to build linework using a bearing and distance.

Job skills will benefit from the use of this command because surveyors routinely draw property and easement linework using distances and bearings.

1. Continue in the same drawing or open the drawing you just saved.
2. Using the **V** for **View** command, highlight the view called **Geom-A**, hit **Restore**, and then hit **OK**. You draw linework using a bearing and distance from where you left off in the previous command. You can also use the line visible on the screen if you did not perform the last example. Of course your results may vary from the figure somewhat.
3. `Command:` L
4. `LINE Specify first point:` Using an Endpoint Osnap pick the end of the last line you drew in the previous example. It is near 1050, 969.

5. `Specify next point or [Undo]:` Type '**BD** or select the second icon on the **Transparent Commands** toolbar. A reminder displays: `Quadrants - NE = 1, SE = 2, SW = 3, NW = 4`
6. `>>Specify quadrant (1-4):` **1**, for Northeast quadrant (Figure 4-30).

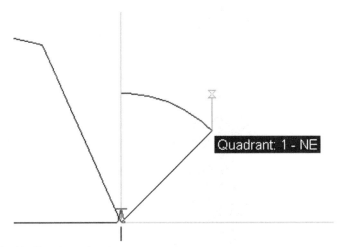

Figure 4-30   The Northeast quadrant

7. A reminder displays: `Current direction unit: degree, Input: DD.MMSSSS (decimal dms)`
8. `>>Specify bearing:` **41.3329**, indicates a bearing angle of 41 degrees, 33 minutes, and 29 seconds.
9. `>>Specify distance:` **75**
10. `Resuming LINE command. Specify next point or [Undo]: (1100.12 1026.09 0.0)`
11. `Specify next point or [Undo]: Quadrants - NE = 1, SE = 2, SW = 3, NW = 4`
12. `>>Specify quadrant (1-4):` **3** `Current direction unit: degree, Input: DD.MMSSSS (decimal dms)`
13. `>>Specify bearing:` **16.1907**
14. `>>Specify distance:` **80**
15. `Resuming LINE command. Specify next point or [Undo]: (1077.64 949.309 0.0)`
16. `Specify next point or [Close/Undo]: Quadrants - NE = 1, SE = 2, SW = 3, NW = 4`
17. `>>Specify quadrant (1-4):` **<ESC>**
18. Save your file with the same name and a suffix of your initials.

Exercise 4-4: The 'ZD Command

This command allows you to build linework using an azimuth (from North) and distance (Figure 4-31).

TIP   If your jurisdiction requires the use of Azimuth (or South Azimuth), this command assists in laying out deed information.

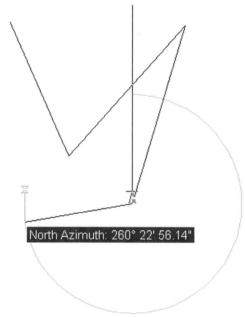

North Azimuth: 260° 22' 56.14"

Figure 4-31   Linework using a North Azimuth

1. Continue in the same drawing or open the drawing you just saved.
2. Using the **V** for **View** command, highlight the view called **Geom-A**, hit **Restore**, and then hit **OK**. You draw linework using a bearing and distance from where you left off in the previous command. You can also use the line visible on the screen if you did not perform the last example. Of course your results may vary from the figure somewhat.
3. Command: **L**
4. LINE Specify first point: Using an Endpoint Osnap pick the end of the last line you drew in the previous example. It is near 1077, 949.
5. Specify next point or [Undo]: Type '**ZD** or select the third icon on the **Transparent Commands** toolbar.
   Current direction unit: degree, Input: DD.MMSSSS (decimal dms)
6. >>Specify azimuth: **260.2256**
7. >>Specify distance: **100**
8. Resuming LINE command. Specify next point or [Undo]: (979.048 932.602 0.0)
9. Specify next point or [Undo]: Current direction unit: degree, Input: DD.MMSSSS (decimal dms)
10. >>Specify azimuth: **<ESC>**
11. Save your file with the same name and a suffix of your initials.

Exercise 4-5: The 'DD Command

This command allows you to build linework using a deflection angle and a distance (Figure 4-32).

1. Continue in the same drawing or open the drawing you just saved.
2. Using the **V** for **View** command, highlight the view called **Geom-A**, hit **Restore**, and then hit **OK**. You draw linework using a bearing and distance from where you left off in the previous command. You can also use the line visible on the screen if you did not perform the last example. Of course your results may vary from the figure somewhat.
3. Command: **L**
4. LINE Specify first point: Using an Endpoint Osnap pick the end of the last line you drew in the previous example. It is near 979, 932.

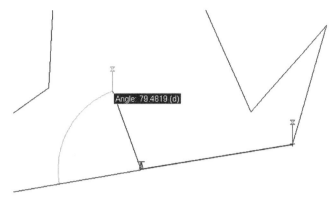

Figure 4-32   Deflection angle and a distance

5. `Specify next point or [Undo]:` Type **'DD** or select the fourth icon on the **Transparent Commands** toolbar.
   `Current direction unit: degree, Input: DD.MMSSSS (decimal dms)`
6. `>>Specify angle or [Counter-clockwise]:` **79.4819**
7. `>>Specify distance:` **75**
8. `>>Specify angle or [Counter-clockwise]:` **<ESC>**
9. Save your file with the same name and a suffix of your initials.

Notice that each of these commands keeps you in the data entry mode until you switch to another to terminate.

Exercise 4-6: The 'NE Command

This command allows you to build linework using a key-in of the Northing and Easting.

TIP   The nice part of this command allows for data entry using the numeric keypad, which expedites the typing.

1. Continue in the same drawing or open the drawing you just saved.
2. Using the **V** for **View** command, highlight the view called **Geom-A**, hit **Restore**, and then hit **OK**. You draw linework using a bearing and distance from where you left off in the previous command. You can also use the line visible on the screen if you did not perform the last example. Of course your results may vary from the figure somewhat.
3. `Command:` **L**
4. `LINE Specify first point:` Using an Endpoint Osnap pick the end of the last line you drew in the previous example. It is near 953,1003.
5. `Specify next point or [Undo]:` Type **'NE** or select the fifth icon on the **Transparent Commands** toolbar.
6. `>>>>Enter northing <0.0000>:` **960**
7. `>>>>Enter easting <0.0000>:` **950**
8. `>>>>Enter northing <0.0000>:` **<ESC>** to terminate.
9. Save your file with the same name and a suffix of your initials.

The 'GN Command

This command allows you to build linework using a key-in of the grid Northing and grid easting. Again, a nice feature allows for data entry using the numeric keypad, which expedites the typing. This is similar to the geodetic calculator used earlier because it reads the geodetic zone when placing points.

The 'LL Command

This command allows you to build linework using a key-in of the latitude and longitude. This is similar to the geodetic calculator used earlier because it reads the geodetic zone when placing points.

Exercise 4-7: The 'PN Command

This command allows you to build linework using a key-in of the point number.

TIP This command is not limited to simply typing a point number; it can also connect multiple points using commas and apostrophes. For instance, you can draw a polyline by starting at a point number of, say, 100 and for the remaining vertices of the polyline you can refer to points in the file by 101-110, 100, which will draw linework through the consecutive points 101 through 110 and connect back to the starting point of 100.

1. Continue in the same drawing or open the drawing you just saved.
2. Using the **V** for **View** command, highlight the view called **Geom-A**, hit **Restore**, and then hit **OK**. You draw linework using a bearing and distance from where you left off in the previous command. You can also use the line visible on the screen if you did not perform the last example. Of course your results may vary from the figure somewhat.
3. Command: **L**
4. LINE Specify first point: Type **'PN** or select the eighth icon on the **Transparent Commands** toolbar.
5. >>Enter point number <1>: **<Enter>**
6. Resuming LINE command. Specify first point: (826.126 943.344 0.0)
7. Specify next point or [Undo]:
8. >>Enter point number <1>: **2**
9. Resuming LINE command. Specify next point or [Undo]: (908.038 1000.71 0.0)
10. Specify next point or [Undo]:
11. >>Enter point number <2>:**<ESC>** to terminate.
12. Save your file with the same name and a suffix of your initials.

Exercise 4-8: The 'PO Command

This command allows you to build linework using a pick of a point on the screen without snapping to it. You can pick any visible portion of the point or its text attributes.

1. Continue in the same drawing or open the drawing you just saved.
2. Using the **V** for **View** command, highlight the view called **Geom-A**, hit **Restore**, and then hit **OK**. You draw linework using a bearing and distance from where you left off in the previous command. You can also use the line visible on the screen if you did not perform the last example. Of course your results may vary from the figure somewhat.
3. Command: **L**
4. LINE Specify first point: Type **'PO** or select the tenth icon on the **Transparent Commands** toolbar.
5. Select point object: Simply pick on the word **UNIT** for Point 2.
6. Resuming LINE command. Specify first point: (908.038 1000.71 0.0)
7. Specify next point or [Undo]:
8. >> Select point object: Hit **<Enter>**

9. `>> Specify next point or [Undo]:` Type in **950,960**
10. `Resuming LINE command.`
11. `>> Specify next point or [Undo]:` **<ESC>** to terminate.
12. Save your file with the same name and a suffix of your initials.

Exercise 4-9: The 'SO Command

This command allows you to build linework using a station and offset (positive for right and negative for left) from an alignment.

> **TIP** This routine is useful because it also works inside other AutoCAD commands and references the alignment in question. It can be used to draw manholes at a station offset from an alignment.

1. Continue in the same drawing or open the drawing you just saved.
2. Using the **V** for **View** command, highlight the view called **Geom-B**, hit **Restore**, and then hit **OK**. You see a shot alignment in the view. Draw linework using a station and offset of this alignment.
3. Type **CIRCLE** or pick it from the **Draw** toolbar.
4. `CIRCLE Specify center point for circle or [3P/2P/Ttr (tan tan radius)]:` Type **'SO** or select the eleventh icon on the **Transparent Commands** toolbar.
5. `>>Select alignment:` Pick the alignment in the view.
6. `>>Specify station:` **75**
7. `>>Specify station offset:` **25**
8. `Resuming CIRCLE command. Specify center point for circle or [3P/2P/Ttr (tan tan radius)]: (920.909 655.768 0.0)`
9. `Specify radius of circle or [Diameter] <5.0000>:`
10. `>>Specify station:` **<ESC>** to terminate the station, offset part of the routine.
11. `Resuming CIRCLE command. Specify radius of circle or [Diameter] <5.0000>:` **1.5**
12. Save your file with the same name and a suffix of your initials.

Exercise 4-10: The 'SS Command

This command allows you to build linework using side shot data. You pick an occupied point on which to establish your station and then a second point representing your backsight. Following that you provide an angle and distance to locate your object.

1. Continue in the same drawing or open the drawing you just saved.
2. Using the **V** for **View** command, highlight the view called **Geom-B**, hit **Restore**, and then hit **OK**. You see an alignment in the view. You draw linework using a station and offset of this alignment.
3. Type **L** for **Line** or pick it from the **Draw** toolbar.
4. `Command: l LINE Specify first point:` Using a Center Osnap pick the center of the circle you drew in the last exercise.
5. `Specify next point or [Undo]:` Type **'SS** or select the twelfth icon on the **Transparent Commands** toolbar.
6. `>>Specify ending point or [.P/.N/.G]: Current angle unit: degree, Input: DD.DDDDDD (decimal)`
7. `>>Specify angle or [Counter-clockwise/Bearing/Deflection/aZimuth]:` **60.9261**, as in Figure 4-33.

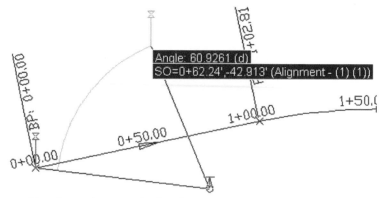

Figure 4-33   Building linework using side shot data

8. >>Specify distance: **100**
9. Line Specify first point: <**ESC**> to terminate.
10. Save your file with the same name and a suffix of your initials.

Exercise 4-11: The 'PSE Command

This command allows you to build linework using profile, station, and elevation criteria.

TIP   When using this routine notice that graphics appear to show you where the station you type is on the profile, and then a graphic appears to show you the elevation that you type.

1. Open the drawing called Chapter-4-Profile.Dwg.
2. Using the **V** for **View** command, highlight the view called **Access Road-Profile View**, hit **Restore**, and then hit **OK**. You see a shot profile in the view. You draw a bridge crossing over the road using a station and elevation data from this profile.
3. Type **L** for **Line** or pick it from the **Draw** toolbar.
4. Command: l LINE Specify first point: Type '**PSE** or select the fourteenth icon on the **Transparent Commands** toolbar.
5. >>Select a Profile View: Pick a grid line in the profile.
6. >>Specify station: **300**
7. >>Specify elevation: **364**
8. Resuming LINE command. Specify first point: (18300.0 21640.0 0.0)
9. >>Specify station: **300**
10. >>Specify elevation: **355**
11. Resuming LINE command. Specify next point or [Undo]: (18300.0 21550.0 0.0)
12. Specify next point or [Undo]:
13. >>Specify station: **350**
14. >>Specify elevation: **355**
15. Resuming LINE command. Specify next point or [Undo]: (18350.0 21550.0 0.0)
16. Specify next point or [Close/Undo]:
17. >>Specify station: **350**
18. >>Specify elevation: **364**
19. Resuming LINE command. Specify next point or [Close/Undo]: (18350.0 21640.0 0.0)
20. Specify next point or [Close/Undo]:
21. >>Specify station: <**ESC**> to terminate.
22. Save your file with the same name and a suffix of your initials.

You see the outline of an overhead bridge in a thick linestyle, Figure 4-34.

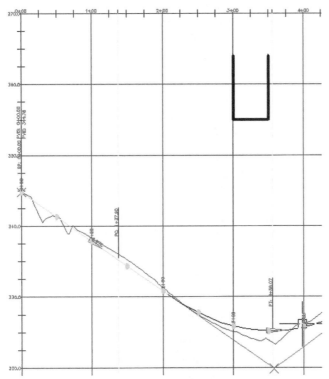

Figure 4-34   Using profile, station, and elevation Criteria

Exercise 4-12: The 'PGS Command

This command allows you to build linework using profile, grade, and station criteria (Figure 4-35).

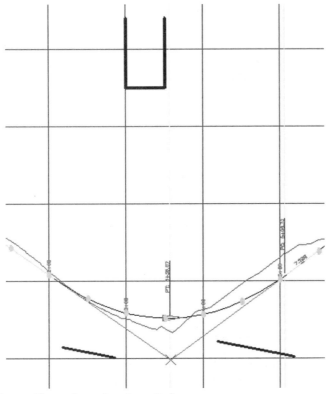

Figure 4-35   Using profile, grade, and station criteria

Note that these commands can be used for almost any AutoCAD command. For instance, you can draw a pipe in profile while maintaining the grade of the pipe. You can draw the invert of the pipe and the other invert can be established using a grade and station.

Student Files

1. Continue on the same drawing, or open the drawing called Chapter-4-Profile.Dwg.
2. Using the **V** for **View** command, highlight the view called **Access Road-Profile View**, hit **Restore**, and then hit **OK**. You see a shot profile in the view. You draw a culvert from a point at −2% using the profile data.
3. Type **L** for **Line** or pick it from the **Draw** toolbar.
4. `Command: l LINE Specify first point:` **18217,21215**
5. `Specify next point or [Undo]:` Type '**PGS** or select the fifteenth icon on the **Transparent Commands** toolbar.
6. `>>Select a Profile View:` Pick a grid line in the profile.
7. `>>Specify grade <0.00>:` **-2**
8. `>>Specify station:` **285**
9. `Resuming LINE command. Specify next point or [Undo]:` (18285.0 21201.4 0.0)
10. `Specify next point or [Undo]:`
11. `>>Specify grade <-2.00>:` **<ESC>** to terminate.
12. Save your file with the same name and a suffix of your initials.

Exercise 4-13: The 'PGL Command

This command allows you to build linework using profile, grade, and length criteria.

Student Files

1. Continue on the same drawing, or open the drawing called Chapter-4-Profile.Dwg.
2. Using the **V** for **View** command, highlight the view called **Access Road-Profile View**, hit **Restore**, and then hit **OK**. You see a shot profile in the view. You draw a culvert from a point at −2% using the profile data.
3. Type **L** for **Line** or pick it from the **Draw** toolbar.
4. `Command: l LINE Specify first point:` **18418,21222**
5. `Specify next point or [Undo]:` Type '**PGL** or select the sixteenth icon on the **Transparent Commands** toolbar.
6. `>>Select a profile view:` Pick a grid line in the profile.
7. `Current grade input format: percent`
8. `>>Specify grade <0.00>:` -2
9. `>>Specify length:` **100**
10. `Resuming LINE command. Specify next point or [Undo]:` (18518.0 21202.0 0.0)
11. `Specify next point or [Undo]:`
12. `Current grade input format: percent`
13. `>>Specify grade <-2.00>:` **<ESC>** to terminate.
14. Save your file with the same name and a suffix of your initials.

You can continue drawing linework in this command if you have a series of pipe inverts to draw. This command can be used to draw anything in the profile that has a length and grade, such as the bottom of a ditch.

Exercise 4-14: The 'MR and 'ML Commands

These commands allow you to build linework by matching the radius or length of an object.

TIP

These commands are excellent for obtaining data from other criteria either known or within the file somewhere. Remember to use these as you create parcel and alignment data in the future.

Student Files

1. Continue on the same drawing, or open the drawing called Chapter-4-Profile.Dwg.
2. Using the **V** for **View** command, highlight the view called **Geom-D**, hit **Restore**, and then hit **OK**. You see an arc and a line.
3. Type **Circle** or pick it from the **Draw** toolbar.
4. Command: _circle Specify center point for circle or [3P/2P/Ttr (tan tan radius)]: **19135, 20050**
5. Specify radius of circle or [Diameter]: Type **'MR** or select the second from last icon on the **Transparent Commands** toolbar.
6. >>Select entity to match radius: Pick the arc on the screen.
7. Notice a circle shows up with the same radius of the arc.
8. Type **L** for **Line** or pick it from the **Draw** toolbar.
9. Command: l LINE Specify first point: **18530,20100**
10. Specify next point or [Undo]: Type **'ML** or select the second from last icon on the **Transparent Commands** toolbar.
11. >>Select entity to match length: Pick the line on the screen.
12. Notice a line shows up with the same length as the other line.
13. Terminate and save your file with the same name and a suffix of your initials.

PARCELS

This next section explores parcel information. You begin with the Parcel Settings and move toward generating parcels and see how Civil 3D handles them in a state-of-the-art manner.

Parcel Development in Civil 3D

Due to the potential complexities of laying out parcels, the Autodesk Civil 3D includes a wide range of tools to assist in the construction of the primitive components of the parcels, namely the horizontal geometry, the lines and the arcs. We have already seen the concepts in the earlier examples, of the dynamic link that exists between the display styles for an object and the appearance of the object. The act of updating an object's style results in automatic changes to the appearance of the object in AutoCAD. For example, if you update the label style for a parcel, the parcel linework annotation is updated.

Open the **Settings** tab in the **Toolspace**. If you expand **Parcel**, notice Parcel Styles, Label Styles, *Table Styles*, and Commands. See Figure 4-36.

These are discussed in detail, but generally the **Parcel Styles** handle how the parcel appears, what displays, and on which layers. The **Label Styles** control what annotation appears and how it looks for the components of the parcel; in other words, the area and the rectilinear and curvilinear linework. The **Table Styles** setting controls how tables involving parcel data appears and which layers are involved. The **Commands** settings include preestablished settings for the details of building annotation and tables, such as units and precisions.

Table Styles—A library of styles

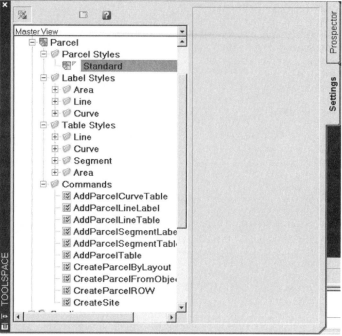

Figure 4-36 Parcel Settings

Exercise 4-15: Creating Parcel Styles

This exercise allows some serious practice in establishing styles for parcels. Civil 3D allows for some new features that are often used to add to the aesthetics of parcel drafting such as placing a hatch pattern around the edge of the parcels. You explore this and annotation styles as well in the next series of exercises.

1. Open the file Chapter-4-Parcel.Dwg.
2. Type **V** for **View** and select and restore the view called **Parcel – A**.
3. Then go to the **Settings** tab in the **Toolspace**; expand the items for **Parcel**.
4. Open **Parcel Styles** and you see **Standard** and **Chapter 4**. These can be edited by right clicking on them and choosing **Edit...** Right click on the **Chapter 4 Style** item and select **Edit...**
5. In the **Information** tab ensure that **Chapter 4** is the name of the style. Choose the **Design** tab in the dialog box. This is where the **Fill** setting is set. It defaults to 5'. This sets a 5' hatch around the perimeter of the parcel, which is often found on engineering and surveying drawings as a highlighting method for parcels. We accept this.
6. Choose the **Display** tab and turn **on** the **Parcel Area fill** layer, so you can see the fill when it occurs. Note that for Parcel Segments under the **Layer** column is where the layer can be set. The layer **V-Prcl-Sgmt** is set to **red** and the layer **V-Prcl-Htch** is set to color **254**. Hit **OK** to exit. Hit **Apply** and **OK**.

You now have a Parcel Style called Chapter 4 with layers set for the parcels, as in Figure 4-37.

Exercise 4-16: Creating an Area Label Style

Now create an Area Label Style for the parcels.

1. In the same drawing, expand **Label Styles**, right click on the **Chapter 4 Area Style** and choose **Edit...**
2. In the **Information** tab ensure that **Chapter 4 Area Style** is the name. Select the **General** tab and observe the settings.

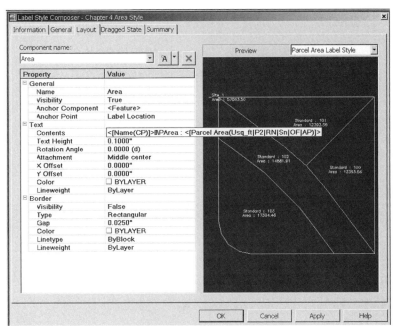

Figure 4-37   Parcel Label Style Composer

3. Click in the Value column for **Text Style** and you see a button with three dots (ellipses) on it. Click this button to see the available styles.
4. Select **Romans** and hit **OK**.
5. Select the **Layout** tab next. Note in the Value column for the **Property** of **Text** Contents. Click in the Value column for this item, and take note how the field and the related appearance of the area text can be altered. The **Preview** window on the right of Figure 4-38 can be edited by the user and prefixes or suffixes can be typed in if needed.
6. Now select the **Dragged State** tab and observe the settings in here that control the properties for labels when they are dragged away from their initial insertion points.

TIP The **Summary** tab provides a single location to review and set all of the settings at once. This is a great destination for experts.

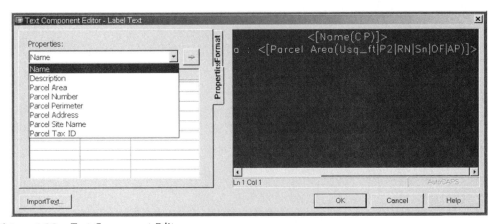

Figure 4-38   Text Component Editor

7. You use the **Standard** Style for Lines and Curves, but you may want to take a moment to review these by selecting either **Lines** or **Curves** and right clicking and choosing **Edit**... Hit **OK** to depart the editing.

Although most of the settings are fairly routine and easy to understand, let us pause for a moment to explore some settings that may not be so self-evident. Expand the item for **Command Settings**. Right click on the **Create-ParcelbyLayout** setting and choose **Edit Command Settings**... An **Edit Command Settings** dialog box displays, shown in Figure 4-39. The interesting settings to take note of are the blue icons. These are expanded in Figure 4-39 and specify the default options for **Parcel** commands.

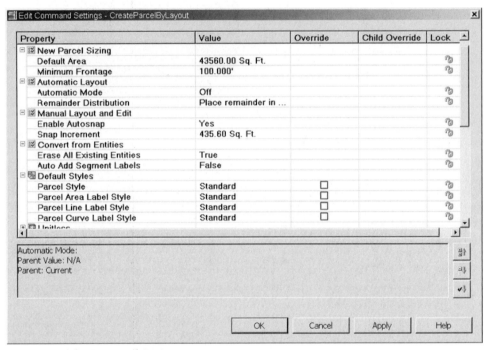

Figure 4-39   Edit Command Settings

Another area of note is at the bottom of this same dialog tab. It is the **Transparent Commands** settings, shown in Figure 4-40.

Property	Value	Override	Child Override	Lock
⊞ Coordinate				
⊞ Grid Coordinate				
⊞ Elevation				
⊞ Area				
⊞ Volume				
⊞ Angle				
⊞ Direction				
⊞ Lat Long				
⊞ Grade/Slope				
⊞ Station				
⊟ Transparent Commands				
Prompt for 3D Points	False	☐		⊘
Prompt for Y before X	False	☐		⊘
Prompt for Easting then Northing	False	☐		⊘
Prompt for Longitude then Latitude	False	☐		⊘
Grade Input Format	percent	☐		⊘
Slope Input Format	run:rise	☐		⊘

Figure 4-40   Transparent Commands settings

There is a myriad of settings here that can control whether you have Nor-things/Eastings or *X/Y* prompts, whether you are prompted for the third dimension, how the order for latitude/longitude will be prompted for, and what units slopes and grades are being asked for in.

> Observe these settings carefully because they are similar to the **Command Settings** in other items.

TIP

Now that you have explored many of the settings controlling parcel development and display, let us create some using the commands provided.

In this same drawing and in the view provided earlier, **Parcel-A**, you create some linework for a parcel.

From the **Parcels** pull-down menu, choose **Create By Layout** and a toolbar displays as shown in Figure 4-41. Note that it can be expanded/retracted by using the arrow at the far right of the toolbar.

Figure 4-41   Parcel Layout tools

The commands available are described from the left and are referred to as Buttons 1 through 10 in their descriptions that follow:

- Button 1:   **Create Parcel**

- Button 2:   **Add Fixed Line—Two Points,** which draws a lot line as a line segment. Choose a starting point and an endpoint to create it.

- Button 3:   **Add Fixed Curve:**

  - **By 3 Points,** which draws a lot line as a curved component. You would define a starting point, a point on curve, and an endpoint.
  - **By 2 Points and a Radius,** which draws a lot line as a curve component. You would define a starting point, a radius value, the curve's direction, and an endpoint.

- Button 4:   **Draw Tangent with No Curves,** which draws a connected series of lot line components. You would identify a series of points.

- Button 5:   This button has several options to it including:

  - **Slide Angle—Create,** creates one or more new lot lines defined with starting and ending points along the frontage of the lot or, at your option, using an angle relative to the frontage. These angles are measured in positive degrees, from 0 degrees (toward the endpoint) through 180 degrees (toward the start point).
  - **Slide Angle—Edit,** moves a lot line. The user can keep or modify the line's frontage angle in this routine.
  - **Slide Direction—Create,** develops one or more new lot lines. You can define starting and ending points along the frontage of the lot and an

absolute direction for the lot line using azimuths, bearings, or any other two points in the drawing.

- **Slide Direction—Edit**, moves a lot line. The user can keep or modify the line's absolute direction.
- **Swing Line—Create**, develops a lot line defined with starting and ending points along the frontage of the lot and a fixed swing point on the opposite side of the parcel. The size of the parcel can be modified by swinging the lot line to intersect a different point along the frontage. This is limited by the user through minimum areas and frontage limits that are established.
- **Swing Line—Edit**, moves a lot line by pivoting it from one end. You can select which end to use as the swing point on the fly.
- **Free Form Create**, develops a new lot line. You can define an attachment point to begin and then use bearings, azimuths, or a snap to a second attachment point to complete the command.

- Button 6: **PI Commands** for:

  - **Inserting**, creates a vertex at the point clicked on a parcel segment.
  - **Deleting**, eliminates a vertex that is selected on a parcel segment. It then redraws the lot line between the vertices on either side to repair the parcel line. This command can delete or merge parcels depending on how the vertex was deleted.
  - **Breaking apart** PIs, separates end points at a vertex selected with a separation distance specified by the user. This command does not delete or merge parcels; it does, however, make them incomplete and the components that are affected become geometry elements. These elements become parcels again once any loose vertices are reconnected to a closed figure.

- Button 7: **Delete Sub-Entity**, eliminates parcel components. If a sub-entity is eliminated when it is not shared by another parcel, the parcel is deleted. If a shared sub-entity is eliminated, the two parcels that shared it are merged together.

- Button 8: **Parcel Union**, merges two adjacent parcels together where the first parcel selected determines the identity and properties of the joined parcel, similar to the AutoCAD **PEDIT** command.

- Button 9: **Pick Sub-Entity**, selects a parcel component for display in the **Parcel Layout Parameters** dialog box. Use the **Sub-Entity Editor** before choosing this command.

- Button 10: **Sub-Entity Editor**, displays the **Parcel Layout Parameters** dialog box that allows for the review or editing of attributes of selected parcel components.

- There is also an **Undo** and a **Redo** button for creating and editing parcel data.

The graphics capabilities that accompany these commands are extraordinary, and a few are used next to explore how they work.

Exercise 4-17: Create Parcel By Layout

1. Remain in the same drawing. In the **Prospector**, expand **Open Drawings**, expand the **Chapter-4-Parcel.dwg**, expand **Sites**, expand **Site 1**.
2. Then right click on **Parcels**. Choose **Properties**.
3. Under the **Composition** tab, set the **Site Parcel Style** to **Chapter 4**, which was created earlier. Set the **Site Area Label Style** to **Chapter 4 Area Style**, also created earlier. Hit **Apply** and **OK**.

4. In the Chapter-4-Parcel.Dwg drawing, select **Create By Layout...** from the **Parcels** pull-down menu.
5. When the **Create By Layout** toolbar displays, select the **Add Fixed Line** command. The dialog box for **Create Parcels Layout** displays. Set the **Parcel Style** to **Chapter 4**. Set the **Area Label Style** to **Chapter 4 Area Style**. Turn **on** the toggle for **Automatically add segment labels**. Hit **OK**.
6. When it asks for a start point, use an Endpoint snap to select the point at **16000,19000**, which is a vertex of the parcel polyline already drawn in the file.
7. Then using Endpoint snaps select the vertices of the polyline in a clockwise fashion. Set the second and third vertex and stop after selecting an endpoint at the third vertex at $X = 16268.9231\ Y = 19285.9656$.

TIP   Note that after you draw one line, you must begin the next line again, unlike an AutoCAD line.

8. While staying in the command, select the third button in the toolbar, **Add Fixed Curve**, **By 3 Points**. Use an Endpoint snap to select the third vertex, use a nearest snap to pick a point on the curve and an endpoint to select the fourth vertex.
9. Then use the second button, **Add Fixed Line**, to select a first point at the endpoint at the fourth vertex and the second point at the beginning of the parcel where you started.
10. When the software asks for a new start point, hit <**Enter**> and then type **X** for **Exit**, as shown in the command prompt. You see a parcel, fully developed with annotation and hatching, as discussed in the setup of the settings earlier.

If you look in the **Prospector** (Figure 4-42), you see this parcel listed there, with a **Preview** of the parcel.

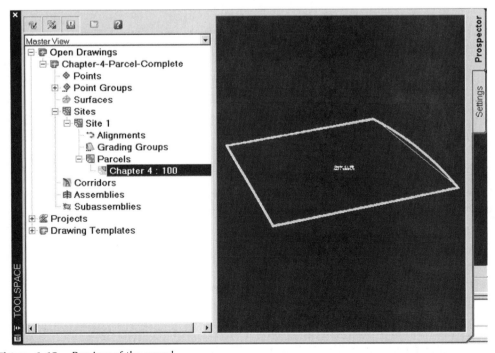

Figure 4-42   Preview of the parcel

Exercise 4-18: Editing Parcel Data

1. To edit the parcel, use the **Create By Layout...** from the **Parcels** pull-down menu again.
2. This time select button 6, **Insert PI**.
3. When prompted, pick the north parcel line. It requests that you pick a point for the new PI. Type **16100, 19300** as the new PI location. Hit **<Enter>** to terminate and **X** to leave the routine.
4. You see the parcel updated with the new PI, new annotation, and a new area. The **Prospector** is updated if you select **Refresh** when right clicking the parcel item. Refer to Figure 4-43.

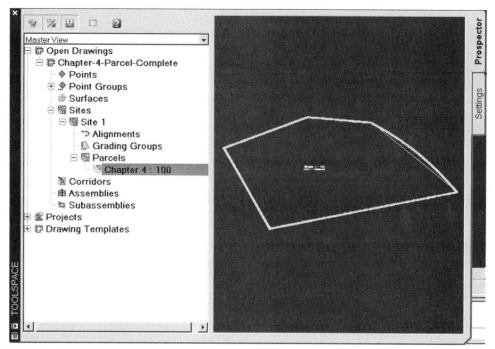

Figure 4-43   Updated Preview of the Parcel

Exercise 4-19: Creating Parcel Data from Objects

A second option for creating parcel data is to select it directly from pre-drawn entities. Let us perform an example of this.

1. Type **V** for **View** and select the view called **Parcel—B**.
2. A copy of the same parcel data displays. Select **Create From Objects** from the **Parcels** pull-down menu. The **Create Parcels** dialog box in Figure 4-44 displays.
3. It asks you to: `Select the Lines, arcs or polylines to convert into parcels:` Select the polyline displayed.
4. The dialog box for **Create Parcels From Objects** displays. Set the **Parcel style** to **Chapter 4**. Set the **Area label style** to **Chapter 4 Area Style**. Turn **on** the toggle for **Automatically add segment labels**. Hit **OK**.
5. Again the parcel is defined, hatched, labeled, and displayed in the **Prospector**. Note that you may need to choose **Refresh** from the right click menu to see both parcels in the **Prospector**.

Exercise 4-20: Creating Rights of Way

Another command within the **Parcels** pull-down menu is the **Create ROW** command. This creates a right of way along an alignment. When a right of way is developed, adjacent parcel boundaries are offset by a user-specified distance from the right of way on each side of the

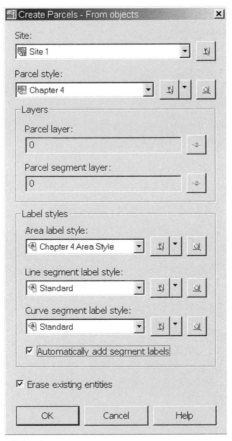

Figure 4-44   Create Parcels dialog box

alignment. Radii can be defined for end returns that exist along the right of way at intersections.

The **Right of Way** command works like a parcel, albeit a narrow variation of a parcel. When the **Create ROW** command is used, it prompts the user to pick parcels. If an alignment is found on one of the edges of the selected parcels, a right of way is created in accordance with the preestablished parameters.

1. Type in **V** for **View** and select the view called **Parcel—C**. This shows an alignment and several parcels along the alignment.
2. Select **Create From Objects** from the **Parcels** pull-down menu.
3. It asks you to: `Select the Lines, arcs or polylines to convert into parcels:` Select the polylines of the three parcels in the view.
4. The dialog box for **Create Parcels From Objects** displays. Set the **Parcel style** to **Chapter 4**. Set the **Area label style** to **Chapter 4 Area Style**. Turn **on** the toggle for **Automatically add segment labels**. Set layers for the parcels to **V-Prcl-Sgmt** and **V-Prcl-Htch**. Hit **OK**.
5. Again the parcels are defined, hatched, labeled, and displayed in the **Prospector**.
6. Now choose the command **Create ROW** from the **Parcels** pull-down menu. When it asks you to select the parcels, choose the three parcels you just created. Then the **Create Right of Way** dialog box, shown in Figure 4-45, displays to obtain the criteria for developing the ROW.
7. Set the value for **Offset from alignment** to **45**. Set the **Cleanup at parcel boundaries** to **35**. Set the **Cleanup at alignment intersections** to **35** as well. Hit **OK**.
8. The configuration should appear as shown in Figure 4-46.

For the next step in this exercise, let us add a line table for the parcel in the **View Parcel—B**.

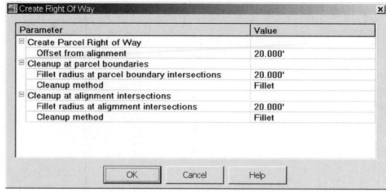

Figure 4-45   Create Right of Way dialog box

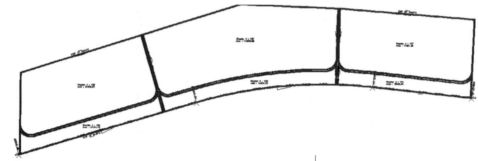

Figure 4-46   Results of parcel layout

9. Type in **V** for **View** and select the view called **Parcel—C**. This shows an alignment and several parcels along the alignment.
10. Select **Tables** from the **Parcels** Pull-down menu. Select **Add Line** table.
11. The **Table Creation** dialog box displays, as shown in Figure 4-47.

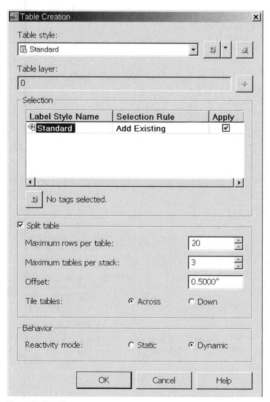

Figure 4-47   Table Creation dialog box

In the selection window in the center of the dialog box, turn on the **Apply** button and make sure it is checked as in Figure 4-47. Then hit **OK**. A table is on your cursor and asks you to pick a point for the table.

Pick a point away from the parcel graphics. It appears as shown in Figure 4-48. Other tables can be brought in as well for curves and areas.

Line Table		
Line #	Length	Direction
L100	388.861	N72° 47′ 09.54″E
L101	165.234	S16° 18′ 58.54″E
L102	461.693	S73° 41′ 01.46″W
L103	79.578	S1° 32′ 20.88″W
L104	153.731	N1° 32′ 20.88″E
L105	412.792	S85° 32′ 05.26″E
L106	79.578	S1° 32′ 20.88″W
L107	153.731	S1° 32′ 20.88″W
L108	250.000	N30° 00′ 00.00″W
L109	239.992	N69° 38′ 27.50″E
L110	169.505	S85° 15′ 02.49″E
L111	250.000	N30° 00′ 00.00″W
L112	400.000	N80° 00′ 00.00″E
L113	509.902	S78° 41′ 24.24″W

Figure 4-48   Completed Table

ALIGNMENTS

Alignments are developed to assist engineers, surveyors, and contractors in their computations and layout of projects. Alignments provide a common structure by which they can refer to the corridor and locate the positions easily and be able to duplicate that location reliably. An alignment consists of "primitive elements" such as lines, arcs, and spirals. Although each of these is an independent component, when strung together they create an alignment and can be thought of and referred to as a single object.

Alignments can be used in all corridor situations whether they are roadways, channels, aqueducts, or utilities such as waterlines.

The notion of "stationing" is what unifies thinking about alignments. Stationing (Figure 4-49) is a mathematical concept whereby the beginning of the alignment commences with a value, say 0, and this number increases by the lengths of the lines, arcs, and spirals until terminating at the end of the alignment. To make it a little easier to understand and work with, stationing also includes the concept of dividing the actual linear value by 100′. In other words, station 2 + 50 would be located 250′ from the beginning of the alignment. Station 18 + 32.56 would be 1,832.56′ from the beginning. The even stations are denoted on the left side of the plus sign and the leftover value is on the right side of the plus sign. Unless otherwise instructed, corridor alignments often begin with the station 10 + 00 or 1000′. This allows for changes to be

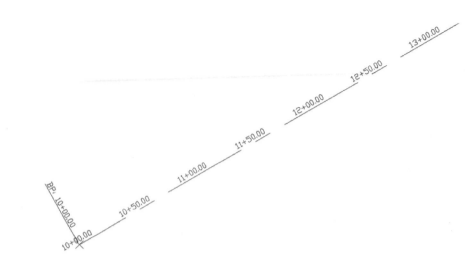

Figure 4-49   Stationing

made to the alignment and within reason you will at least remain in positive numbers when computing stationing. For example, if you began a roadway at station 0+00, it could happen that a superior would want to extend the roadway length 100′ from the beginning, thereby causing the beginning station to be -1+00. This causes potential confusion simply because of the addition of negative numbers. Anything designers can do to simplify the design and construction process pays off in the long run, and this is one of those things it is important to do.

In combination with stationing, a concept of offset values is also used in conjunction with alignments. Typically, a positive offset is to the right of the alignment and a negative value is to the left, when looking forward toward increasing stationing. This additional concept allows users to locate objects, either existing or proposed, using the stationing for the alignment as the baseline for computations. For instance, it would not be difficult to locate a proposed fire hydrant with a station of 18+32.56 and an offset of 55′ Right, or the corner of a drain inlet at station 9+55.32 and an offset of −14.5 (Left).

One last item associated with alignments and their respective stationing is the concept of station equations or equalities. This is a situation in which the roadway begins, say, at station 10+00 and perhaps the alignment is 2000′ long. Thus the ending station is 30+00. Now let us further say that a side road intersects the roadway halfway through it, at station 20+00. The hypothetical State Department of Transportation (DOT) has decided that the intersecting roadway is the predominant roadway and yours is not. Therefore, it decrees that your roadway will have your stationing from the beginning and will maintain your stationing until it intersects with the side road. At that point, the DOT wants the stationing of your road to pick up where the side street leaves off and continue on until your road ends.

This is where a station equation will exist. Our roadway stationing begins at 10+00 and progresses until station 20+00 where the side road enters the project, as in Figure 4-50.

If the side road has an ending station of, say, 11+96.32 when it intersects your road, then your road continues from that point onward accumulating its additional stationing values and adding them to the 11+96.32 from the side road. So your road would end at station 11+96.32 plus 10+00 remaining after the side road intersects your road for a total ending station of 21+96.32. To complicate things just a little further, note that the stationing could be decreasing instead of increasing. If it were, then your road could end at 11+96.32. It could also occur that there would be multiple stations on the alignment of the same value depending on how the station equations are applied to the alignment. See Figure 4-51.

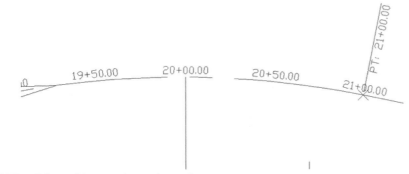

Figure 4-50   Side road intersecting main road

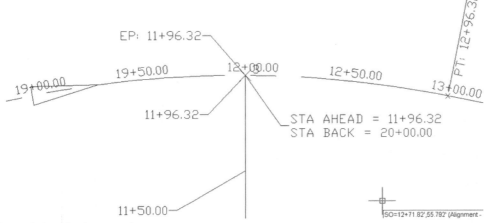

Figure 4-51   Station equation

In a subdivision corridor, items that reference horizontal alignments are gutters, sidewalks, and lot corners; water, sanitary, and storm sewers; catch basins; and manholes. This conversation is limited for the time being to roadway applications. However, they can be generalized to the other aforementioned project types.

The criteria used for the design of horizontal alignments for roadways are represented in terms of number of lanes, design speeds, and minimum curve radii. The minimum radius for curves increases with design speed such that large radius curves are required for high vehicle speeds. Correspondingly, there also tends to be a minimum radius once you go below a certain design speed such as that found in residential subdivisions. A governmental agency, perhaps a DOT, usually specifies horizontal alignment design criteria for highways. For subdivision roads, it may be the county's highway department or the municipality that dictates the design criteria.

Horizontal Alignment Development in Civil 3D

Autodesk Civil 3D includes a robust toolset to assist in the construction of the primitive components of the horizontal alignment, namely the horizontal geometry, the lines, arcs, and spirals. You have already seen the concepts in the previous examples, of the dynamic link that exists between the display styles for an object and the appearance of the object. The act of updating an object's style results in automatic changes to the appearance of the object in AutoCAD. For example, if you update a component within the alignment, a ripple-through effect updates profiles, sections, and annotations as well. There are similar tools for developing alignments as there are for creating parcel data. You can draw the alignment using polylines and convert them to alignment objects, or you can use the **Layout** toolbar to develop the design with alignment tools custom developed for use in design alignments.

Exercise 4-21: Alignment Settings

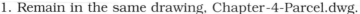

Alignment creation is explored in this section, beginning with the **Alignment Settings.** You see how Civil 3D handles them in a state-of-the-art manner.

1. Remain in the same drawing, Chapter-4-Parcel.dwg.
2. Open the **Settings** tab in the **Toolspace.** If you expand **Alignments** notice **Alignment Styles, Label Styles, Table Styles,** and **Commands.** Each of these is discussed in detail, but generally the **Alignment Styles** handle how the alignment appears, what will display, and on what layers. The **Label Styles** control what annotation appears and how it will look for the components of the alignment; in other words, the stationing, equations, alignment components, station/offsets, and so on. The **Table Styles** setting, shown in Figure 4-52, controls how tables involving alignment data appear and what layers are involved. The **Commands** settings include preestablished settings for the details of building annotation and tables, such as units and precisions.

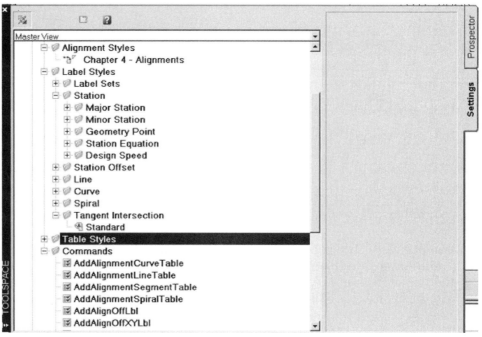

Figure 4-52   Table Styles setting

3. Right click on the **Alignment Styles** item and select **Edit....** The **Information** tab allows for naming and optional descriptions. In the name field type a name of **Chapter 4—Alignments.**
4. The **Design** tab allows you to control the **Grip Edit Behavior.** This can be enabled or disabled as a toggle, and it controls whether the cursor snaps to a specified increment when grip editing the radius of a curve in the alignment. The **Radius Snap Value** specifies the curve's incrementing radius. Enable the **Grip Edit Behavior** and set the value to **25.**
5. In the **Display** tab you can set the layers for individual components of the alignment if you choose to. For this example, set the color for **Lines** to **red** and the color for **Curves** to **green.**
6. Hit **Apply** and **Close.**

7. Now click on the **Label Styles** item, select **Stations,** and then right click on **Major Stations**. Select **New....**

8. Click on the **Information** tab, type in the name of our style as **Chapter 4—Major Stations Label Style.**

9. Then in the **Layout** tab, Figure 4-53, under the **Text Property,** click in the Value column for **Contents**. Notice that a button appears on the right side of the field. Click on the button with the ellipsis (...).

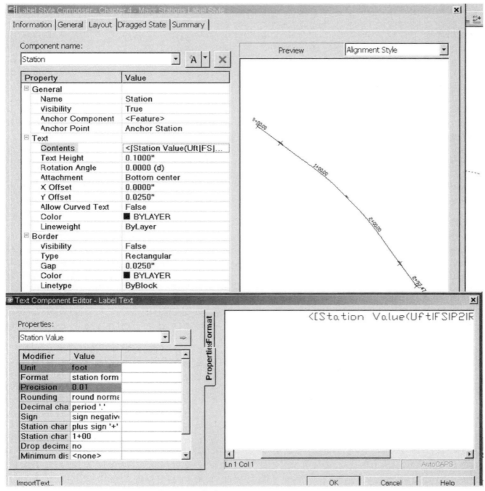

Figure 4-53 Text Component Editor—Label Text

10. A dialog box called **Text Component Editor—Label Text** displays.

11. You see the formula in the window to the right. There is a parameter in the formula, **P2** for Precision to 2 decimal places. Click in there and delete the entire current formula.

12. In the **Modifier** column notice the **Precision** item. Click in the Value column for this modifier. The default is 0.01. Change it to **1**.

13. *This next step is important.* Make sure you click the button with the **blue right arrow** on it to send the parameter change into the formula window to the right.

14. If you look in the formula, notice that the parameter **P** is now set to 0.

15. Hit **OK**. The **Preview** in the **Label Style Composer** shows the stationing with no decimal places now.

16. Hit **Apply** and **OK** to exit.

17. Repeat this for the **Minor Stations** and call the label style **Chapter 4—Minor Stations Label Style**. Repeat steps 8 through 15 for the **Minor Label Styles.**

This exercise should provide you with the idea of how the settings for alignments work. They are intended to be prebuilt into the Civil 3D product and then take effect on the creation of the alignment.

Now that you have gone through these basic examples, note that the dialog boxes for all of the settings work essentially the same way and have the same types of parameters. This is true for the Label Styles for Label Sets; all of the Stationing styles; the styles for lines, curves, spirals, tangent intersections; and Table Styles.

The dialog boxes for the **Commands** section of the **Settings** were discussed under **Parcels, Command Settings**.

Fixed, Floating, and Free Lines and Curves

In getting started developing alignment objects, an important concept must be discussed first. That concept involves "Constraint-based" commands and the creation of fixed, floating, and free lines and curves. These objects can be edited dynamically and the objects retain their tangencies.

A Fixed entity is fixed in its position and is defined by criteria such as a radius or located points. It is not dependent on other entities for geometry development or tangency. A Floating entity is always tangent to one entity and is defined by the parameters provided or is dependent on one other entity to define its geometry. A Free entity is always tangent to an entity before and after it and must have at least two other entities to attach to that define its geometry.

Although a user can always develop polylines to create their alignments, there are also some state-of-the-art tools in the **Create By Layout** toolbar (Figure 4-54). They are called up by choosing the **Alignments** pull-down menu and selecting **Create By Layout**. A dialog box first displays asking you for some

Figure 4-54   Create By Layout toolbar

initial alignment style parameters but then the **Alignment Layout Tools** toolbar displays. The commands available within it are described from the left and are referred to as buttons 1 through 12 in their descriptions below:

- Button 1: This button has three choices. They are **Draw Tangent-Tangent without Curves, Draw Tangent-Tangent with Curves**, and **Curve Settings**. By using the **Tangent-Tangent** command, the user can pick a series of points end-to-end in order to draw a fixed alignment in a fashion similar to drawing AutoCAD lines. The choices provide options to insert free curves automatically at each PI. These commands can be effectively used to create quick layouts of an alignment. The lines and curves can be modified by the user, and as the linework is edited it always maintains tangency.

- Button 2: **Insert PI**.

- Button 3: **Delete PI**.

- Button 4: **Break-Apart PI**.

- Button 5: [icon] This button allows for creating **Fixed, Floating, and Free lines**. This concept is defined by the characteristics inherent within each type of line. These commands have several options tool, including:

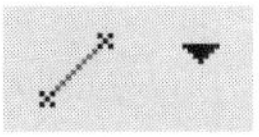

  - **Draw Fixed Line and Draw More Fixed Lines**—These commands allow you to enter a fixed line through two points, or enter a fixed line through two points where the line is tangent to the end of the selected fixed or floating entity.
  - **Draw Floating Line and Draw More Floating Lines**—These commands allow you to enter floating lines through points that are always tangent to fixed or floating curves, or you can enter a floating line tangent through a point. The lines drawn are always tangent to the end of fixed or floating curves.
  - **Draw Free Lines Between Two Curves**—Allows you to enter a free line that is always tangent to a Fixed or Floating curve, either before or after the line.

- Button 6: [icon] This button has several options to it including:

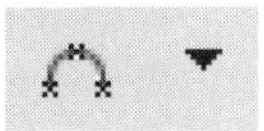

  - **Draw Fixed Curve and Draw More Fixed Curves**—These commands, in Figure 4-55, allow you to enter a fixed curve using 3 points, or you can

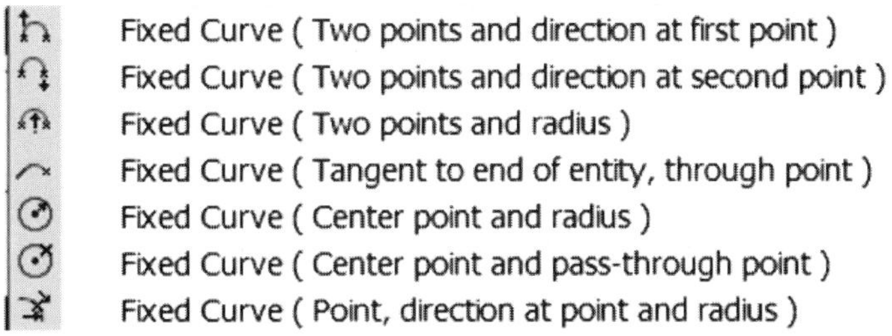

Fixed Curve (Two points and direction at first point)
Fixed Curve (Two points and direction at second point)
Fixed Curve (Two points and radius)
Fixed Curve (Tangent to end of entity, through point)
Fixed Curve (Center point and radius)
Fixed Curve (Center point and pass-through point)
Fixed Curve (Point, direction at point and radius)

Figure 4-55   Draw fixed curve and draw more fixed curves

enter a fixed curve using the following criteria to define the curves as shown in the figure.
- The *first three commands* are performed by specifying two points and some additional criteria. They all achieve a fixed curve through three points.
- The next command develops a *fixed curve tangent to the end of an entity and through a chosen point*. This achieves a fixed curve through three points that is also tangent to the end of the entity you selected.
- Following that is the *fixed Curve developed by selecting a center point, providing a radius, and a curve direction*, which achieves a fixed circle.
- The next command allows entering a *fixed curve by specifying a center point, a pass-through point, and the curve's direction*, which achieves a fixed curve that passes through the specified point.
- The last command enters a *fixed curve by specifying a point, the curve's direction at the point, and a radius*, which achieves a fixed circle with a center point and a radius.
  - **Draw Floating Curve and Draw More Floating Curves**—These commands (Figure 4-56) allow you to enter floating curves using a variety of criteria that will be maintained on modification of those curves. The **Draw Floating Curve** allows you to enter a floating curve that is attached to a fixed or floating line or curve entity, by specifying a radius,

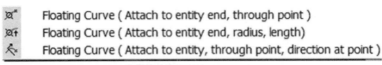

Figure 4-56   Draw Floating Curve and Draw More Floating Curves

a pass-through point, and an angle range. This achieves a floating curve, which is always tangent to the entity to which it is attached. The submenu **Draw More Floating Curves** is shown in the figure and differs from the **Fixed** commands in that these entities are not pinned to any specific location in the drawing. These commands are described as follows:

- **Draw Floating Curve—Attach to Entity End** and specify a pass-through point achieves a floating curve that is always tangent to the entity to which it is attached.
- **Draw Floating Curve—Attach to Entity End** and specify a radius and length, which achieves a floating curve entity that always starts at the end of the entity to which it is attached.
- **Draw Floating Curve—Attach to Entity** by specifying a pass-through point and a direction at that point, which achieves a floating curve that remains tangent to the entity to which it was attached.

- **Draw Free Curves and Draw More Floating Curves**—The first command, **Draw Free Curves Attached to Two Entities**, allows you to enter a free curve that is attached to two objects by specifying a radius and an angle range, which achieves a free curve that remains tangent to the two entities to which it is attached. The command **Draw More Floating Curves** allows entering a free curve that is attached to two objects by specifying a pass-through point, which achieves a free curve that remains tangent to the entities to which it is attached.

These next three button tools were added to 2007.

- Button 7: **Floating Spirals:**
  - Floating line with Spiral: (From Curve through point)—adds a floating spiral from an entity with a line at its end.
  - Floating line with Spiral: (From Curve end, length)—adds a floating spiral from an entity with a line at its end.

- Button 8: (More) **Floating Spirals:**
  - Floating curve with Spiral: (From entity end, radius, length)—adds a floating spiral from an entity based on the radius and length provided.
  - Floating curve with Spiral: (From entity, radius, through point)—adds a floating spiral from an entity based on the radius and a point provided.
  - Free spiral—curve—spiral: between two entities. This will prompt to select the incoming and outgoing tangents and then prompt for spiral lengths and a radius for the curve.

- Button 9: This button allows for a large number of sophisticated spiral geometry commands. These routines are summarized in the commands which follow.
  - Fixed Spiral: develops a spiral with a spiral in length and a curve radius.
  - Free spiral: (between two entities)—requests length and radius information and then shows extraordinary graphics in the placement of the object.
  - Free compound Spiral-spiral (between two curves)—develops two spirals back to back
  - Free Reverse spiral (between two curves)—develops two spirals back to back but in a reverse curve scenario.
  - Free compound Spiral-Line-spiral (between two curves, spiral lengths)—develops two spirals back to back

- Free Reverse spiral-line-spiral (between two curves, spiral lengths)—develops two spirals back to back but in a reverse curve scenario.
- Free compound Spiral-line-spiral (between two curves, line lengths)—develops two spirals back to back
- Free Reverse spiral-line-spiral (between two curves, line lengths)—develops two spirals back to back but in a reverse curve scenario.

- Button 10: **Delete Sub-Entity**, eliminates alignment components.

- Button 11: **Pick Sub-Entity**, selects an alignment component for

  display in the **Alignment Layout Parameters** dialog box. Use the **Sub-Entity Editor** before choosing this command.

- Button 12: **Sub-Entity Editor**, displays the **Alignment Layout**

  Parameters dialog that allows for the review or editing of attributes of selected alignment components.

- Button 13: **Alignment Grid View**, brings up the **Panorama** view for

  reviewing the alignment.

- There is also an **Undo** and a **Redo** button for creating and editing Parcel data.

How to Design an Alignment

Now let us use some of the alignment development routines to check out the advantages that Civil 3D provides to its users. Begin with the first **Layout** tool in the toolbar.

Exercise 4-22: Tangent-Tangent (With Curves)

1. Open the file called Chapter 4 Alignments-2.Dwg.
2. Type **V** for **View** and select **Align-A** and restore it.
3. Set your current layer to **C-Road-Cntr**.
4. Turn **off** the surface's contour layers on **C-Topo-Existing Ground**.
5. From the **Alignments** pull-down menu select, **Create By Layout**.
6. Accept the defaults except choose the **Alignment Style** of **Chapter 4—Alignments**. Figure 4-57 shows this dialog box. Hit **OK** when finished.
7. The **Create By Layout** toolbar now displays.
8. Using the **down arrow** next to the first button, select the command shown in Figure 4-58, called **Curve Settings**....
9. In the **Properties** dialog box that displays, use the Clothoid type, which is the most commonly used, and hold the **200′** radius.
10. Now choose the command **Tangent-Tangent (With Curves)**. At the command prompt when asked to provide the first point, type in the following coordinates: **14685,18545**. Then follow these instructions.
11. Specify next point: Enter **15175,18510**
    Specify next point: **15250,18275**
    Specify next point: **15475,18175**
    Specify next point: <Enter> to terminate.
12. Zoom in on the **Alignment** and notice it has the component coloring as described earlier. Curves are already developed. Annotation is evident as well.
13. Now pick the **Alignment** and grips appear. Select a grip on the PI for the first curve. Drag it slightly to a new location while watching the curve.

Student Files

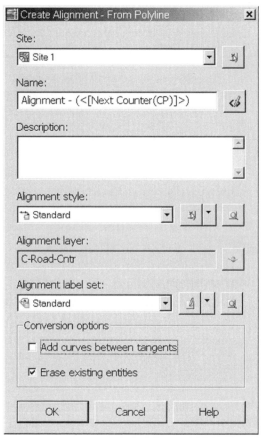

Figure 4-57   Create By Layout

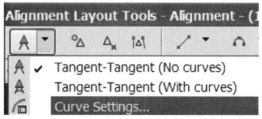

Figure 4-58   Curve Settings...

Do not move the PI so far that the curve cannot be computed and causes it to disappear. Cancel when finished moving the PI.

14. Now go to the **Alignments** pull-down menu and choose **Edit** *Alignment Geometry*.... Select the alignment and the **Edit** toolbar appears. Choose the command on the right of the toolbar called **Alignment Grid View** and the **Panorama** displays. Scroll through the Panorama and notice that the alignment held the radii of 200′.

Exercise 4-23: Insert PI

Now use the second tool called **Insert PI**.

1. Again, go to the **Alignments** pull-down menu and choose **Edit**.... Select the **Alignment** in Figure 4-59 and the **Edit** toolbar appears.
2. Choose the second command, **Insert PI**.

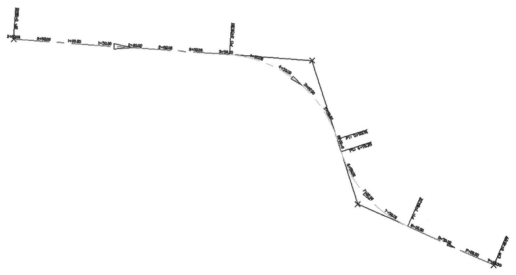

Figure 4-59   Select the alignment

3. When asked to **Pick a point near PI** to insert, use the Midpoint snap and pick the first tangent. It places a new PI at 1+77.90 and a label appears.
4. Hit <**Enter**> to leave the editing facility.
5. Now pick the **Alignment** and notice that grips appear at keypoints on the alignment.
6. There are different types of grips denoted by their shape and include square, triangular and circular grips. The Square grips are on alignment endpoints and allow the entire tangent component to be moved. The curves are the connecting segments and hold their radius as the tangent is relocated. The circular grips identify the curve geometry in the alignment. These can be slid along the tangents to lengthen or shorten the curve. The triangular grips are located at the PIs and can be moved to any other location as needed. When they are moved, they drag the curve along all the while holding the radius of the curves. Refer to Figure 4-60.

Figure 4-60   Edit with grips

7. Experiment with moving and relocating these grips. Move the PI grip for the PI you just added and observe how the alignment recomputes and redisplays.
8. Move that PI to a coordinate at or near an *X,Y* of **14865,18560**.
9. Then pick the **Alignment**, right click, and select **Edit Alignment...** to redisplay the **Layout** toolbar.
10. Choose the **down arrow** on the sixth button from the left in the toolbar. When the menu drops down, select the command **Draw Free Curves Attached to Two Entities, Specify Radius.**
11. The software responds and asks you to: `Select entity before to attach to:` Pick the tangent before the PI you just created.
12. `Select entity after to attach to:` Pick the tangent after the PI you just created.
13. `Is curve solution angle [Greaterthan180/Lessthan180] <Lessthan180>:` It is, so hit **<Enter>**.
14. `Specify radius in or [curveLen/Tanlen/Chordlen/midOrd/External]:` Type **250**.
15. `Select entity before to attach to:` **<Enter>** to terminate. Refer to Figure 4-61.

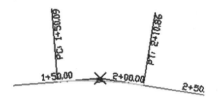

Figure 4-61   Specifying a radius

Notice that you now have a new curve in the alignment with a radius of 250′.

Exercise 4-24: Grips

Because you have seen several types of grips, let us take a moment to further experiment with these grips.

1. Now pick the **Alignment** and notice that grips appear at keypoints on the alignment.
2. Move the circular grip for a curve and slide it along the tangent. Observe how the curve gets shorter or longer.
3. Grab a rectangular grip for an internal tangent segment and move it to a new location. Watch how the tangents and curves attached to it become relocated as well, while maintaining the curve radius.

Exercise 4-25: Using Spirals in Alignments

The software also supports spirals within the alignments. State DOTs use spirals in different ways. In some states, entrance and exit ramps for highways may be designed using spiral mathematics. Remember from the earlier description in this chapter of spirals that a spiral is a constantly changing radius. This differentiates spirals from simple arcs, which have a constant radius. You often need more physical space to build spiral ramps than those with arcs; hence you might find that the Texas DOT uses spirals more often than does the DOT in the Washington, DC, region, which has less available space. You may find that the DOT uses arcs for urban design but requires spirals for rural design for the same reasons.

What you do here for the alignment is add a spiral to the end of the alignment.

1. Pick the **Alignment**, right click, and select **Edit Alignment**... and the **Edit** toolbar appears.
2. Choose the eighth icon from the left called Floating curve with Spiral: (From entity end, radius, length).
3. The following prompts appear; provide the suggested responses shown.
4. `Select entity for start point and direction:` Use the Endpoint snap and snap to the right end of the alignment.
5. `Select spiral type [Compound/Incurve/Outcurve] <Incurve>:` Hit **<Enter>**.
6. `Specify curve direction [Clockwise/counterclockwise] <Clockwise>:` Hit **<Enter>**.
7. `Specify length:` **200**
8. `Specify radius:` **100**
9. `Select entity for start point and direction:` Hit **<Enter>** to terminate.

Observe that a spiral curve has been added to the end of the alignment as expected; refer to Figure 4-62. Notice that the stationing is automatically continued from the previous length of the alignment.

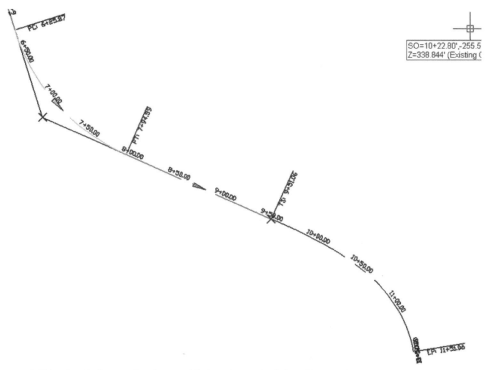

Figure 4-62   A spiral curve has been added to the end of the alignment

When an alignment is being constructed, you may want to know detailed information about any of the components.

Exercise 4-26: Spiral Information

Run through the following routine to obtain information about the spiral, which as you know can become quite complex in its mathematics.

1. Pick the **Alignment**, right click, and select **Edit Alignment Geometry**... and the **Edit** toolbar appears.

2. Pick the **Sub-Entity Editor**, the fourth icon in the toolbar from the right. Then select the command **Pick Sub-Entity**, the fifth icon from the right.
3. Select the **Spiral Component**.
4. A **Criteria** window is populated with all of the spiral criteria for your inspection. See Figure 4-63.

Parameter	Value
Spiral In	
Number	4
Curve Group I...	
Curve Group S...	
Type	Spiral-Curve
Constraint1	Float
Constraint2	Spiral length, Radius and Curve L...
Spiral type	Simple
Length	**100.000'**
A	**141.421'**
Delta angle	14.3239 (d)
Start Station	4+82.12'
End Station	5+82.12'
Start Direction	S66° 02' 15.04"E
End Direction	S51° 42' 48.84"E
Start Point	(15475.0000',18175.0000',0.0...
End Point	(15562.4423',18127.0581',0.0...
Incurve	Incurve
Compound	false
Radius in	Infinity'
Radius out	200.000'
Total X	99.377'
Total Y	8.296'
Short tan	33.533'
Long tan	66.886'
P	2.079'
K	49.896'
Spiral Definition	Clothoid
SPI Station	5+49.01'
SPI Northing	18147.8349'
SPI Easting	15536.1214'
SPI Included A...	165.6761 (d)
Radial Point No...	17970.0735'
Radial Point Ea...	15438.5236'
Curve	
Number	4
Curve Group I...	
Curve Group S...	
Type	Spiral-Curve
Constraint1	Float
Constraint2	Spiral length, Radius and Curve L...
Length	**100.000'**
Radius	**200.000'**
Delta angle	28.6479 (d)
Start Station	5+82.12'
End Station	6+82.12'
Start Direction	S51° 42' 48.84"E
End Direction	S23° 03' 56.44"E
Start Point	(15562.4423',18127.0581',0.0...
End Point	(15622.5349',18048.4307',0.0...
Center Point	(15438.5236',17970.0735',0.0...
Pass Through ...	
Pass Through ...	
Pass Through ...	

Figure 4-63   Spiral Criteria

Notice that you can pick any sub-entity to obtain its respective data as well.

Exercise 4-27: Fixed Lines and Curves

The next routine explores creating fixed lines and curves and shows what these features offer.

1. In the same file called Chapter 4 Alignments-2.Dwg, type **V** for **View** and select **Align-A** and restore it.

2. Set your current layer to **C-Road-Cntr.**
3. Turn off the surface's contour layers on **C-Topo-Existing Ground.**
4. From the **Alignments** pull-down menu, select **Create By Layout.**
5. Select the black **down arrow** for the fifth icon from the left, and select the command called **Fixed Line (Two Points).** Enter the following data to construct that line and stay in the **Layout** toolbar when you are finished creating the line.
6. The prompt responds: `Specify start point:` **15200,18700**
7. `Specify next point:` **@200<30**
8. Now select the black **down arrow** for the sixth icon from the left, and select the command called **More Fixed Curves.** In the submenu choose the **Fixed Curve (Tangent to end of entity, through point).**
9. The prompts requests: `Select entity for start point and direction:`
10. Pick near the end of the line you just drew.
11. The prompt requests: `Specify end point:` **15900,18800** and notice the curve emerges from the end of the line and is also tangent. The alignment expands with the associated stationing.
12. Once again, stay in the **Layout** toolbar.
13. Now use a command to add another tangent off of the end of the curve.
14. Select the black **down arrow** for the fifth icon from the left, and select the command called **More Fixed Lines** and in the submenu choose the **Fixed Line (Two Points, Tangent to end of entity).** Enter the following data to construct that line.
15. `Select entity for start point and direction:` Pick near the end of the curve you just drew.
16. `Specify length:` **400.**

Figure 4-64   Line emerges that is tangent to curve

17. Notice that a line emerges from the end of the curve that is also tangent to the end of the curve. See Figure 4-64.
18. Hit <**Enter**> to terminate the routine and leave the **Layout** toolbar.

If at some point you accidentally left the **Layout** toolbar, no problem. Just select the **Alignment,** right click, and choose **Edit Alignment...** to reenter the toolbar.

TIP   Do not pick **Create From Layout** when attempting to edit an alignment because this creates a **New alignment.**

CHAPTER TEST QUESTIONS

Multiple Choice

1. Tie-in methods include which of the following?
   a. Rectangular tie-in
   b. Polar tie-in
   c. Intersection tie-in
   d. All of the above

2. The total number of degrees within a closed polygon depends on the number of sides and is computed using the following formula:
   a. ANGULAR CLOSURE = $(N) \times 180$
   b. ANGULAR CLOSURE = $(N - 2) \times 180$
   c. ANGULAR CLOSURE = $180/\pi$
   d. ANGULAR CLOSURE = $(N - 2)/180$

3. The Arc method of curve calculation occurs where:
   a. The degree of curve is measured over a chord length of 100′
   b. The degree of curve is measured at a radius of 100′
   c. The degree of curve is measured over an arc length of 100′
   d. None of the above

4. The following terms are associated with horizontal arcs:
   a. PVI, PVT, PVC
   b. ST, TS, SC, CS
   c. PI, POC, PRC, PC, PT
   d. All of the above

5. Some of the types of important angles to understand for civil design are:
   a. Interior angles, azimuths, bearings
   b. Cardinal directions
   c. Deflection angles, turned angles
   d. All of the above

6. For constraint-based design, there are:
   a. Fixed objects
   b. Free objects
   c. Floating objects
   d. All of the above

7. Constraint-based design includes the following objects:
   a. Splines and spirals
   b. Polylines
   c. Terrain models
   d. Civil 3D lines and curves

8. Civil 3D will:
   a. Automatically label alignments on creation
   b. Draft alignments based on styles
   c. Annotate stationing equations or equalities
   d. All of the above

9. Civil 3D **Parcel** commands will:
   a. Locate buildings on lots
   b. Fit structures into tight BRL line limits
   c. Warn when BRL lines are violated
   d. None of the above

10. In a subdivision corridor, items that reference horizontal alignments comprise:
    a. Gutters and sidewalks and lot corners
    b. Utilities such as water, sanitary
    c. Storm drainage structures such as catch basins and manholes
    d. All of the above

True or False

1. True or False: **Transparent** commands occur without the user knowing about them.

2. True or False: A bearing has two quadrants, North and South.

3. True or False: Intersection tie-in uses two angles from a baseline to locate a position.

4. True or False: Distances along baselines are called stations.

5. True or False: The distance along the station and at a right angle to it is called an offset.

6. True or False: Arcs have a constantly changing radius.

7. True or False: Spirals and parabolas fall into the rectilinear category.

8. True or False: There are two type of arcs and they are defined using the Roadway and Railway methods.

9. True or False: A PI for an arc is an abbreviation of Point of Interest.

10. True or False: A POC for an arc is an abbreviation of Point of Curvature.

11. True or False: Interior angles are measured only in buildings.

12. True or False: An azimuth angle can be measured from North, South, East, or West and cannot exceed 90 degrees.

13. True or False: Constraint-based design allows designers to embed certain parameters into objects that will be retained during modifications.

14. True or False: Free objects must have both in-coming and outgoing objects to define them.

15. True or False: Floating objects must have an incoming object to define them correctly.

STUDENT EXERCISES, ACTIVITIES, AND REVIEW QUESTIONS

1. What is the slope of a line with coordinates of 5000,5000 and 5100,5050?

2. What is the bearing of that same line in Problem 1?

3. What is a slope distance?

4. Describe the term *stationing* as used in alignment design.

5. How are station/offset values used?

6. Use the **Layout** toolbar to create fixed, free, and floating lines and fixed, free, and floating curves. Draw two unconnected fixed lines. Then add a floating curve to the two lines and note how it reacts after you leave the **Layout** toolbar and edit the grips.

7. Use the **Layout** toolbar to create fixed, free, and floating lines and fixed, free, and floating curves. Draw two unconnected fixed curves. Then add a floating line to the two curves and note how it reacts after you leave the **Layout** toolbar and edit the grips.

8. Name the parts of the arc shown in Figure 4-65.

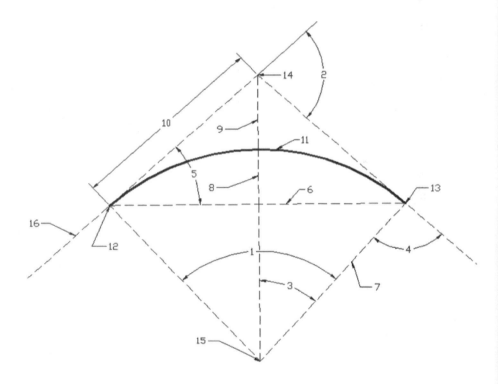

Figure 4-65

What is 1?	What is 2?
What is 3?	What is 4?
What is 5?	What is 6?
What is 7?	What is 8?
What is 9?	What is 10?
What is 11?	What is 12?
What is 13?	What is 14?
What is 15?	What is 16?

9. Compute the missing values and the degree of curve for the data below. Use the formulas discussed in the chapter to solve these. Show your work.

   Curve 1: R = 93.55, L = XXX.XX, Tan = 44.46, Included angle (Delta) = 50-50-38, Chord Length = 80.32
   Curve 1: L = _____   Degree of Curve: _____

   Curve 2: R = 52.10, L = 46.01, Tan = XX.XX, Included angle (Delta) = 50-35-50, Chord Length = 44.53
   Curve 2: T = _____   Degree of Curve: _____

10. Fill in the missing blanks for the parts of the alignment (i.e., PI, PC, etc.) in Figure 4-66. Note: A spiral, curve, spiral begins at 14+86.68.

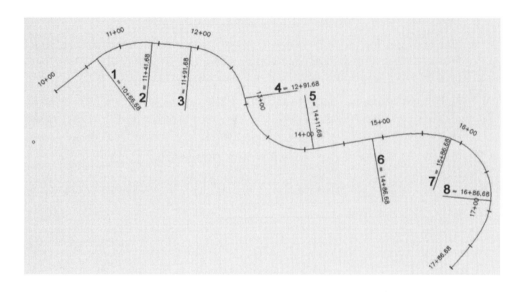

Figure 4-66

What is 1? _____   What is 2? _____   What is 3? _____

What is 4? _____   What is 5? _____   What is 6? _____

What is 7? _____   What is 8? _____

What is the total length of this alignment?

11. Plot the following alignment into your CAD file. Plot the drawing at a scale of 1″ = 50′.

Horizontal Alignment Station and Curve Report—Alignment

Desc. Station	Spiral/Curve Data	Northing	Easting
PI	10+00	17405.34	14810.53
	Length: 102.81	Course: N 34-23-28 E	
PI	11+02.81	17490.18	14868.60
	Length: 163.49	Course: N 68-46-06 E	
	Delta: 34-22-39		

Circular Curve Data

PC	10+71.88	17464.65	14851.13	
RP		17408.17	14933.65	
PT	11+31.88	17501.38	14897.44	
	Delta: 34-22-39	Type: RIGHT	Radius: 100.00	Es: 4.68
	DOC: 57-17-45	Length: 60.00	Tangent: 30.93	
	Mid-Ord: 4.47	External: 4.68	Chord: 59.10	Course: N 51-34-47 E

PI	12+64.43	17549.39	15021.00	
	Length: 157.56	Course: S 15-44-19 E		
	Delta: 95-29-35			

Circular Curve Data

PC	11+81.88	17519.49	14944.04	
RP		17449.58	14971.20	
PT	13+06.88	17469.92	15043.39	
	Delta: 95-29-35	Type: RIGHT	Radius: 75.00	Es: 36.54
	DOC: 76-23-40	Length: 125.00	Tangent: 82.56	
	Mid-Ord: 24.57	External: 36.54	Chord: 111.03	Course: S 63-29-06 E
PI	13+81.88	17397.74	15063.74	

Advanced 3D Surface Modeling

5

Chapter Objectives

After reading this chapter, the student should know/be able to:
- Understand the mathematics needed for Civil 3D terrain modeling.
- Compute information about terrain models and contours by hand using mathematical interpolation.
- Use Civil 3D to accomplish the same functions automatically and much faster than by hand.
- Understand the TIN process and the components of the TIN data.
- Understand how to analyze surface data.

INTRODUCTION

This chapter introduces Terrain modeling by discussing the various algorithms and methodologies used to develop Digital Terrain Models. A Digital Terrain Model is commonly known as a *DTM*. The DTM is the basis of most of the computations that occur in Civil 3D when it comes to the third dimension. These activities include road and utility profile creation, cross sections, earthwork calculations, contour representation, and other functions. The DTM may be one of the most important components to any project.

DTM—Digital Terrain Model

Several advanced terrain problems and related solutions are also explored. Terrain models are developed and edited. Several advanced techniques are deployed to analyze the models in 3D. Surfaces, contours, breaklines, and spot shots are discussed in detail.

To begin, when working on an engineering project, whether it is a road design job or a site development project, it almost always starts with an evaluation of the existing conditions. This can be done in a number of ways; for instance, by using a traditional survey crew with total stations or by commissioning a photogrammetist to obtain aerial photographs and stereodigitizing them into a CAD file. Control would be established in either case using local monumentation or Geographic Positioning Systems (GPS). The existing ground data are collected, processed, analyzed, and certified for use. In addition to the horizontal aspects of that data the DTM would likely be developed as well.

REVIEW OF FUNCTIONS AND FEATURES TO BE USED

This chapter procedurizes how to create surfaces, how to create and modify Surface Styles to control the appearance of the surface, and how to display the surface data in a variety of different ways. Surface data are computed, the surface is analyzed, and a variety of display mechanisms are taught to view the site in innovative ways.

DTM Theory

In many cases vertical geometry is developed from an original ground surface. In the "old" days engineers would hand interpolate contours that represented certain elevations on the ground surface.

In Figure 5-1, to plot a contour for every 1 foot of constant elevation change, you would locate where the contour enters the data set and plot that point. You would then locate where that same contour would then travel as it meanders through the data set. The following sequence illustrates how to do this by hand. The method parallels that of Civil 3D in that it also uses linear interpolation to compute contour locations.

Example 5-1: Manually Compute Contour Locations

Let us identify where the 100′ contour would be. There are no data indicating that a 100′ contour exists in this data set because the lowest value is 100.3. See Figure 5-1. Therefore, locate the 101′ contour. It must be near the 100.3 ground shot, right? It appears that the 101′ contour will enter between Points 1 and 2 and exit between Points 1 and 5 (note that the direction of the contour is of no consequence to these calculations).

First compute the distance between Points 1 and 2, or because they may be on paper, scale it off the paper. This distance is 51′. Then the distance between Points 1 and 5 are scaled off. This distance is 75.4′.

Then compute the difference between the elevations of Points 1 and 2, which is 2.9′. The difference between the elevations of Points 1 and 5 is 1.5′. You now need to compute the difference from the 101′ contour to the elevation of Point 1, which is 0.7′. Compute the ratio of these elevations so that you can prorate the horizontal distance from Point 1 where the 101′ contour begins. Compute the ratio of the 101′ contour from the elevation at Point 1 with the elevation difference of Points 1 and 2. That would be 0.7/2.9, which equals 0.24. Therefore the 101′ contour begins at 24% of the total distance away from Point 1 toward Point 2. Therefore, 75.4*.24 equals 18.2′. You should scale and plot a point 18.2 from Point 1 toward Point 2.

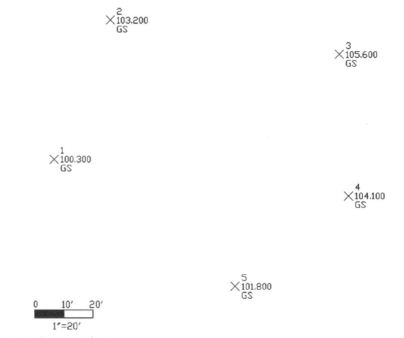

Figure 5-1   Surface spot shots

Let us compute the distance where the 101' contour exits the data set. This must occur between Points 1 and 5, because no elevations exist anywhere else that would allow a 101' contour to exit. Scale the distance from Point 1 to Point 5. You should get 75.3'. The elevation difference is 101.8 − 100.3 = 1.5. The elevation difference between Point 1 and the contour elevation of 101' is 101 − 100.3 = 0.7. The ratio of distance that the 101' contour is from the 100.3 spot shot is 0.7/1.5 = 0.467, which means that the 101' contour is 46.67% of the distance from Point 1 to Point 5, which is 75.3*46.67 = 35.14'.

Repeat this for any locations where the 101' contour must pass related to the data you possess. In other words, the 101' contour must pass between Points 1 and 3 and Points 1 and 4 on its way toward exiting between Points 1 and 5.

So, continuing on with the computations in the same manner as just described, you need to be able to plot where the 101' contour is relative to Points 1 and 3 and 1 and 4. Then you can sketch the probable path that the contour takes as it winds through the data set.

Distance from Point 1 to Point 3 = 103.48.
Positive Elevation difference between Points 1 and 3 = 105.6 − 100.3 = 5.3.
The elevation difference between Point 1 and the 101' contour = 0.7.
Therefore, the ratio of 0.7/5.3 = 0.13, or 13% of the distance from Point 1 to Point 3, is established as the location of the 101' contour, which is 0.13*103.48 = 13.67'.

Distance from Point 1 to Point 4 = 101.18.
Positive Elevation difference between Points 1 and 4 = 104.1 − 100.3 = 3.8.
The elevation difference between Point 1 and the 101' contour = 0.7.
Therefore, the ratio of 0.7/3.8 = 0.184, or 18.4% of the distance from Point 1 to Point 4, is established as the location of the 101' contour, which is 0.184*101.18 = 18.64'.

Now compute the location for the 102' contour.

Distance from Point 1 to Point 2 = 51.
Positive Elevation difference between Points 1 and 2 = 103.2 − 100.3 = 2.9.
The elevation difference between Point 1 and the 102' contour = 1.7.
Therefore, the ratio of 1.7/2.9 = 0.586, or 58.6% of the distance from Point 1 to Point 2, is established as the location of the 102' contour, which is 0.586*51 = 29.89'.

It can be seen that the 102' contour does not exit between Points 1 and 5 because the elevation of point is less than 102'. It must exist between Points 5 and 4.

Distance from Point 5 to Point 4 = 49.1.
Positive Elevation difference between Points 5 and 4 = 104.1 − 101.8 = 2.3.
The elevation difference between Point 5 and the 102' contour = 0.2.
Therefore, the ratio of 0.2/2.3 = 0.086, or 8.6% of the distance from Point 5 to Point 4, is established as the location of the 102' contour, which is 0.086*49.1 = 4.26'.

Distance from Point 1 to Point 3 = 103.48.
Positive Elevation difference between Points 1 and 3 = 105.6 − 100.3 = 5.3.
The elevation difference between Point 1 and the 102' contour = 1.7.
Therefore, the ratio of 1.7/5.3 = 0.32, or 3.2% of the distance from Point 1 to Point 3, is established as the location of the 102' contour, which is 0.32*103.48 = 33.19'.

Distance from Point 1 to Point 4 = 101.18.
Positive Elevation difference between Points 1 and 4 = 104.1 − 100.3 = 3.8.
The elevation difference between Point 1 and the 102' contour = 1.7.
Therefore, the ratio of 1.7/3.8 = 0.447, or 44.7% of the distance from

Point 1 to Point 4, is established as the location of the 102' contour, which is 0.447*101.18 = 45.26'.

Now compute the location for the 103' contour.

Distance from Point 1 to Point 2 = 51.
Positive Elevation difference between Points 1 and 2 = 103.2 − 100.3 = 2.9.
The elevation difference between Point 1 and the 102' contour = 2.7.
Therefore, the ratio of 2.7/2.9 = 0.93, or 93% of the distance from Point 1 to Point 2, is established as the location of the 102' contour, which is 0.93*51 = 47.48'.

It can be seen that the 103' contour exits between Points 5 and 4.

Distance from Point 5 to Point 4 = 49.1.
Positive Elevation difference between Points 5 and 4 = 104.1 − 101.8 = 2.3.
The elevation difference between Point 5 and the 103' contour = 1.2.
Therefore, the ratio of 1.2/2.3 = 0.52, or 52% of the distance from Point 5 to Point 4, is established as the location of the 103' contour, which is 0.52*49.1 = 25.6'.

Distance from Point 1 to Point 3 = 103.48.
Positive Elevation difference between Points 1 and 3 = 105.6 − 100.3 = 5.3.
The elevation difference between Point 1 and the 103' contour = 2.7.
Therefore, the ratio of 2.7/5.3 = 0.509, or 50.9% of the distance from Point 1 to Point 3, is established as the location of the 103' contour, which is 0.509*103.48 = 52.7'.

Distance from Point 1 to Point 4 = 101.18.
Positive Elevation difference between Points 1 and 4 = 104.1 − 100.3 = 3.8.
The elevation difference between Point 1 and the 102' contour = 2.7.
Therefore, the ratio of 2.7/3.8 = 0.71, or 71% of the distance from Point 1 to Point 4, is established as the location of the 103' contour, which is 0.71*101.18 = 71.89'.

Exercise 5-1: Manually Compute Contours

Compute the locations for where the 104' and 105' contours exist in the data set.

ANSWER TO EXERCISE 5-1

Distance from Point 2 to Point 3 = 78.8.
Positive Elevation difference between Points 2 and 3 = 103.2 − 105.6 = 2.4.
The elevation difference between Point 2 and the 104' contour = 0.8.
Therefore, the ratio of 0.8/2.4 = 0.33, or 33% of the distance from Point 2 to Point 3, is established as the location of the 104' contour, which is 0.33*78.8 = 26.26'.

Distance from Point 1 to Point 3 = 103.48.
Positive Elevation difference between Points 1 and 3 = 105.6 − 100.3 = 5.3.
The elevation difference between Point 1 and the 104' contour = 3.7.
Therefore, the ratio of 3.7/5.3 = 0.698, or 69.8% of the distance from Point 1 to Point 3, is established as the location of the 104' contour, which is 0.698*103.48 = 72.2'.

Distance from Point 1 to Point 4 = 101.18.
Positive Elevation difference between Points 1 and 4 = 104.1 − 100.3 = 3.8.
The elevation difference between Point 1 and the 104' contour = 3.7.
Therefore, the ratio of 3.7/3.8 = 0.97, or 97% of the distance from Point 1 to Point 4, is established as the location of the 104' contour, which is 0.97*101.18 = 98.14'.

It can be seen that the 104' contour exits between Points 5 and 4.

Distance from Point 5 to Point 4 = 49.1.
Positive Elevation difference between Points 5 and 4 = 104.1 − 101.8 = 2.3.
The elevation difference between Point 5 and the 104' contour = 2.2.
Therefore, the ratio of 2.2/2.3 = 0.956, or 95.6% of the distance from
Point 5 to Point 4, is established as the location of the 104' contour,
which is 0.956*49.1 = 46.9'.

Distance from Point 2 to Point 3 = 78.8.
Positive Elevation difference between Points 2 and 3 = 103.2 − 105.6 = 2.4.
The elevation difference between Point 2 and the 105' contour = 1.8.
Therefore, the ratio of 1.8/2.4 = 0.75, or 75% of the distance from
Point 2 to Point 3, is established as the location of the 105' contour,
which is 0.75*78.8 = 59.1'.

Distance from Point 3 to Point 4 = 47.8.
Positive Elevation difference between Points 3 and 4 = 105.6 − 104.1 = 1.5.
The elevation difference between Point 3 and the 105' contour = 0.6.
Therefore, the ratio of 0.6/1.5 = 0.4, or 40% of the distance from
Point 3 to Point 4, is established as the location of the 105' contour,
which is 0.40*47.8 = 19.1'. See Figure 5-2.

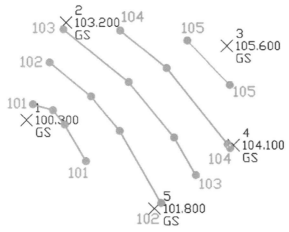

Figure 5-2   Hand-drawn contours

Today, many engineers use a computer and its software to perform this
interpolation. Computers create and use Digital Terrain Models (DTM). The
source data for this DTM can be composed of various types of data ranging from
spot shots or ground shots, breaklines, and contours. A DTM is a digital approximation (i.e., a model) of the data collected to represent the 3D conditions of
the site. At various stages assumptions, interpolations, and computations occur
to prepare the "terrain surface." The user must study and understand what the
software is trying to do in order to interact with the model. The user changes may
make to override interpolations and assumptions made by the system, if desired.

The DTM TIN algorithm is based on Delauney's criteria, which create triangular planes that define the terrain surface. This is typically the method
most manufacturers use and is one in which a triangle is created between the
closest three points in the site. The triangle is a planar face from which computations can be performed. For instance, a profile of a roadway centerline can
be computed from a TIN using linear interpolation. A new and increasingly
used methodology for developing DTMs is topologically triangulated networks,
or TTNs. In this algorithm there is a vehicle for segregating the data used as
source data for the TIN processing.

Once a TIN is prepared, then, computations can be "sampled" or calculated from it. These might produce spot elevations interpolated from the triangle face, contours, a profile, or a cross section. A profile is the graphical intersection of a vertical plane, along the route in question, with the Earth's surface. Profiles are covered in detail in Chapter 6.

Cross sections are shorter profiles made perpendicular to the route in question. There are often two types of cross sections, the first is for roadways or corridors and the second is for borrow pits, landfills, and detention ponds.

The TIN Process

Figure 5-3 shows four spot shots.

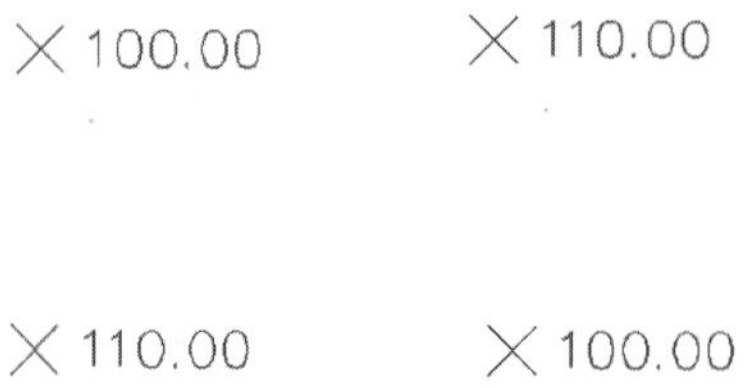

100.00
110.00
110.00
100.00

Figure 5-3 Four simple spot shots

Figure 5-4 shows a resulting TIN.

100.00
110.00
110.00
100.00

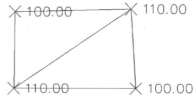

Figure 5-4 A TIN from the four shots

Figure 5-5 shows contouring and depicts a hill or a rise. How do you know this is not a stream or ditch? If the algorithm had placed the interconnecting TIN line such that it connected the 100′ elevations, there would be a different condition to solve for.

1
100.00
2
110.00
3
100.00
4
110.00
100
102
104
106
108
110

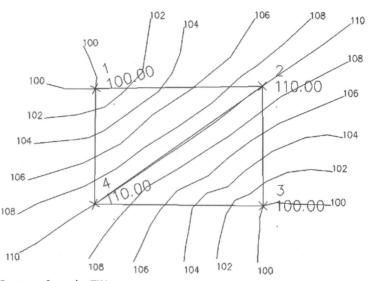

Figure 5-5 Contours from the TIN

This example shows that the software can easily produce a seriously incorrect interpretation of the data, if controls are not put into place to prevent it. The contours generated in Figure 5-6 differed by almost 90 degrees from those in Figure 5-5! Only a technician who knew that the area contained a swale would be able to estimate contours properly for the data given. Of course, the software had no way of knowing whether to connect the points to produce a swale or a highpoint (ridge) condition. That information was not available from the data set of points alone. This situation typically occurs in low areas, hilltops, and ridges. The answer is to place breaklines at the tops of ridges and in the centerlines of swales, at the tops of banks and in the toes of slopes, and every place a "break" exists that must be retained.

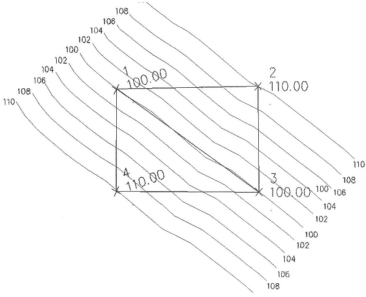

Figure 5-6   Same TIN, different contours

Ground information

A contour is a linestring that represents a single elevation along a surface. It is a coplanar object that always has a consistent elevation at each vertex of the object. It is an imaginary linestring that connects points of equal elevation.

Contours usually have a primary interval and a secondary interval. Some call these index and intermediate, whereas others call them high and normal contour intervals, respectively. Some typical intervals might be 2' and 10' intervals where the 10' intervals' linework is designated as a more dominant lineweight and perhaps a different linestyle than the 2' contours. In high terrain areas the interval might jump to 5' and 25' intervals. In flat areas the interval might be 1' and 5' intervals, whereas metric projects may use .5 m and 2.5 m intervals.

The interval is important so that enough data are shown to convey an understanding of the site.

Some interesting facts about contours follow. Refer to Figure 5-7.

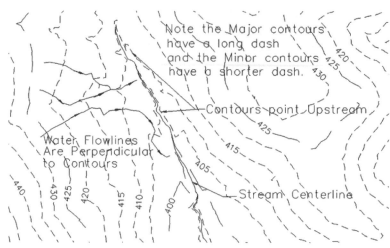

Figure 5-7   Contours speak to us

- Contours were first introduced in 1729 by a Dutch surveyor named Cruquius in connection with depth soundings of the sea. Laplace then used the concept to represent terrain in 1816.
- Water flows perpendicular to the contour.
- Contours bend or point upstream when they cross creeks or rivers.
- Evenly spaced contours represent a fairly uniform or constant slope condition.
- Closed contours that end on an intermediate elevation should be identified with hachures or spot shots to indicate whether the ground is rising or dropping.

WHAT METHOD IS USED TO CALCULATE THE CONTOURS FOR THE SITE?

Let us review the contouring processing. In a particular set of data, the contours may be drawn as shown next. Please note that the algorithms for calculating contour lines vary from manufacturer to manufacturer. One general means for generating contours is straight-line interpolation from the TIN surface.

The calculations interpolate where the desired contour crosses the TIN line. In this manner you can find out where the contour enters and leaves the triangle. As can be seen in Figure 5-8, this kind of processing is quite exact. Notice the 105′ contour. It must touch the 105′ spot shot.

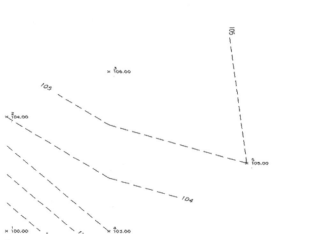

Figure 5-8   An accurate contour

However, because the ground tends to roll in natural conditions, the contour must usually be "smoothed" to represent this rolling. When the software smoothes a contour, it adds error to it. See Figure 5-9.

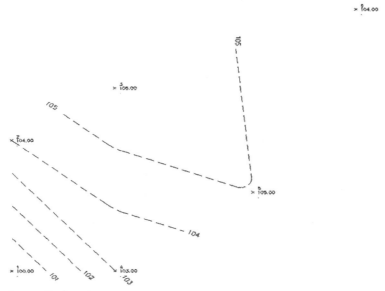

Figure 5-9   A smoothed contour

This error can be noticed in Figure 5-9 because the 105′ contour no longer touches the 105′ spot shot. The smoothing process sacrifices the sharp vertices while trying to retain the shape and location of the remaining contour. This is only one of several variations that can occur in contour generation.

The data in Figure 5-10 show a 140 acre site with breaklines and mass points. Breaklines represent a break in the surface where one plane changes

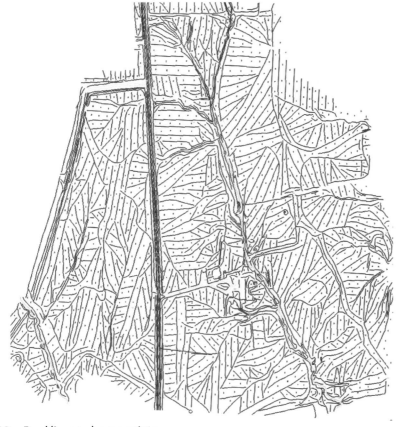

Figure 5-10   Breaklines and mass points

slope from the plane adjacent to it. Mathematically, the breakline serves the TIN creation algorithm in that no triangle will ever be allowed to cross a breakline. This rule allows the user to control how a TIN is developed. This is more efficient than depicting it with contour data. The contour file for this project was around 20 MB whereas this file is less than 1MB. The resulting contours for this site are shown in Figure 5-11.

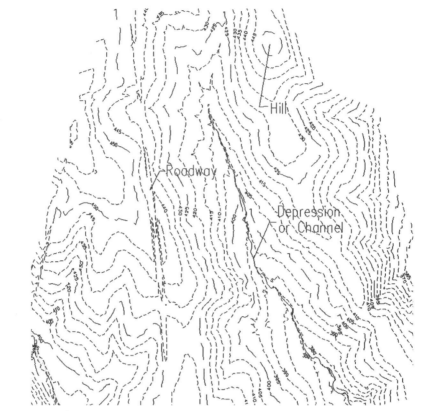

Figure 5-11   Resulting site contours

With the accuracies attained by breaklines and mass point data, one can compute profiles and sections with more precision than that which is possible with contours.

The Civil 3D has the industry's leading terrain modeling capabilities. Now that the theories and algorithms involving DTMs have been discussed, you now explore how Civil 3D handles these computations.

SURFACES IN CIVIL 3D

Next you run through an exercise that sets some basic settings and then you create an Existing Ground Style and generate a surface. The **Surfaces** pull-down menu, Figure 5-12, contains the following commands:

> **Create Surface**... allows for the user to create and name a new surface and choose its surface type, TIN, Grid, Volume....
>
> **Create From DEM**... allows for creating a surface from a **Digital Elevation Model**. These are available from several sources, including the USGS at usgs.org.
>
> **Import TIN**... allows for importing a TIN from Land Desktop. It requires a .tin and its associated .pnt file.

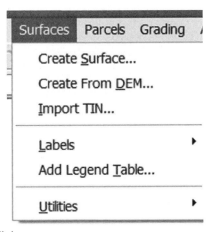

Figure 5-12   The Surface pull-down menu

Labels allows for labeling the surface or its contours.

Add Legend Table... allows for bringing in a legend table for a variety of surface analyses that can be generated in Civil 3D.

Utilities is where you perform volume computations and the **Water drop** command to check for drainage flow directions.

Exercise 5-2: Open Drawing and Inspect Source Data

1. Launch Civil 3D and open the file called Chapter-5-DTM.dwg. Type **V** for **View** and restore the view called **DTM-A**.
2. Notice that there are breaklines and mass points in the file. The mass points are actually AutoCAD blocks floating at the correct elevation. The breaklines were developed by an aerial photogrammetist using stereo-digitizing techniques. The breaklines are three dimensional and have elevations at each vertex. The mass points are placed in the data set to densify the surface data. Together the breaklines and mass points represent the existing ground conditions.
3. Click on a breakline somewhere in the file and notice the grips. Place your cursor in one of the grips, but do not click it. Simply look at the lower left corner of the screen and observe that the *X, Y, Z* coordinates are shown for the grip. Notice the *Z* coordinate or elevation has a value representing the elevation of the surface at that location.

Now that you have observed the data set, prepare a Civil 3D surface for existing ground. First you must inspect some important surface-related settings. The first one you change is the **Precision of the Contour** label. Because you are labeling even contours that are multiples of 2, there is no need to have thousandths place precision on the annotation when whole numbers suffice. The precision is unneeded and the extra decimal places tend to clutter up the drawing. Additionally, you create a layer for the annotation to reside on.

Exercise 5-3: Develop Surface Settings

1. Go to the **Toolspace**. Click on the **Settings** tab.
2. In the **Master View**, expand **Drawings**. Then expand **Surface**. Then expand **Label Styles**.
3. Choose the **Contour** label style, and note that one exists called **Standard**.

4. Click on the **Standard** label style, right click the mouse, and choose **Edit**....
5. The **Label Style Composer** dialog box displays. Click on the **Information** tab and observe the contents of the panel. Click on the **General** tab and observe the contents of the panel.
6. Click the **Layout** tab. Several properties exist in the **Properties** column.
7. Click on **Text**; expand **Text**. Choose **Contents** and then refer to the **Values** column. Click on the value for **Contents**. Choose the button with the ellipsis (...).
8. This brings up the **Text Component Editor—Label Text**.
9. Click into the **Preview** window on the right side of the dialog box; highlight the data there and erase them so that the window is blank.
10. Click on **Precision** in the **Modify** column.
11. The value for the **Precision** is **0.001**, which is not appropriate for your contour label because you do not wish to have decimal points on even contours. Click on 0.001 and select **1** from the pop-down window in the **Values** column.
12. Then click the **blue arrow** at the top of the dialog box to populate the **Preview** window with the new data settings.
13. Look on the **Preview** window and see some code indicating that new settings exist. Hit **OK**.
14. In the **General** tab, for the **Layer** property click on the value where it shows **Layer** zero and notice that a button with an ellipsis ... exists. Choose the button and a **Layer** dialog box appears. Choose **New**. In the **Create Layer** dialog box, click on the value for **Layer** property where it says **Layer 1** and type in a new layer name of **C-Topo-Labl**.
15. In the **Label Style Composer** for **Standard** under the **General** tab, there is a property for **Plan Readability**. Set **Flip anchors** with text; set it to **True**.
16. Click **Apply** and say **OK**.

You need to set one more setting. When executing the procedure to label contours, several technicalities are established by the software that will provide benefits later on. One of those has to do with a Contour Label Control Line. This linework appears in the center of the contour label and can be placed on its own layer so as to be nonintrusive to the user when plotting or negotiating through the drawing.

Exercise 5-4: Set Layer

You need to set the layer on which this object will fall.

1. Go to the **Toolspace**. Click on the **Settings** tab.
2. In the **Master View**, expand **Drawings**. Then expand **Surface**. Then expand the **Commands** item.
3. Click on the item called **CreateSurfCntrLabelLine**. Then right click on the item and choose **Edit Command Settings**.... See Figure 5-13.
4. A dialog box displays, as shown in Figure 5-14, called **Edit Command Settings—CreateSurfCntrLabelLine**. It has many parameters within it, but the one of interest at present is the **Default Layer**.
5. Select the **Default Layer** under the **Properties** column. Click in the Value column and hit the button with the ellipsis (...).
6. A **Layer Selection** dialog box displays. Click **New**....
7. Click in the **Values** column for the **Property** of layer and type in **C-Topo-Ctrl-Line**.
8. Click the value of **Yes** for the **Property Plot** and set it to **No**.
9. Then hit **OK**. Hit **OK** for the **Layer Selection** dialog box. Hit **Apply** and **OK** for the **Edit Command Settings—CreateSurfCntrLabelLine** dialog box.

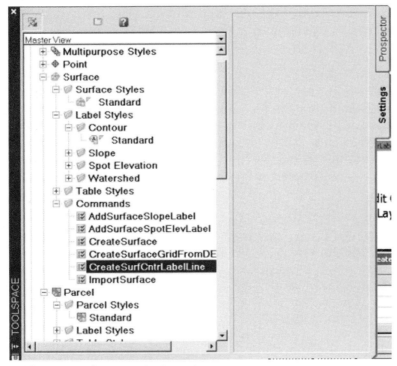

Figure 5-13   Edit Command Settings, Settings tab

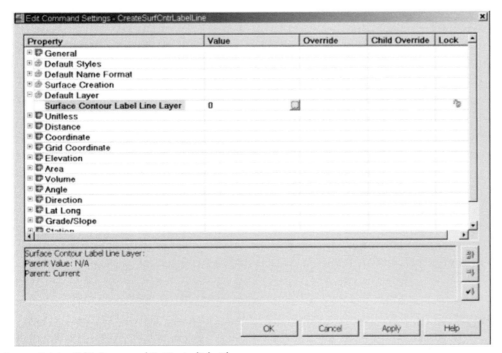

Figure 5-14   Edit Command Settings dialog box

Exercise 5-5: Set Up a Surface

Now allocate a surface for the project. Develop a surface name and set the type of surface you will have.

1. From the Civil 3D pull-down menu choose **Surfaces**.
2. Select **Create Surface**....
3. The **Create Surface** dialog box appears.
4. Several types of surfaces can be created: a TIN surface, a Grid surface, and two types of Volume surfaces. The Volume surfaces are created

during earthwork take-off computations. The Grid surface is for when a surface of evenly spaced rectangular data whose elevations are interpolated from the data is desired.

5. We want a **TIN** surface. In the window is the informational item that includes the name of the surface. Click on the Value column for the **Surface** name.
6. The **Surface** name is incremental so the first surface will be Surface 1, the second surface is Surface 2, and so on. Click in this field and type **Existing Ground** as your name.
7. To the right of the name is the **Layer** field. Pick on the button on the right side of the dialog box for **Surface Layer**. An **Object Layer** dialog box appears. Where it says **Base Layer Name**, choose the button on the right for **Layer** and the **Layer Selection** dialog box opens.
8. Hit the button for **New** layer and enter **C-Topo** in the field for Layer 1. Then where it says **Modifier**, select **Suffix** from the pop-down window.
9. When the field to the right becomes available, type in a dash and an asterisk (-*). This creates a layer with the name of the surface as a suffix to the layer name. We should get a layer called **C-Topo-Existing Ground**, as can be seen in the **Preview**.
10. Type **Existing Ground** for the **Description** also.
11. Leave all else as default and click **OK**.
12. Next go to the **Toolspace** and choose the **Prospector** tab.
13. Look under **Open Drawings** for your drawing. Expand **Surfaces** to show the new surface you just created for Existing Ground.

Exercise 5-6: Build a Surface

The next step is to identify the 3D surface data and build the surface.

1. Expand the **Surfaces** item. You see an item for **Definition**.
2. Expand **Definition** and you see all of the data types that can go into the preparation of the surface. These include **Breaklines, Contours, DEM Files**, and **Drawing Objects**.
3. Your data set contains breaklines and blocks, as discussed earlier. Therefore, use the commands for adding **Breaklines** and use **Drawing Objects** to add the blocks.
4. Begin by clicking on **Breaklines**, then right click on **Breaklines** and select **Add**....
5. A dialog box for **Add Breaklines** displays. Add a **Description** of **EG**.
6. Use **Standard** breaklines as the type of breakline but notice that other choices exist in the pop-down list. The choices allow for **Proximity breaklines, Wall breaklines**, and **Breaklines from a file**.

TIP

Proximity breaklines are 2D polylines that have points with elevations at each vertex in the polyline. The software uses the point elevations in creating 3D breaklines from the 2D polylines. The Wall breakline option is used to represent vertical components of surfaces. In actuality they are not truly vertical as this would cause issues in the software relating to a divide by zero in the denominator for various slope computations. Therefore, the software allows for Wall breaklines that contain wall height data but the top of the wall is slightly offset from the bottom of the wall so as to avoid the vertical computational issue. The last choice is to bring in breaklines from an external text file.

You use **Standard** breaklines, which are the 3D polylines in the file. The last setting is a **Mid-ordinate distance** field, which is available for creating 3D breaklines from 2D polylines that have curves in them. In this case if you use the default of 1', a 3D polyline would be created such that small chords simulate the curve but do not deviate from the arc by more than 1'.

7. Say **OK** and the command prompt instructs you to `Select objects:` Place a window around the data and do not worry if there are points mixed in with the selection set of breaklines because the software filters anything that is not a breakline.
8. A **Panorama** dialog box may appear that indicates these are two crossing breaklines. Coordinates are displayed so the user can check the severity of the issue. In this case you have determined that this will not harm your surface integrity and so continue on. Click the **green arrow** in the top right of the **Panorama** to continue. On completion of this function, the software creates the surface as can be seen by a yellow boundary.
9. Now in the **Prospector**, choose **Drawing Objects under Surface >> Definitions** area. Right click and choose **Add**....
10. For the **Object type**, choose **Blocks**.
11. Type **EG** as the **Description** field in the dialog box and say **OK**.
12. Say **OK** and the command prompt says `Select objects:` Place a window around the data and do not worry if there are breaklines mixed in with the selection set of points because the software filters out the unnecessary data.
13. On completion the yellow surface boundary is updated to include the breaklines.

Now that the surface is created, you typically want to view contours representing the surface.

Exercise 5-7: Develop Contours for the Surface

1. In the **Toolspace**, choose the **Settings** tab. Choose **Surface >> Surface Styles >> Standard**. Right click on **Standard** and choose **Edit**....
2. In the **Edit Surface Styles** dialog box, choose the **Display** tab.
3. With the view direction in the pop-down window set to **2D**, click **on** the lightbulb to turn on the **Major** and **Minor contours** and turn **off** the lightbulb for **Boundary**. Note the colors are preset and can be changed here if so desired.
4. On leaving the dialog box, the contours in Civil 3D are automatically updated to show the contours as defined in the Style.
5. The next step is to create contour labels on the contours. From the Civil 3D pull-down menu, choose **Surfaces >> Labels >> Add Contour labels**.
6. A toolbar shown in Figure 5-15 for performing this function appears and allows for last-minute or on-the-fly changes to the criteria for labeling. Choose the fourth button from the left and click the **down arrow**. Choose **Label Multiple Group Interior**. On doing this, a field opens to the right of the **down arrow**; type **500** in this field. This instructs the software that a label is to be placed every 500' along the contour.

Figure 5-15   Create Contour Labels toolbar

7. The command prompt then says to `Identify a start point:` for the labeling to begin. Place a point at the top of the site and place a second point to the bottom and outside of the site.

8. Hit **<Enter>** to terminate the command. On completion of the command, zoom in and inspect the data and the contour labels every 500' across the site.
9. Turn off the layer created in the settings called **C-Topo-Labl-Line**.
10. Save your file to a name with your initials as the suffix of the name, Chapter-5-DTM-YourInitials.dwg.

TIP

Use the contour labeling as often as is needed to obtain coverage across the site for all of the desired contours that you wish to be labeled.

Job skills will benefit in that the drawing labels can be customized. If you have duplicate labels, notice that you can delete the **Contour Label Control Line** to delete the label. Notice too that you can move this **Contour Label Control Line** to relocate the label!

Congratulations! You have created a surface, displayed its contours, and labeled them. Note that you can also add other labels, as shown in Figure 5-16.

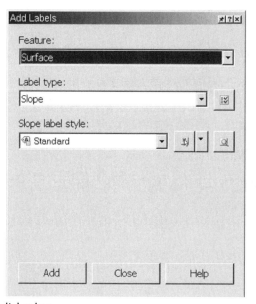

Figure 5-16   Add Labels dialog box

Exercise 5-8: Create a Surface from DEM Data

Now use another Civil 3D command to build a surface based on DEM data. A DEM is a Digital Elevation Model data set. DEM data can be found in a number of locations but one source is the USGS.org website. This exercise uses a DEM data set for Norfolk-E, Virginia.

1. Launch Civil 3D and open the file called Chapter-5-DEM.dwg.
2. Under the **Surfaces** pull-down menu, choose the option for **Create From DEM....**
3. When prompted, select the file called **Norfolk-E.dem**. This file was downloaded from the USGS.org website with the filename Norfolk-E.gz, which is a WinZip file that produces a file called Norfolk-E. It was renamed to a .DEM file prior to using it in this exercise. No scaling or geoedetics have been applied to the file, and it is being used directly as it was downloaded from the USGS.org.

4. At that point, the software reads the file, prepares the surfaces, and generates the contours for the area.
5. Use **ZOOM EXTENTS** to see the result on completion.

Exercise 5-9: Importing a Surface from Land Desktop

Now use another Civil 3D command to import a TIN surface from a Land Desktop project. This file, specifically the .TIN file, can be found in the DTM folder of a Land Desktop project.

1. Launch Civil 3D and open the file called Chapter-5-importtin.dwg.
2. Under the **Surfaces** pull-down menu, choose the option for **Import Surface**....
3. When prompted, select the file called **OSI-INTX.TIN**.
4. At that point, the software reads the file, prepares the surface for Civil 3D usage, and displays it based on the Style in effect.
5. Use **ZOOM EXTENTS** to see the result on completion.

Student Files

Exercise 5-10: Surface Labeling

This exercise explores some of the surface labeling features in Civil 3D. Zoom into the side road in the file you just imported. Let us evaluate the slopes of the roadway. See Figure 5-17.

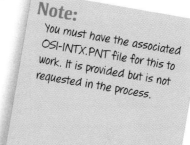
Note:
You must have the associated OSI-INTX.PNT file for this to work. It is provided but is not requested in the process.

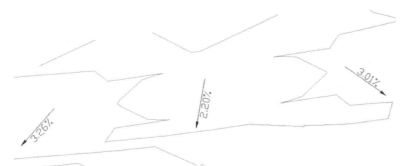

Figure 5-17   Roadway Contours with Labels

1. Under the **Surfaces** pull-down menu, choose the option for **Labels >> Add Surface Labels**....
2. A dialog box appears requesting information on the type of label you want to create.
3. Use a **Surface Feature**; the **Label Type** will be a slope (as opposed to a spot elevation label), and you use a **Standard** slope label style. Hit the **Add** button.
4. The prompt requests you to choose whether you want a one point or two point label.

TIP
A one point label computes the slope at the point you pick based on the triangle slope of the TIN triangle. The two point option allows you to pick two points where it computes the elevations of the two points, the distance between the two points, and the average slope between them.

5. Hit <Enter> for **One Point** and pick points inside the surface of the side road. Slope labels appear and you can repeat this as often as needed.

Surface Analysis

The next item of discussion is how to perform analyses on the surface. Civil 3D has many types of analyses that can be computed. They include:

Contour renders surface triangles according to the range into which a contour falls

Directions renders surface triangles according to the direction that they face

Elevations renders surface triangles according to the elevation range they fall into

Slopes renders surface triangles according to the slopes that have

Slope Arrows places directional vector arrows based on the flow direction that a triangle has

User-Defined Contours draws the location for user-specified contour elevations

Watersheds draws the perimeters of drainage divides

Exercise 5-11: Performing Slope Analysis

Quite often there is concern with the slopes occurring across a site, so it is important to perform a slope analysis. Perform a slope analysis of the roadway as in Figure 5-18.

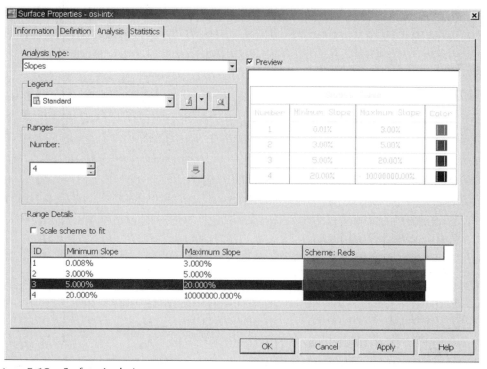

Figure 5-18   Surface Analysis

1. Select the surface in the file. Right click and choose **Surface Properties**....
2. A **Surface Properties** dialog box appears. Set the Analysis type to Slopes.
3. Set the **Number** of **Ranges** to 4.
4. Turn off the **Scale schema to fit** option. Click the Blue Down a menu.
5. For ID1, set the **Maximum Slope** to 3%.
6. For ID2, set the **Minimum Slope** to 3% and the maximum slope to 5%.
7. For ID3, set the **Minimum Slope** to 5% and the maximum slope to 20%.
8. For ID4, set the **Minimum Slope** to 20% and the maximum slope to 10000000%. Hit **Apply** and **OK**.

9. The system sets the surface to display these data once the Style is set to do so.
10. Click on a contour and right click the mouse. Choose **Edit Surface Style**....
11. Under the **Display** tab, turn off the **Contour** components and turn on the **Slopes**. Hit **OK**.
12. The surface should display with red colored faces.
13. Now from the **Surfaces** pull-down menu, choose **Add Legend Table**....
14. Select the surface when requested. The software responds with:

    ```
    Enter table type:
    Directions/Elevations/Slopes/slopeArrows/Contours/Usercontours/
    Watersheds]. Type S for Slope.
    ```

15. Then it asks `Behavior [Static/<Dynamic>]` Hit **<Enter>** for **Dynamic**.
16. `Select Upper left corner:` Pick a point below and out of the way of the surface data and a table of the statistics appears. See Figure 5-19.

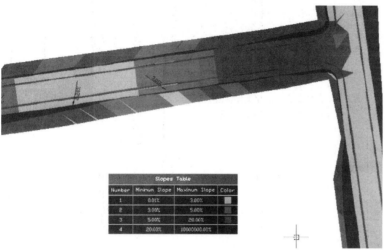

Figure 5-19   A Legend

Exercise 5-12: How to Edit a Surface

Now explore surface editing a little. You have identified that at the intersection of these roads there needs to be a low spot for an inlet. Add that low spot by editing the surface and dropping the desired location by 1 foot.

1. Start by selecting the surface, right click, and choose **Edit Surface Style**....
2. In the **Display** tab turn on the **Contours** and turn off the **Slopes** from the last exercise.
3. When you say **OK**, you again see the contours for the surface.
4. Type **V** for **View** and **Restore a View** called **DTM-B**. This places your view near the intersection of the two roads.
5. In the **Prospector**, expand the drawing file you are in. Expand the **Surfaces** item. Expand the surface called **osi-intx**. Expand **Definition**. Right click on **Edits** and a menu displays surface editing commands.
6. Choose **Modify Point**, shown in Figure 5-20. The prompt returns with `Select point:`.
7. Type **15700,18843.5** and **<Enter>**.
8. The prompt then asks for you to `Enter the new elevation`. Type **441.8**.
9. Notice that a new contour appears immediately at the intersection of the roads, indicating that the point has dropped 1 foot, thereby creating a new contour. That new contour shows up just on top of the cursor in Figure 5-21.

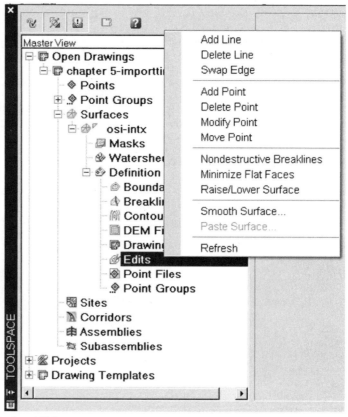

Figure 5-20   Surface Editing

Figure 5-21   Contour updating

Exercise 5-13: Adding Surface Boundaries

In the last surface-related exercise, let us explore the concept of surface boundaries. Boundaries allow you to curtail surface computations based on the boundaries that you define.

1. Zoom into the view of the same file where you imported the **osi-intx** surface of the two roads. Zoom into the side road.
2. Draw a polyline around the outside edge of the side roadway, as shown in Figure 5-23.

3. In the **Prospector**, expand the drawing file you are in. Expand the **Surfaces** item. Expand the surface called **osi-intx**. Expand **Definition**. Right click on **Boundaries** and select **Add...** as in Figure 5-22.

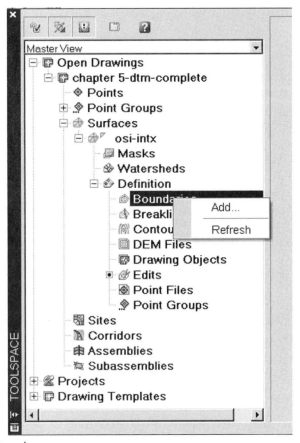

Figure 5-22   Add Boundary

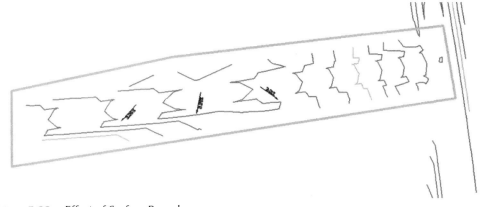

Figure 5-23   Effect of Surface Boundary

4. When the **Add Boundary** dialog box appears, type **B1** for the name of the boundary.
5. Choose **Outer** for the type of boundary. Leave the mid-ordinate distance at 1.0'. Hit **OK**.
6. The prompt asks you to `Select object:` Pick the polyline you drew.
7. The surface is instantly reduced to exist within the boundary drawn. Notice that the adjacent road is now gone. See Figure 5-23.

To continue exploring with the surfaces features, take some time to change the **Surface Style** and turn on or off various items to change how the surface can display itself.

CHAPTER TEST QUESTIONS

Multiple Choice

1. Which of the following acronyms pertain to surface models?
   a. TIN
   b. DTM
   c. TTN
   d. All of the above

2. A DTM can represent which of the following?
   a. Caves and escarpments
   b. Existing and finished ground surfaces
   c. Roadway profiles
   d. None of the above

3. DTMs can be created by using the following as source data:
   a. Points, breaklines
   b. Roadway profiles and alignments
   c. DEMs combined with utility data
   d. All of the above

4. The following computations usually use surface data:
   a. Profiling
   b. Earthworks
   c. Cross sections
   d. All of the above

5. Contour intervals can be which of the following?
   a. 1′ and 5′ intervals
   b. 2′ and 10′ intervals
   c. 5′ and 25′ intervals
   d. All of the above

6. The following information items are attributed to contours:
   a. Laplace used the concept to represent terrain in 1816.

   b. Water flows perpendicular to the contour.
   c. Contours bend or point upstream when the cross creeks or rivers.
   d. All of the above.

7. What method is used to calculate the contours for a site?
   a. Straight-line interpolation from the TIN surface
   b. Slope analysis
   c. Delauney's method
   d. None of the above

8. Surfaces in Civil 3D can be:
   a. TIN surfaces
   b. Grid surfaces
   c. Volume surfaces
   d. All of the above

9. Proximity breaklines are:
   a. Breaklines near other breaklines
   b. Polylines that have points with elevations near their vertices
   c. Breaklines that are in the proximity of mass points
   d. None of the above

10. Civil 3D can analyze surfaces according to which of the following:
    a. Triangle direction
    b. Slope
    c. Elevation
    d. All of the above

True or False

1. True or False: A DTM stands for a Digital Terrain Methodology.

2. True or False: A contour with a value of 101.5 can never be a legitimate contour elevation.

3. True or False: A DEM is a legitimate source data type for a surface.

4. True or False: Contours can be used in conjunction with spot shots to develop surfaces.

5. True or False: The TIN theory is based on Descartes' methods.

6. True or False: Most of the computations for the third dimension emanate from the surface model.

7. True or False: Cross sections are usually computed orthogonally from the centerline.

8. True or False: The TIN algorithm computes triangle vertices based on the closest three points.

9. True or False: A contour is a linestring that represents a single elevation along a surface.

10. True or False: Contours may have primary and secondary intervals.

STUDENT EXERCISES, ACTIVITIES, AND REVIEW QUESTIONS

1. Draw the 5′ contours where they fall on this map of spot elevations shot by a surveyor in the field (Figure 5-24). The tick marks designate where the actual spot shot is located. Use manual interpolated techniques to determine where the contours fall.

X 427.52 X 422.27 X 430.02 X 429.44 X 428.53 X 430.36 X 431.70 X 425.28 X 414.89 X 407.92 X 402.45

X 422.40 X 418.86 X 428.16 X 427.08 X 424.56 X 427.98 X 430.27 X 426.69 X 417.42 X 406.51 X 401.28

X 417.91 X 420.07 X 425.48 X 422.56 X 422.33 X 426.26 X 428.90 X 425.74 X 416.24 X 407.08 X 401.63

X 415.39 X 412.61 X 420.19 X 417.10 X 419.57 X 425.70 X 428.31 X 427.93 X 421.14 X 410.72 X 401.71

X 410.87 X 409.28 X 410.60 X 413.67 X 421.79 X 426.95 X 431.14 X 427.78 X 417.31 X 407.79 X 406.03

X 405.64 405.48 X 407.95 X 415.10 X 421.30 X 426.62 X 431.22 X 430.56 X 421.27 X 411.25 X 411.57

X 399.67 X 409.32 X 415.52 X 420.04 X 424.90 X 429.92 X 433.73 X 429.64 X 419.44 X 415.39 415.49

X 404.79 409.59 X 415.65 X 420.97 X 426.32 X 425.35 X 432.04 X 431.25 X 424.36 X 422.29 X 419.74

Figure 5-24   Surface Data—spot shots

2. Show the drainage for the site in Problem 1. Draw six arrows representing the drainage across the site.

3. Circle the high and the low points on the site in Problem 1.

4. What is the slope in Figure 5-25 from the 434 contour to the 424 contour, the 426 to the 424 contour, and the 422 contour to the 420 contour?

5. How many 2′ contours would exist on the side sloping of a retention pond if the top of the pond elevation were at 500′ elevation, the side sloping were at 3:1 down, and the existing ground tie-out were at 485?

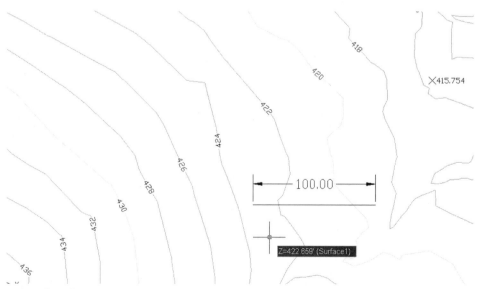

Figure 5-25   Computing slope from contours

Advanced Profiles and Sections

6

Chapter Objectives

After reading this chapter, the student should know/be able to:
- Understand the algorithms and formulas involved in computing existing and proposed profiles.
- Understand how profiles and sections build on what was learned in the previous chapter on surfaces.
- Understand how existing ground profiles and cross sectional data are associated with horizontal alignments.
- Design, label, and edit a proposed vertical alignment.
- Produce, label, and edit sections.

INTRODUCTION

This chapter begins with some of the theory involved in designing roads and moving the student into the third dimension with creating roadway profiling; developing vertical tangents, vertical curves, vertical alignments; and the editing and labeling of same. Roadway vertical tangents, vertical curves, and vertical alignments (profiles) and profile views are discussed in detail. Cross sections are also covered in this chapter because they are closely related to profiles.

REVIEW OF FUNCTIONS AND FEATURES TO BE USED

There are three main requirements for designing a roadway. The first is the development of an alignment, covered in Chapter 4. The second is the roadway profile, covered in this chapter. The third is the road's cross sectional *template*, to be discussed in Chapter 7 on corridor design. Here you learn how to produce existing ground profiles, how to place them into profile views, and how to develop *proposed profiles* and annotate them. This chapter explores the link between horizontal alignments, existing ground profiles, and cross sections and how they update when modifications are made to the data. Similar material will be procedurized for section-based views of the roadway as well.

In some locations across the country, road design can be accomplished with the alignment, the profile, and a simple cross sectional detail outlining the typical cross section for the road. The contractor follows the alignment for horizontal control, the profile for vertical control, and then inputs the cross sectional data into the paving machinery.

Template—A Land Desktop term for Assembly

Proposed Profile—The proposed vertical alignment

THEORY OF VERTICAL ALIGNMENTS

Before getting into the Civil 3D approach for developing profiles, some of the rudimentary mathematics involved with the vertical alignments for roadways must be discussed.

Profiles are created by sampling the elevation at intervals along a horizontal alignment (see Figure 6-1). The existing profile is shown as the heavy white line in Figure 6-2. For tangent sections, this is often accomplished by computing the elevation wherever the alignment crosses the TIN lines in the terrain model. In curved locations, many times a chord-based system is used to simulate the curve. The location of the chord is where the computation occurs as it crosses TIN lines in the model. Figure 6-2 shows an alignment lying on the terrain model.

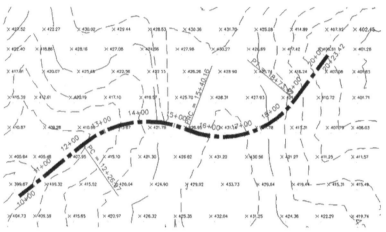

Figure 6-1   A horizontal alignment

The proposed vertical alignment, shown as the light linework in Figure 6-2, is comprised of vertical tangents and vertical curves. Two tangents meet at a PVI. Usually if there is a grade break between the two longitudinal tangents that is considered slight, a vertical curve is not required. Slight differences between agencies but may range from 0.5% to 2.5%. Anything beyond these values requires a vertical curve to smooth out the transition between the tangents. Figure 6-2 shows a dark, thick existing ground line a light thin line for proposed tangents and vertical curves. There are elevations on the vertical axis and it is usually exaggerated to show vertical detail. There are stations along the horizontal

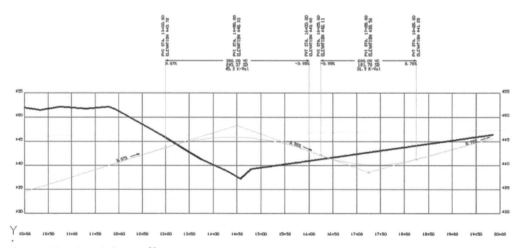

Figure 6-2   An existing profile

axis that match the alignment. There are three proposed tangents and two vertical curves. The tangents are denoted by a beginning station and elevation and a grade. It ends at the **PVC**, where the vertical curve begins. The curve ends at the **PVT**. A short tangent exists and then another PVC begins the next curve.

PVC—The Point of Vertical Curvature

In Figure 6-3, a vertical curve is shown emboldened with annotation for the PVC and the PVT. The PVI is shown as a triangle. The back tangent is before the curve and a forward tangent occurs after the curve. Once these are known, the only other criterion is the length of the curve because a parabolic equation governs the remaining computations.

PVT—The Point of Vertical Tangency

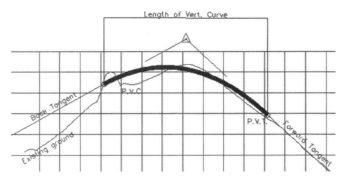

Figure 6-3   A vertical curve

In Figure 6-4, criteria are identified with the curve data. A completed proposed profile is where the incoming tangent slope is 5.44%, the outgoing is −9.15%, and the vertical curve length is 400′.

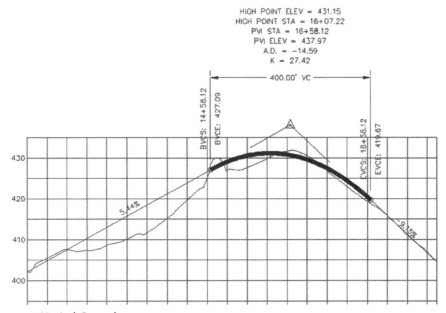

Figure 6-4   Vertical Curve data

THE PROPERTIES OF PARABOLIC CURVES

1. A curve's elevation at the midpoint of the curve is halfway from the elevation at the PVI to the elevation at the midpoint of a straight line from the PVC to the PVT.

2. The tangent offsets vary as the square of the distance from the points of tangency.

3. For points spaced at equal horizontal distances, the second differences are equal, which is useful when checking the vertical curve computations. The differences (or subtractions) between the elevations at equally spaced stations are called first differences. The differences between the first differences are called the second differences.

The design of a vertical alignment is usually specified in terms of a maximum grade and the minimum length for the vertical curve. The maximum allowable grade is a parameter that considers whether the service vehicles can negotiate the grade easily and safely. In other words, trucks would have considerable difficulty in climbing a road with a 12% grade. It would be clearly unsafe for trucks to be descending this grade should the brakes fail. So a highway might limit the maximum grade to 8%, so that the service vehicles (trucks and cars) could negotiate the grade safely. Because the vertical curves are parabolic, their criteria are represented a K factor. The K factor is computed by dividing the length of the curve by the grade break at the Point of Vertical Intersection (PVI). The relationship here is that a smaller K factor means a sharper curve and a larger K factor designates a flatter curve.

Cross sectional design criteria depend on the service classification for the roadway. Subdivision roads may have four lanes for high-density areas and two lanes for less dense areas. They may be bounded by curb and gutter and sidewalks for drainage and pedestrian traffic. Highways, on the other hand, usually have several lanes and are bounded by shoulders. Often there is a center median to divide the bidirectional traffic.

FORMULAS USED FOR COMPUTING PARABOLIC VERTICAL CURVE DATA

Where L = the Length of curve, the PVC station = PVI station − $(L/2)$.

The PVT = PVI + $(L/2)$. The PVT = PVC + L.

The general equation for a vertical curve is a parabolic equation, which derives as follows:

$Y_p = a + bX_p + cX_p^2$, where Y_p is any point on the parabola located X_p from the origin. And a, b, and c are constants.

For a vertical curve,

$a = Y_{BVC}$
$bX_p = g_1X$
$c = (g_2 − g_1)/2L$

Therefore:

$Y = Y_{BVC} + g_1X + ((g_2 − g_1)/2L)*X^2$

A useful term is the rate of change of grade. This is defined as r. Therefore:

$r = (g_2 − g_1)/L$

Combining the expressions results in the formula for a vertical curve of a roadway:

$Y = Y_{BVC} + g_1X + (r/2)*X^2$

Example 6-1: A Manually Computed Vertical Curve Problem

With a back tangent of 2%, a forward tangent of −2.5%, and a vertical curve length of 800′, compute the starting and ending stations of this vertical curve and its accompanying table if the PVI station is 36+70 and the PVI elevation is 813.53.

$$r = (g_2 - g_1)/L \rightarrow (-2.5 - 2)/8 = -.5625\%/\text{station}$$

$$\text{VPI} = 36+70$$
$$-L/2 = 4+00$$
$$\text{BVC} = 32+70$$
$$+L = 8+00$$
$$\text{EVC} = 40+70$$

The Elevation at BVC = $813.53 - (2.00)(4) = 805.53$
The Elevation at EVC = $Y_{\text{BVC}} + g_1X + (r/2)X^2 = 805.53$
  $+ (2)(8) + (-.5625)(64)/2 = 805.53 + 16.0 + (-18.054) = 803.53$
The Elevation at the midpoint of the chord of the curve = (805.53
  + 803.53)/2 = 804.53
The Elevation at midpoint of curve = $(813.53 + 804.53)/2 = 809.03$

A table, shown in Table 6-1, can be established to compute the elevations along the vertical curve. You may select an interval that you are interested in computing the information for, say, every 20 or 25 feet. Use the third vertical curve property listed earlier to ensure that you have computed the correct answers. Simply break down the formula for the vertical curve into its parts and make each part a column in the spreadsheet. The second to last column will compute the first difference between the elevations. The last column will compute the second difference (or the difference between the first differences). Note that it equals r.

Sta	X(Sta)	g_1X	$rX^2/2$	Curve Elev.	1st Diff.	2nd Diff.
44+70	8	16	−18	803.53		
44+00	7.3	14.6	−14.99	805.14		
43+00	6.3	12.6	−11.16	806.97	−1.83	
42+00	5.3	10.6	−7.9	808.23	−1.26	−.57
41+00	4.3	8.6	−5.2	808.93	−.7	−.56
40+00	3.3	6.6	−3.063	809.067	−.137	−.56
39+00	2.3	4.6	−1.488	808.642	.425	−.56
38+00	1.3	2.6	−.475	807.655	.987	−.56
37+00	.3	.6	−.025	806.105	1.55	−.56
36+70	0	0		805.53	.575	

Table 6-1   Table for Computing Elevations Every 100′ Along a Vertical Curve

The High/Low points of a vertical curve are computed as such:

$X = g_1L/g_1 - g_2 \rightarrow -g_1/r \rightarrow -2/-.5625 = 3.555$ stations = X
The High point = $36+70 + (3.555) = 40+25.5$
The Elevation = $805.53 + 2(3.555) + (-.5625/2)(3.555)^2$
  $= 805.53 + 7.11 - 3.554 = \textbf{809.086}$

Another method for computing vertical curves for roads is by using the K value, where $K = L/A$. L is the length of the vertical curve and A is the algebraic difference between the outgoing and incoming vertical tangents. Therefore the length of vertical curve can be computed based on a required K minimum value such that $L = KA$.

Table 6-2 and Table 6-3 provide examples of the American Association of State Highway Transportation Officials (AASHTO) criteria for vertical curves for both crest and sag conditions with regard to being able to stop the vehicle. There must be enough sight distance for the driver to realize a problem and then apply the brakes. This information can be found in the AASHTO publication *A Policy on Geometric Design of Highways and Streets.*

Very often designers need to design a vertical curve so that it meets a minimum K value as established by the transportation authority with jurisdiction over the design. The curve can exceed the minimum K value but must not be less than the required value. Engineers usually compute the length of curve needed to meet this condition and then round it up to the next even 50' value to make it easier for others to check the work and stake it out for construction.

Design Speed	Minimum Stopping Distance (ft)	K value for Crest Curve
30	200	19
35	250	29
40	305	44
45	360	61
50	425	84
55	495	114
60	570	151
65	645	193
70	730	247

Table 6-2   AASHTO Stopping Sight Distances for Crest Curves

Design Speed	Minimum Stopping Distance (ft)	K value for Sag Curve
30	200	37
35	250	49
40	305	64
45	360	79
50	425	96
55	495	115
60	570	136
65	645	157
70	730	181

Table 6-3   AASHTO Stopping Sight Distances for Sag Curves

Another method for determining lengths of vertical curves has to do with passing sight distance. When a driver is in a sag curve, oncoming traffic can be observed easily because both cars are in the concave portion of the curve. However, when the driver is approaching a crest curve, eyesight is obscured by the crest of the hill. Therefore, there must be enough forward sight distance for the driver to see an oncoming vehicle before making a decision to pass another vehicle in his or her lane. Table 6-4 provides an indication of the AASHTO passing sight distances required for a roadway.

Sections

Roadway sections (Figure 6-5, showing SL-1) are similar to profiles in that they represent conditions before and after design. They are three dimensional in what they are representing and are denoted with axes similar to profiles.

There are elevations on the vertical axis, which is usually exaggerated to show vertical detail, and there are distances along the horizontal axis that equate to the length of the cross section to the left and to the right of the centerline. Typically where the centerline crosses the section is denoted with zero (0) whereas the distances are shown positive leaving the centerline to the right and negative to the left.

The section information is shown in Figure 6-6 for the existing ground section.

Design Speed	Minimum Passing Distance (ft)	K value for Crest Curve
30	1090	424
35	1280	585
40	1470	772
45	1625	943
50	1835	1203
55	1985	1407
60	2135	1628
65	2285	1865
70	2480	2197

Table 6-4   AASHTO Passing Sight Distances

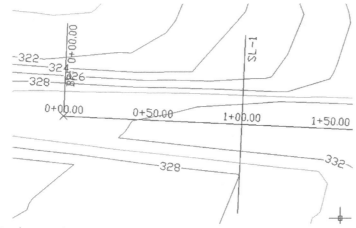

Figure 6-5   Roadway sections

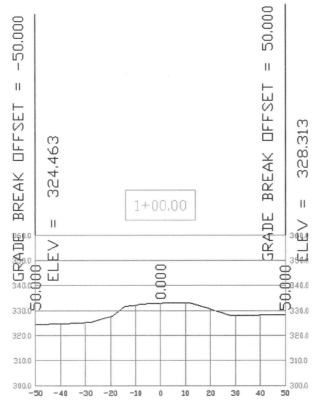

Figure 6-6   An existing ground section

APPLICATIONS USING CIVIL 3D FOR PROFILING

Profiles and sections in Civil 3D are intelligent objects, similar to the point groups, parcels, terrain models, and alignments already discussed. Profiles and cross sections are dynamically linked to horizontal alignments and surfaces. When the horizontal alignment is edited, the profile and cross section associated to it automatically update. If the surface is modified, the profile and cross section data also update automatically. The related annotation updates as well, thereby saving time in drafting functions. Styles in Civil 3D are defined to control the display of this profile and cross sectional data as well as the related annotation. These Styles represent your company or agency's or your client's standards. They are developed and stored in prototype Civil 3D drawings and within the project drawings.

In Civil 3D, existing ground profiles and cross sections are related to horizontal alignments. Changes to the alignment result in automatic updates to the profile and to the sections. This is referred to as a *dynamic model*. I have referred to it as the *ripple-through effect*.

A step-by-step procedure for designing a road right through profiles and sections and then sending it to construction would involve the following tasks:

1. Create and, if needed, edit the horizontal alignment.
2. Compute and plot the existing ground profile in the **Profile View**.
3. Develop and, if needed, edit the vertical alignment.
4. Compute and analyze the existing ground cross sections.
5. Develop roadway template(s) or assemblies.
6. Process the Corridor.
7. Plot the proposed roadway design cross sections. Repeat Steps 1 through 7 as needed.
8. Develop construction stakeout information for traditional staking, or 3D digital data for 3D Machine Control using Trimble Link, Carlson Connect, or Leica XChange. Compute earthwork takeoffs.

In order to focus on profiling and sections in the next project, that is undertaken just us open a drawing that already has an alignment defined. The process to be followed in the next steps is:

1. Compute the data and plot existing ground profile data in a **Profile View**. The **Profile View Style** controls the display of the profile grid, the title, and the horizontal and vertical axes annotation for stations and elevations. The **Profile View Label Styles** control the display of the annotation inside the **Profile View**.

2. Develop the finished ground profile using vertical tangents and vertical curves.

3. For the cross sections, compute and plot the existing ground cross section data.

4. Civil 3D simplifies the modification process for existing ground profiles and cross sections due to the dynamic link to the horizontal alignment. To experience this, modify the source data and again observe the ripple-through effect.

Begin by opening a drawing that has a surface and an alignment prepared so that your profiling activities can be expedited. Then you will use the **Profiles** pull-down menu, shown in Figure 6-7, with its commands described.

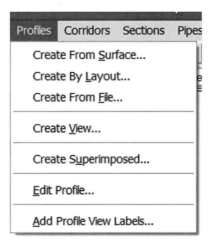

Figure 6-7   The Profiles pull-down menu

Create From Surface... allows you to develop a profile of an alignment from the surface object, normally Existing ground.

Create By Layout... opens the **Layout** toolbar for designing profiles using criteria.

Create From File... allows the creation of a profile from an external file with data in it.

Create View... allows for creating a **Profile View** for use later.

Create Superimposed... allows for superimposing multiple profiles into a **Profile View**.

Edit Profile... invokes the **Layout** toolbar for editing design profiles criteria.

Add Profile View Labels... allows for adding specific profile labels.

Exercise 6-1: Create Profile from Surface Data

1. Launch Civil 3D and open the file called Chapter-6-a.dwg.

Student Files

TIP

Go to the Prospector and expand Sites >> Site 1 >> Alignments >> Alignment–Chapter 6 >> Profiles. You will see 3 existing ground profiles. Right click on them and select Delete... for each. This is how to delete unwanted profiles.

2. Type **V** for **View** and restore the view called **Plan-A**.
3. From the **Profiles** pull-down menu, select the command **Create From Surface...**. See Figure 6-8.
4. The **Alignment** of **Alignment—Chapter 6** should be set and now select the **Existing Ground** under the **Select surfaces** window and hit the **Add>>** button.
5. Then turn on the **Sample offsets** and type **45** into the **Sample offsets** field, which opens, and hit **Add>>**. In the **Profile list** window at the bottom, click on the second row where the offset of **45** exists and double-click on the value in the **Style** column. From the selection options, choose **Existing Ground BRL Right** for the style of this offset profile (because 45 is positive and therefore to the right). You want to differentiate it from the centerline so as not to confuse it with the centerline profile.

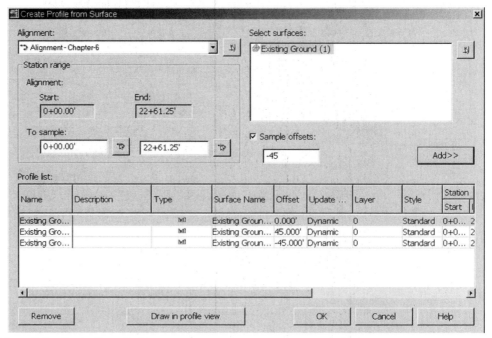

Figure 6-8   Create From Surface

Job skills benefit in that it is often desired to compute and draw the profile for the BRL, or the Building Restriction Line, so the reviewer can see what the ground is doing where a structure might be built.

6. Then type **−45** into the **Sample offsets** field and hit **Add>>**. In the **Profile list** window at the bottom, click on the third row where the offset of **−45** exists and double-click **Standard** in the **Style** column. From the selection options, choose **Existing Ground BRL Left** for the style of this offset profile. Again, you want to differentiate it from the centerline and the other BRL linework so as not to confuse it. The values should appear as indicated in Figure 6-8 for the dialog box.

7. Click on the button that says **Draw in profile view**. Hit **OK**.

8. Then the **Create View** dialog box displays, as shown in Figure 6-9.

9. In the dialog box, check the name of the **Profile view name** field. Accept the automatic naming **PV- (<[Next Counter(CP)]>)**, which is 1. Provide a description of Chapter 6. From the **Alignment** pop-down window, **Alignment—Chapter 6** should be selected. Notice the profiles you computed in the previous box are shown in the window at the bottom of the box, all of which should be checked to draw.

10. As you have done several times in the previous chapters, create a layer for the **Profile View** called **C-Road-Prof-View**. Hit **OK** and place the location of the view to the right of all of the plan view data, say, at a coordinate of **18800,19500**.

11. To create the Styles for the left and right profiles, to go the **Settings** tab of the **Toolspace** and expanded **Profiles** item and the **Profile Styles** item. Then copy the Existing ground style and paste it, creating a Copy of Existing ground. By right clicking on that style and editing it, the **Information** tab is selected and the name is changed to **Existing Ground BRL Left (or Right)**. Then the **Display** tab is selected and the colors are changed to **green** and **cyan** to differentiate the linework for the profiles.

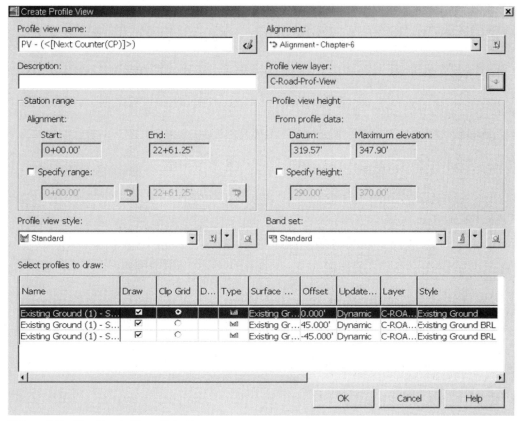

Figure 6-9   The Create View dialog box

Take a moment to inspect these to ensure that you could accomplish the same result if needed.

Now you have an existing ground profile for your alignment. It also shows left and right offset profiles, which are often required by reviewing jurisdictions so they can observe the ground activity near where buildings might be constructed. The centerline profile should be red, the left profile is green, and the right profile is blue. The engineer also uses this information to analyze the impact that a proposed roadway or corridor might have near the BRL line.

Exercise 6-2: Develop a Finished Grade Profile

The next task is to develop a proposed profile for the roadway. Use the Civil 3D commands to execute the formulas discussed earlier in this chapter. Create vertical tangents with PVIs between them and vertical curves at these PVIs. To begin, open a drawing containing the existing ground profile.

1. Launch Civil 3D and open the file called Chapter-6-b.dwg.
2. Type **V** for **View** and restore the view called **Profile-A**.
3. From the **Profiles** pull-down menu, select the command **Create By Layout**....
4. The prompt asks you to select the **Profile View** in which you will be working. Select the grid of the profile view shown in the view you just restored.
5. A **Create Profile** dialog box displays, requesting information about your profile. Set it up as shown in Figure 6-10, with a **Description**, a new layer for the proposed profile, and a style called **Design Style**. Use the **Complete Label Set** for the **Profile label set**. Hit **OK**.

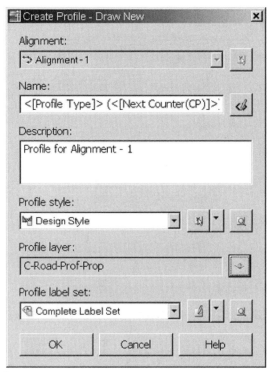

Figure 6-10 Profile information

PROFILE LAYOUT TOOLBAR

On hitting **OK**, the **Profile Layout** toolbar displays. It is shown in Figure 6-11 and a description of the commands follows.

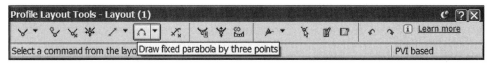

Figure 6-11 The Profile Layout toolbar

This command has a pop-down arrow that *allows you to draw vertical tangents with or without curves*. A **Curve Settings** icon allows you to set the curve settings that will be used for the **Curves** option.

This command allows for the *insertion of a PVI*.

This command allows for the *deletion of a PVI*.

This command allows for *moving a PVI*.

Use the pop-down arrow to obtain three icons to add a curve at a PVI, to set the curve parameters in the process, and to delete a curve.

These two new commands introduce Constraint Based Design to vertical geometry. These are similar to what was discussed in Chapter 4 except these control vertical computations. These two commands allow for Fixed, Free, and Floating Lines and Vertical Curves.

NEW
to AutoCAD
2007

This command *deletes a curve.*

 This command allows for the *insertion of a PVI using numeric table input.* It requires station, elevation, and curve values if they are to be added here as well.

 This command allows for the *raising and lowering of a PVI.*

 This command allows for *copying a profile.*

 Another new function is the ability to display PVI-based data in a pop-up window.

NEW
to AutoCAD
2007

The first icon is a sub-entity selector, the second icon is a command that opens or closes the **Profile Layout Parameters** dialog box. The third icon opens or closes the **Profile Grid View** dialog box.

NEW
to AutoCAD
2007

Exercise 6-3: Create a Proposed Profile

So, continuing on using the **Profile Layout** toolbar that appears, create a proposed profile. The software prompts you to Select a command from the **Layout** tools.

1. Using the pop-down arrow for the first icon, select the **Curve Settings** command shown in Figure 6-12.

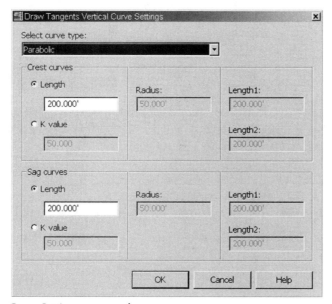

Figure 6-12   The Curve Settings command

2. The **Curve Settings** dialog box displays. In the **Select curve type** drop-down window, use **parabolic**, although circular and asymmetrical curves are available. Use a curve length of **400′**, so set the sag and crest curve lengths to **400**. Hit **OK**.

A crest curve goes over a hill whereas a sag curve goes into a depression.

3. The software prompts you to select a start point for the vertical tangents with curves. `Specify start point:` Use the Endpoint snap to pick the left edge of the red existing ground centerline profile. It should be around 0+00 and elevation 324.5.

4. It then says to `Specify end point:` looking for the end of the tangent segment.

> Notice that as you move the cursor around the profile, the cursor tip will appear, telling you the SZ of the point you are at. The SZ is the station and elevation of your current location.

5. Locate the station of 9+00 and an elevation near 339.0 by using these cursor tips. When you find that approximate location, use the Nearpoint snap and pick it on the vertical gridline.

6. It then says to `Specify end point:` Locate the station of 15+00 and an elevation near 327.0 by using these cursor tips. When you find that approximate location, use the Nearpoint snap and pick it on the gridline.

7. It then says to `Specify end point:` Locate the end of the red existing ground profile and use the Endpoint snap to pick it, thereby completing the development of the vertical tangents. This should be close to 21+53 and elevation 343.0. Feel free to zoom in when needed to snap to these locations.

> It is usually a smart move to place your PVIs at even stations to make it easier for surveyors or contractors to build the road without having to worry about odd station numbers. Nice, even 25, 50, or 100′ stations accommodate this.

8. Zoom out to view the profile and notice that curves exist automatically at the PVIs. Guess what the curve lengths are?

9. Take a moment to inspect the profile. Notice that the vertical tangents are one color whereas the curves are another color. This is to help you visualize when lines end and curves begin.

Exercise 6-4: Re-create Profile with Different Styles

Now let us assume that this profile is ready for plotting. Re-create this profile view with a different set of styles to make it more appropriate for a plot.

1. From the **Profiles** pull-down menu, choose **Create View....** The **Create Profile View** dialog box appears, as shown in Figure 6-13. Provide the **Description** shown in the figure.

2. Change the **Band set** to the **Profile Data & Geometry**.

3. In the **Select profiles to draw** window, change the **Style** for the **Layout (1)** to **Finished Grade**. This is the one highlighted at the bottom of the graphic in Figure 6-14.

> Notice that if you do not want to draw a particular profile, you can turn off the check mark in the **Draw** column of the window. You will draw them all. Another important item to notice here is the column called **Update Mode**. The **Existing Ground profiles** are set to **Dynamic**. This indicates that when the alignment or the terrain data are modified, these profiles are automatically updated for the designer.

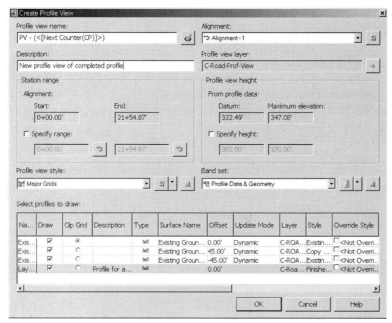

Figure 6-13   The Create Profile View dialog box

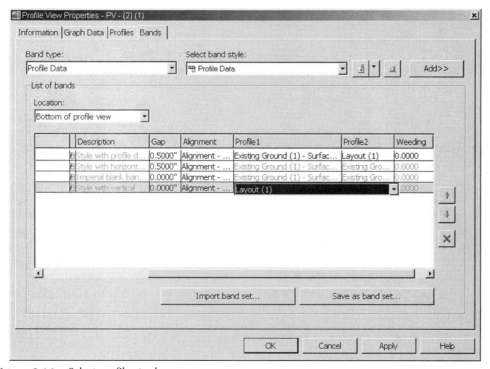

Figure 6-14   Select profiles to draw

4. When you say **OK**, the software prompts for a point to place the new profile. Pick a point somewhere well above the original profile.
5. Inspect the profile when it comes in and notice that curves are labeled and data bands come in at the bottom. Notice that one shows the horizontal geometry relative to the profile. This is a wonderful analysis tool to compare the horizontal and vertical geometries.
6. In order to acquire some additional experience in editing these Styles, notice that there is a band of information at the bottom representing

the existing ground that is too verbose and unneeded. Now if this represented the finished ground, that could be useful in seeing the vertical geometry is a single rapid observation.

7. Click on a grid for the **Profile View**. Right click and select **Profile View Properties**.... You can also access the **Profile View Properties** from the **Prospector**, as shown in Figure 6-15.

8. Choose the **Bands** tab and select the last item in the **Band Type** column, vertical geometry.

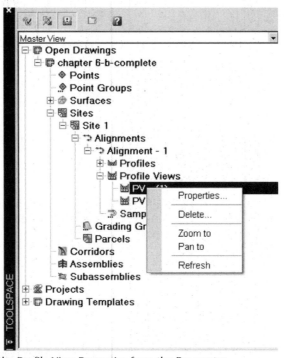

Figure 6-15   Access the Profile View Properties from the Prospector

9. In the column labeled **Profile 1** set the value to **Layout (1)** so it knows to display graphics for the finished ground profile.

10. The last step is to set the annotation in the **Band Type** column for the Profile Data to reflect the existing ground AND the Proposed profile. For the **Profile Data Band Type**, change the column for **Profile 2** to **Layout (1)** so it will automatically annotate the proposed ground. The **Profile 1** cell is already set to **Existing ground**. Hit **OK**.

11. Next, delete the first **Profile View** used to construct the profile. Open the **Toolspace** and go into the **Prospector**.

12. Expand the items for the Current Drawing, **Sites**, **Site 1**, **Alignments**, **Alignment—1**, and **Profile Views**. Right click on **PV – 1** and select **Delete**... to remove it from the project and drawing. Click **Yes** to the confirmation and notice that when it is complete the second "final" **Profile View** remains while the first is removed.

You can also just simply delete the graphics of the **Profile View** in AutoCAD.

This exercise illustrates how to develop an Existing Ground profile, what the **Profile Views** offer, how to alter the **Profile View Styles**, and how to develop a proposed roadway profile and alter its appearances and display capabilities.

Job skills will benefit from reviewing the bands showing the horizontal and vertical geometry so that an engineer or manager can easily review the design data without having to hunt through the profile or alignment data.

The next exercise explores what happens in Civil 3D when modifications are made.

Exercise 6-5: Modify the Label Style

You may have noticed that the vertical tangents have no labeling on them. Add these now as they and other labels can be affected on the fly or from the start.

1. Launch Civil 3D and open the file called Chapter-6-c.dwg.
2. Notice that the previously developed plan and profile are visible.
3. Zoom into the **Profile** and notice that there are no grades on the vertical tangents. You want to add those to the **Profile**. Modify the tangent labels.
4. In the **Toolspace**, click on the **Settings** tab. Expand the **Profiles** item, the **Label Styles** item, and the **Line** item.
5. Right click on the label style **Finished Ground** and choose **Edit**. See Figure 6-16.

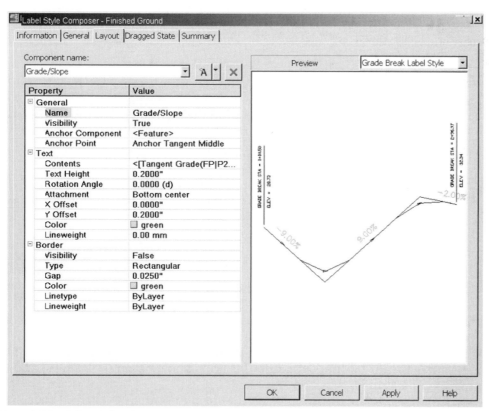

Figure 6-16 Edit Label Styles

6. In the **Layout** tab, change the **Text Height** to **0.2** and the **Y Offset** to **0.2**. Set the color to **green**. Hit **OK** to exit.

7. Next, select the proposed profile linework in the **Profile View**, right click, and choose **Edit Labels...**

8. The **Profile Labels** dialog box appears.

9. Under the **Type** pop-down, choose **Lines**.

10. Under **Profile Tangent Label Style**, choose **Finished Ground**.

11. Then hit the **Add>>** button. It should appear as shown in Figure 6-17.

12. Hit **OK** and notice that you now have grade labels on the vertical tangents.

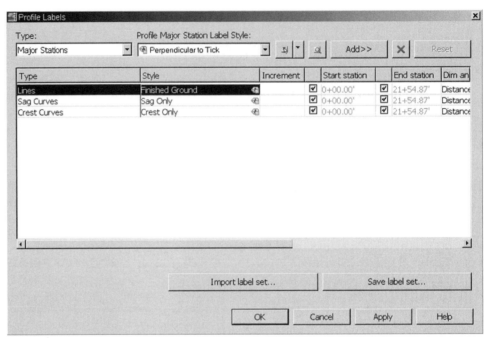

Figure 6-17   Profile Labels

THE RIPPLE-THROUGH EFFECT

Civil 3D has state-of-the-art capabilities in updating the design model. Because many of the objects are linked, they sense whether modifications have been to the data that affect their display and computations.

Exercise 6-6: Try the Ripple-Through Effect

Next you open a drawing and make some changes to the data to observe how the ripple-through effect works and its potential to save enormous time and on the project.

1. Continuing on in the same drawing Chapter-6-c.dwg, discussion will concern making modifications.

2. Zoom into the beginning of the profile and notice that the existing ground elevation is around 324.5.

3. Now type **V** for **View** and restore the view called **Plan-A**. Notice some contour activity and the alignment.

4. Select the alignment with the mouse. Notice that grips show up. Square grips appear on tangents, circular grips show up on the curves, and triangular grips appear on the PIs.

5. Using **Grip-edit** features, select the grip at station 0+00 and move it to another location within the contoured surface, say, near an X, Y of

14600, 18750. The alignment relocates, recomputes, and redrafts its annotation.

6. Now type **V** for **View** and restore the view called **Profile-A**. Notice that the red Existing ground linework has changed up to around elevation 320, which is where the alignment now begins.

7. If you zoom out to see the whole profile, notice that it actually became longer to compensate for the increased length created when the beginning of the alignment was moved. Normally, the proposed profile would need to be "stretched" to relocate its beginning and endpoints. This would be accomplished using the grip-edit features of the profile in exactly the same manner that you just moved the alignment.

In order to expand your skills, an optional exercise for the reader is to relocate the proposed profile using this method.

8. When you are finished experimenting, either open this drawing up again, without saving it, or type **U** for **Undo** and repeat until the alignment returns to its original location.

9. Now type **V** for **View** and restore the view called **Profile-A**. Notice that the red Existing ground linework has changed back to what it was, around elevation 324.5.

10. This is a major feature of the Civil 3D software and now you have experienced what I call the ripple-through effect.

11. Feel free to experiment with this file and save it as another name so you can return to it in the future.

Job skills benefit from this feature in that "value engineering" can be performed in identifying the best design for the issues at hand. By experimenting with various design alternatives, the engineer can focus in on the most cost-effective solution. Many firms are now offering value engineering to their clients as an added benefit to their services.

Exercise 6-7: Numerical Editing of Profiles

The next function you explore is the numerical editing of a profile. Sometimes designers feel it is advantageous to edit a profile using a tabular view that allows them access to the elevations, slopes, and grades. Civil 3D allows for this in addition to visual or graphical edits.

1. Launch Civil 3D and open the file called Chapter-6-d.dwg. It opens with the profile in the view.

2. From the **Profiles** pull-down menu, choose **Edit Profile**. When prompted select somewhere on the blue proposed profile. The **Profile Layout** toolbar displays.

Student Files

TIP

Note that you can also simply grip the proposed profile, right click the mouse, and select **Edit** to achieve the same result.

3. Pick the command near the end of the toolbar called **Profile Grid View**. This will display the **Panorama** with all of the profile data in it, shown in Figure 6-18.

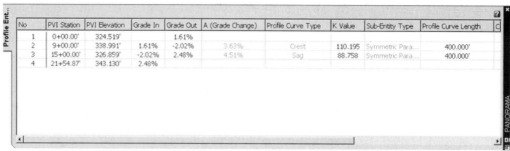

Figure 6-18   Profile Panorama

4. The fields that cannot be edited are screened back and inaccessible. Fields that can be altered are accessible by double-clicking the mouse in the cell.
5. For PVI 2, change the **Profile Curve Length** to **500**. Hit **<Enter>** and close the **Panorama**.
6. Then zoom into PVI 2 and notice the annotated curve length is now 500. The graphics and text have been automatically updated.

Exercise 6-8: Delete a Profile Component

Sometimes it is necessary to use an asymmetrical vertical curve that occurs when there is not enough incoming or outgoing tangent length to accommodate a symmetrical curve. For instance, you might be required to have a 500′ curve; however, there is only 200′ of incoming tangent available. Therefore, an asymmetrical curve where the first portion of the curve is 200′ and the second portion of the curve is 300′ would satisfy the overall length requirement. Let us perform an example of this. Begin by deleting the second curve and notice how the Civil 3D automatically drafts it up. Follow that up by adding an asymmetrical curve to that same PVI.

1. In the same file called Chapter-6-d.dwg, choose **Edit Profile** from the **Profiles** pull-down menu. When prompted select somewhere on the blue proposed profile. The **Profile Layout** toolbar again displays.
2. The software says to: `Select a command from the layout tools:`
3. Choose the command to **Delete** a vertical curve. It prompts you to `Pick point near curve to delete:` Pick a point near the PVI for the second curve at station 15+00.
4. The software responds with **Curve successfully deleted.** Hit **<Enter>** to complete the sequence.
5. Notice that the curve is gone but the PVI remains.

Exercise 6-9: Create an Asymmetrical Curve

Now add in an asymmetrical curve.

1. In the same file called Chapter-6-d.dwg, choose **Edit Profile** from the **Profiles** pull-down menu. When prompted select somewhere on the blue proposed profile. The **Profile Layout** toolbar again displays.
2. The software says to: `Select a command from the layout tools:`
3. Choose the command **More Free Vertical Curves >> Free Asymmetrical Parabola** (PVI Based) from the pop-down arrow in the toolbar. It is associated with the sixth button from the left and prompts you to: `Pick point near PVI or curve to add curve:` Pick a point near the PVI at station 15+00.
4. The software asks you to provide **Length 1** of the asymmetrical curve. Type **200**.
5. The software asks you to provide **Length 2** of the asymmetrical curve. Type **300**.

6. Hit <**Enter**> to complete, and notice the new curve and the related annotation, shown in Figure 6-19. Notice the center of the curve does not exist at the PVI as it would for a symmetrical curve. This asymmetrical curve can be very useful in road design.

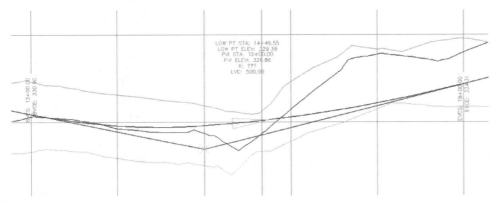

Figure 6-19   Curve Data

Asymmetrical vertical curves are often used as a last resort in road design. They might be used in situations in which a standard symmetrical vertical curve just does not work. For example, let us say you are designing a curve near an intersection and there is a PVI that requires a vertical curve approaching that intersection. The prevailing DOT code may require a vertical curve length of, say, 600'. It could be that you only have 200' to the left of the PVI before you enter the intersection but you have plenty of space to the right side of the PVI. This is where you might place an asymmetrical curve such that 150' occurs to the left side of the PVI and 450' occurs to the right side. In this fashion you have accommodated the requirement to establish a 600' curve without encroaching on the intersection.

SUPERIMPOSING PROFILES

Another powerful feature within the profiling functions of Civil 3D is the ability to superimpose one profile into another. This is useful when you wish to compare how one profile relates to another.

Exercise 6-10: Superimpose One Profile onto Another

1. Launch Civil 3D and open the file called Chapter-6-e.dwg. There is an additional alignment and another profile in this file. They were prepared in the same manner as the one you just completed and are included here in order to expedite this command usage.
2. Zoom into the two profiles so they can be seen.
3. From the **Profiles** pull-down menu, choose **Create Superimpose**... When prompted to `Select the Source profile:` select the red linework in the shorter profile.
4. When asked, `Select the Destination Profile View`, select a gridline in the longer **Profile View**.
5. The **Superimpose Profile Options** dialog box appears. You can restrict the length of the profile to be computed if you wish or just hit **OK**.
6. The superimposed profile appears in the longer profile.
7. Notice that it is longer than it was in the source profile. That is because the profile is projected into the destination orthogonally. The source alignment is not parallel to the destination alignment.
8. Zoom into the plan view and inspect how the source alignment is drawn. It is on the **C-Pipe-Cntr** layer.

CIVIL 3D SECTIONS

The last part of this chapter delves into Sections. As discussed earlier, sections are typically, but not always, orthogonal to the alignment and are created at intervals, as opposed to profiles, travel along the alignment. Sections provide users with more information as to what is occurring adjacent to our alignment. Civil 3D develops cross sectional data in Sample Line Groups and as we have seen already, Styles control the appearance of the Sample Line Groups. Next you use the **Sample Line Tools** toolbar, Figure 6-20.

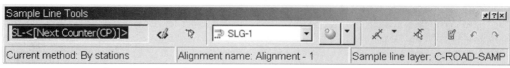

Figure 6-20   Sample Line Tools toolbar

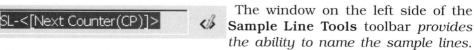

The window on the left side of the **Sample Line Tools** toolbar *provides the ability to name the sample lines.* The icon to its right brings up a dialog box that allows you to change the template naming, which is automatic and sequentially increments the name.

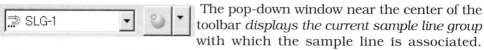

The pop-down window near the center of the toolbar *displays the current sample line group* with which the sample line is associated. The icon to its right allows you to create or modify the sample line group. The options within this icon include **Create** a new sample line group, **Edit** the currently selected sample line group, **Delete** the currently selected sample line group, or you can be prompted to pick a sample line group from the drawing, which then becomes the current sample line group.

The fifth button from the right *displays the various sample line creation methods* used to create the sample line(s). The pop-down arrow allows for other actions such as **Create** sample lines by choosing individual stations along the alignment. In this case a perpendicular transient graphic appears in Civil 3D that you can move along the alignment to witness the current location. The other methods are **Create** sample lines by allowing you to pick points in the drawing, **Create** sample lines by picking existing polylines in the drawing, **Open** a dialog box for **Creating Sample Lines—By Station Range** and **Open** a dialog box for **Create Sample Lines—From Corridor Stations**.

The fourth button from the right *prompts you to select a sample line from the screen.*

The third button from the right *toggles the select or edit sample line functionality.* When the button is down, the **Edit Sample Line** dialog box is displayed.

The last two buttons are **Undo** and **Redo**.

Exercise 6-11: Create Sample Lines for Sections

Next you explore the abilities of sections.

1. Open a file called Chapter-6-f.dwg.
2. From the **Sections** pull-down menu, select **Create Sample Lines**. When prompted select the **Alignment** in the plan view.

3. The **Create Sample Line Group** dialog box opens, as shown in Figure 6-21. Hit **OK**.

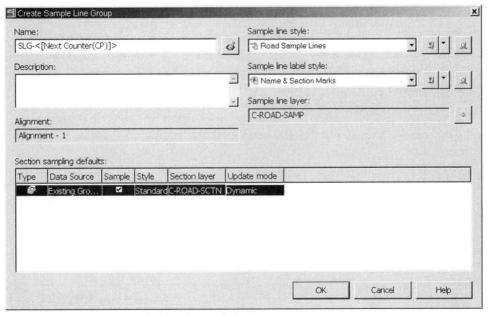

Figure 6-21   The Create Sample Line Group dialog box

4. The **Sample Line Tools** toolbar displays.
5. Using the pop-down arrow associated with the fifth button from the right, select the **By Station Range** icon. The **Create Sample Lines—By Range of Stations** dialog box appears. See Figure 6-22.
6. Notice that by default the cross sectional data are sampled 30′ to the left and right of the alignment and data are sampled at 50′ increments along the alignment.

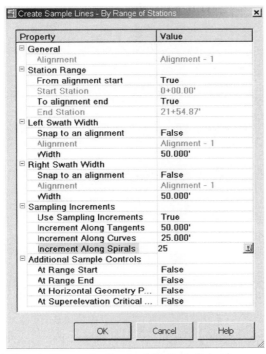

Figure 6-22   The Create Sample Lines—By Range of Stations dialog box

7. Change the **Left** and **Right Swath Width** to **50′**.
8. Change the **Increment Along Curves** and **Spirals** to **25′**. Hit **OK**. Notice that sample lines appear along the alignment with 50′ lines on tangents and 25′ lines on curves.
9. For the **Additional Sample Controls**, turn all of the options to **True**. So, at range start and end, at horizontal geometry points, and at **superelevation** critical points are all marked **True**. This way you obtain additional sections at these locations as well.
10. Hit **OK**.
11. The prompt requests you to `Specify Station`, hit **<Enter>**.
12. You will notice in the **Prospector**, under **Sites**, **Site 1**, **Alignments**, and **Sample Line Group** that a number of sections now appear.

Note that in the plan view linework has shown up illustrating where the sections were cut.

Exercise 6-12: Create Sections

In the next steps, you create existing ground cross section views for each of these locations. In order to view these cross sections to evaluate what is occurring to the side of your alignment, you need to develop cross section views, similar to profile views.

1. From the **Sections** menu, click **Create Multiple Views**, which displays the **Create Multiple Section Views** dialog box, as shown in Figure 6-23. Hit **OK**.

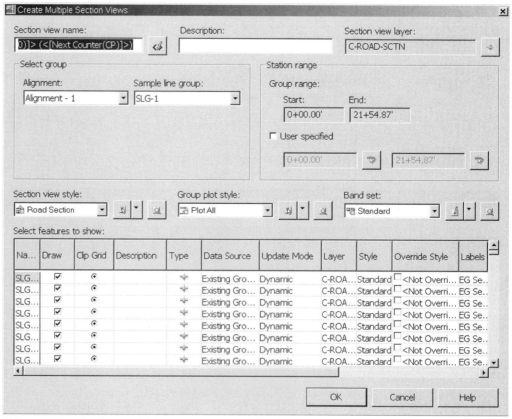

Figure 6-23   The Create Multiple Section Views dialog box

2. When prompted to identify the **section view** origin, pick a point to the right of the profile for the cross sections to be placed, say at **21000,21000**. Zoom in and notice that Section sample lines are shown on the alignment in Figure 6-24 and sections should be in your drawing in Figure 6-25.

Section View—A Civil 3D object

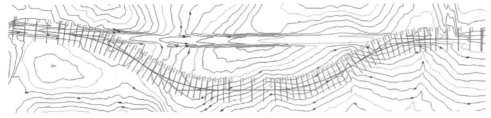

Figure 6-24 Section sample lines are shown on the alignment

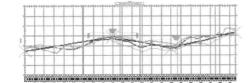

Figure 6-25 Plotted Sections

These cross section views are objects that are dynamically linked to the horizontal alignment in much the same manner that the profile view was when you were making modifications. If the horizontal alignment is relocated, the existing ground cross section views automatically update to reflect those modifications.

3. Zoom into one section (Figure 6-26) and notice that there is too much information for the grade breaks of the existing ground. You do not really need this density of data, so turn it off.

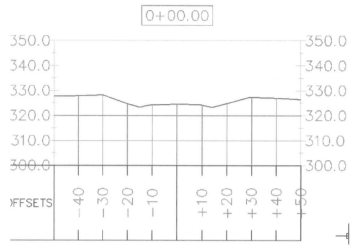

Figure 6-26   One section

4. Click on the gridline for a section and right click and select **Section View Properties** to invoke the dialog box.
5. In the **Labels** column, select the cell for the grade breaks where it says <**Edit...**>.
6. A **Section Label** dialog appears. Click on the grade breaks row and hit the button with the red **X** to delete. Hit **OK**. Hit **OK**. The labels are gone and the section is easier to read.

TIP   To turn these off prior to computing the sections and to effect this change on all sections, you can access the settings in the **Toolspace**. Go to the **Section View** parameter and right click. Select **Edit Feature Settings** to access the **AeCccSectionView** dialog box. Expand **Default Styles**, click on **Section Label Set**, click on the value of **EG Section Labels**. This allows you to select the button for the **Value**. Select **Edit** from the pop-down arrow. The same dialog box appears and you can hit the same red **X** to delete the grade breaks as you did earlier.

CHAPTER TEST QUESTIONS

Multiple Choice

1. There are three main requirements for designing a roadway. They are:

   a. Alignments, sections, and templates (or assemblies)
   b. Profiles, sections, and earthworks
   c. Alignments, corridors, and parcels
   d. Alignments, profiles, and a proposed section

2. Proposed vertical alignments can be developed from:

   a. Horizontal tangents, curves, and spirals
   b. Vertical tangents and vertical curves that do not exceed 8%
   c. Vertical tangents and vertical curves
   d. None of the above

3. Which of the following vertical curve properties are true?

   a. A tangent ends at the PVC and a vertical curve begins at a PVC
   b. A vertical curve begins at a PVT and a vertical tangent begins at a PVT
   c. Two tangents meet at a PVC
   d. None of the above

4. Which of the following are properties of a vertical curve?

   a. The tangent offsets vary as the square of the distance from the points of tangency.
   b. For points spaced at equal horizontal distances, the second differences are equal.

c. A curve's elevation at the midpoint of the curve will be halfway from the elevation at the PVI to the elevation at the midpoint of a straight line from the PVC to the PVT.

d. All of the above

5. The design of a vertical alignment is usually specified in terms of:

   a. A maximum grade and the minimum length for the vertical curve
   b. A minimum grade and the minimum length for the vertical curve
   c. A maximum grade and the maximum length for the vertical curve
   d. Always being dictated by the State Department of Transportation

6. Civil 3D has a ripple-through effect:

   a. Where profiles and cross sections are dynamically linked to horizontal alignments and surfaces
   b. If the surface is modified, the profile and cross section data also update automatically
   c. Where related annotation updates as changes are made
   d. All of the above

7. A proposed profile can be edited using which of the following features?

   a. By graphically selecting grips for PVIs

b. By using the numerical editor in the **Edit Profile** toolbar
c. By graphically selecting grips for vertical tangents
d. All of the above

8. Which of the following modifications can be accomplished by using the **Edit Profile** commands?

   a. Inserting and deleting PVIs
   b. Inserting and deleting alignment segments
   c. Inserting and moving PVIs, but not deleting them
   d. All of the above

9. Which of the following vertical curves are supported?

   a. Parabolic
   b. Circular
   c. Asymmetrical
   d. All of the above

10. Cross sections can be created in the AutoCAD drawing in which if the following ways?

   a. Individually
   b. By sheet
   c. All sections can be imported at once
   d. All of the above

True or False

1. True or False: Profiles are created by sampling the elevation at intervals along a horizontal alignment.

2. True or False: Profiles computed along tangent sections usually sample the elevation wherever the alignment crosses the TIN lines in the terrain model.

3. True or False: A profile cannot be developed if the alignment has horizontal curves or spirals.

4. True or False: A PVI is the intersection of two vertical tangents.

5. True or False: A PVC is the intersection of two vertical curves.

6. True or False: Sometimes a grade break may occur that does not require any vertical curve at all.

7. True or False: A vertical curve is usually parabolic with an ever-changing radius.

8. True or False: The K value for profile computations is derived from empirical results of roadway friction factors.

9. True or False: Roadways are designed for safety and never use subjective parameters such as passing sight distances or stopping sight distances.

10. True or False: In Civil 3D, profiles are linked to the alignment and can react to changes made to the alignment.

STUDENT EXERCISES, ACTIVITIES, AND REVIEW QUESTIONS

1. With a back tangent of −2%, a forward tangent of 1.6%, and a vertical curve length of 800′, compute the starting and ending stations of this vertical curve and its accompanying table if the PVI station is 87+00 and the PVI elevation is 743.24.

Advanced Corridor Development

7

Chapter Objectives

After reading this chapter, the student should know/be able to:
- Understand "corridor design" in which the proposed section or template for the roadway is created.
- Develop assemblies.
- Customize subassemblies.
- Create and edit simple roadways.
- Create multiple assembly roadways.
- Understand a safety technique used by engineers to assist motorists in handling curves called superelevating the roadway.
- Create multiple baseline roadways.
- Perform rehabilitation and reconstruction design for roads that need maintenance.

INTRODUCTION

This chapter commences with discussions and algorithms for roadway super-elevations and then moves into corridor development. Civil 3D offers state-of the-art capabilities for developing roadways and corridors in general.

At this point the text has discussed developing alignments, existing, and proposed profiles, and cross sections. This chapter covers "corridor design," which is creating the proposed section or template for the roadway. The objectives should be generalized because a corridor can include not only roadways but many other projects as well. If a project has a beginning, an end, limitations to the right and left, and standardized cross sectional components, it may be a corridor. Think of stream improvements and railroads as examples of this, perhaps even tunnels and aqueducts. The chapter also delves into a safety technique used by engineers to assist motorists in handling curves called superelevating the roadway.

REVIEW OF FUNCTIONS AND FEATURES TO BE USED

This chapter provides instruction in the functions for corridor design. These include using **Tool Palettes** to access subassemblies and how to configure them for use in almost any scenario. These subassemblies are used to build assemblies for the sections to be designed. The discussion then walks through examples of how to build both simple and sophisticated corridors. The chapter also covers viewing the resulting cross sections of the design and various outputting strategies.

CORRIDOR THEORY

Once the section is developed, considerations are then made for safety as the rider negotiates curves in the road. According to the laws of mechanics, when a vehicle travels on a horizontal curve, it is forced outward by centrifugal force. Therefore, engineers use a method called superelevation by which they tilt the pavement lanes to counteract the centrifugal force. The two figures presented here indicate a template in normal conditions (along the tangent of a road) in Figure 7-1 and in the superelevated condition (around a curve, in this case, turning to the left) in Figure 7-2.

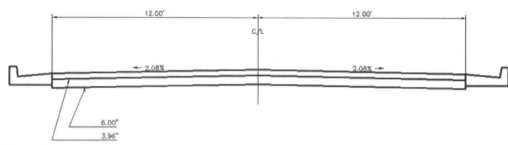

Figure 7-1   A normal template

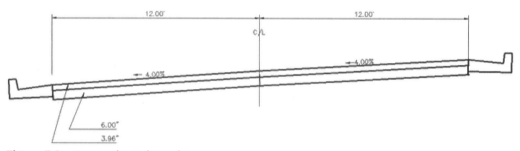

Figure 7-2   A superelevated template

Civil 3D performs DOT-level highway design because it allows for corridors with multiple baselines and multiple assemblies, all operating under a variety of conditions. One item in particular that should be of interest to many engineers is that Civil 3D also performs roadway rehabilitation and reconstruction design. Many subassemblies are included for roadway, curb, or sidewalk maintenance.

SUPERELEVATION OF ROADWAYS—BASIC CRITERIA

Rate of Superelevation—The rate of banking a roadway

On a superelevated highway, this centrifugal force is resisted by the vehicle's weight parallel to the superelevated surface and the friction between the tires and pavement. Centrifugal force cannot be balanced by **superelevation** alone, because for each curve radius a specific superelevation rate must be selected for a particular driving speed. In other words, the *rate of superelevation* varies based on the radius of the curve. Any side thrust must be offset by side friction. If the vehicle is not skidding, these forces are in equilibrium. Figure 7-3 illustrates the trigonometry involved in computing superelevation problems. Refer to the *AASHTO Green Book: A Policy on Geometric Design of Highways and Streets*, 5th Edition for an authoritative reference on this subject.

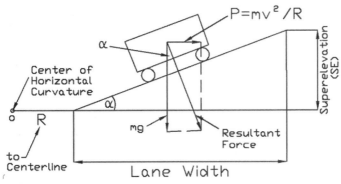

Figure 7-3   Superelevation trigonometry

Variables and terms for superelevations include:

e = Superelevation slope

e_{max} = Maximum superelevation rate for a given condition

R = Curve radius

v = Velocity

$P = mv^2/R$

B = Horizontal lane width

A maximum superelevation (SE) is required when the radius (*R*) is a minimum:

$$\tan (\alpha) = P/mg \quad \tan (\alpha) = (mv^2/R)/mg \quad \tan (\alpha) = v^2/gR$$

$$SE = B*\tan (\alpha) \quad SEmax = Bv^2/gR$$

This model is theoretical and is usually guided by the state DOT or AASHTO.

Superelevation transitions are used to define the pavement rotation up to a maximum superelevation and are also used to bring a superelevated section back to the normal crown state. At a reverse curve the section is often flat at the (PRC) as the superelevation in one direction transitions to the other direction.

Two components make up the total transition for a superelevated section, the Tangent runout and runoff, shown in Figure 7-4. The Tangent runout defines the length of highway needed to bring a normal crown section to a section where the outside lane has a zero percent slope instead of normal crown. The superelevation runoff defines the distance needed to bring a section from flat to fully superelevated, or vice versa.

Daylighting—An intersect with the existing ground surface.

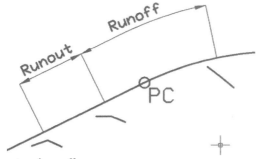

Figure 7-4   Tangent runout and runoff

TIP   The superelevation criteria for entering a curve are usually mirrored as the driver leaves a curve.

Figure 7-5 shows how the edge of pavement rises and drops based on the tilting of the road before and after the (PC) or (PT) of a curve.

Standard superelevation rates are usually provided by the Department of Transportation that oversees roadway design in the jurisdiction. A typical chart

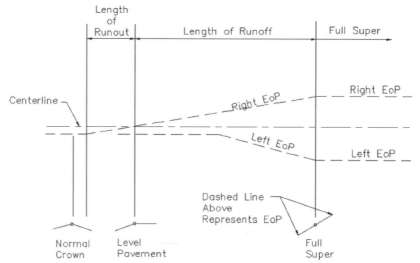

Figure 7-5   The edge of pavement in superelevation

from which to obtain these values, shown in Figure 7-6, comes from the Virginia Department of Transportation (VDOT). For a specific curve radius and a design speed, the superelevation rate can be easily determined. For example, for a 3000′ radius curve and a design speed of 40 mph, the state requires a minimum of 2.5% superelevation rate.

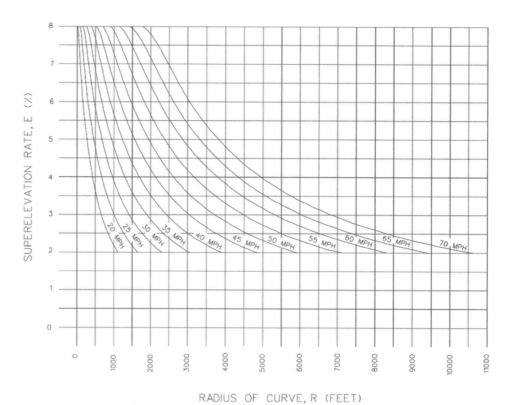

Figure 7-6   Superelevation rates for VDOT

For more information on design standards for roadway design, the American Association of State Highway and Transportation Officials (AASHTO) studies and recommends criteria to engineers. It can be located at www.aashto.org. Consult with your local state, county, or city DOT for its respective requirements as well.

Now that sections and superelevations have been discussed, let us get into Civil 3D and develop a corridor.

CREATE A CORRIDOR FOR A ROAD

Remember that a corridor is a Civil 3D object and as such is linked to the horizontal and vertical alignments, surfaces, and assemblies that were used to create it. If one or more of these components are altered, the corridor is correspondingly modified.

TIP   For the rebuild to be automatic, the user must right click on the corridor and select **Rebuild-Automatic**. Otherwise a rebuild can be executed at the user's choice by right clicking on the corridor and selecting **Rebuild**.

The next step is to modify subassembly properties for design criteria, create an assembly from the subassemblies, and then create the Corridor model.

Subassemblies are the primitive components of assemblies, and they instruct the software in how to handle lanes, curbs, guardrails or shoulders, related side sloping, and tie-out. Civil 3D provides many subassemblies to begin with and the user accesses them from **Tool Palettes**. The assemblies, on the other hand, are stored in the drawing and can be accessed from the **Project Toolspace** under **Assemblies**. They contain the definition of the typical section being designed and are developed by piecing together subassemblies.

This corridor has the following design requirements: The road will be a crowned section with a pavement width of 12′. The cross slope of the lane is −2.0% from the crown. A 2′ curb and 6″ gutter are used, located at the edge of pavement. Add on a 2′ grass strip, a 4′ sidewalk, and another 2′ grass strip, respectively. Then the side sloping will be 3:1 until it hits existing ground.

The **Corridors** pull-down menu is shown in Figure 7-7.

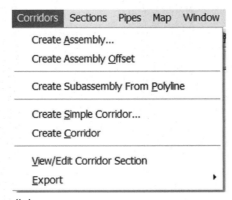

Corridors	Sections	Pipes	Map	Window

Create Assembly...
Create Assembly Offset

Create Subassembly From Polyline

Create Simple Corridor...
Create Corridor

View/Edit Corridor Section
Export   ▶

Figure 7-7   The Corridors pull-down menu

Create Assembly... allows you to develop a roadway template, now called an assembly. It is constructed from subassemblies selected from the **Catalog** or **Tool Palettes**.

Create Assembly Offset allows you to develop a controlling offset on an assembly. This is used for roads that have other paths that will define the design, such as a service road.

Create Subassembly From Polyline allows you to develop new subassemblies from polylines that you draw in AutoCAD.

Create Simple Corridor... designs the corridor from an alignment, a profile, and an assembly.

Create Corridor designs a corridor for a sophisticated roadway including such items as station frequency and controlling offsets, multiple baselines, and regions.

View/Edit Corridor Section allows you to view the corridor sections and edit them interactively.

Export allows for placing portions of the corridor into AutoCAD as polylines, feature lines, or points.

Exercise 7-1: Set Up Toolspace and Subassembly Settings

Student Files

1. Open the drawing called Chapter-7a.dwg.
2. Under the **General** pull-down menu, there is a **Tool Palettes** window command that should be checked **on**. See Figure 7-8 for an example of the **Tool Palettes**. If it is not, click it to set it **on**. A **Tool Palette** should be open in Civil 3D that says **Civil 3D Imperial**. Note that you can switch

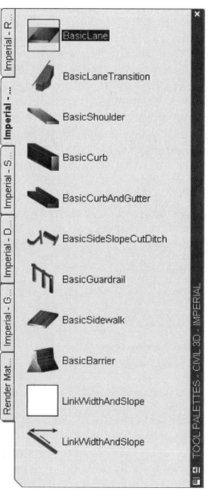

Figure 7-8   The Tool Palettes window

from **Imperial** tool palettes to **Metric** palettes by right clicking on the spine of the palette and choosing **Metric** from the options.

3. As an exercise, browse through the **Catalogs** and look for **Rehabilitation** subassemblies. Civil 3D is very strong at assisting in road reconstruction design. The catalog component can be i-dropped into a new palette that you create called **Rehab**.

MODIFY SUBASSEMBLIES FOR THE ROAD

This **Tool** palette can be used to develop the roadway corridor and it will remain in place for future jobs as well.

Let us now modify the properties of the subassemblies needed for the road. The modifications occur as the subassemblies are placed into the assembly. These include the **BasicLane**, the **BasicCurbandGutter**, the **BasicSidewalk**, and the **DaylightStandard** subassemblies.

You can create the assembly from the subassemblies on the **Tool** palette. The assembly contains the design control for the proposed typical section of the road.

Exercise 7-2: Create the Assembly

1. Type **Z** for **Zoom** at the command prompt and hit **<Enter>**. Type **C** for **Center** and type **17000,20000**. When asked for a height, type **50**.
2. This has placed you in a location where you can create the assembly. There is nothing special about this location; it is simply out of the way and zoomed to a height of 50′.
3. From the **Corridors** pull-down menu, select **Create Assembly**. The **Create Assembly** dialog box opens.
4. Accept the default name of **Assembly 1** and set the name of the layer to **C-Road-Assm**. Leave the remaining settings as default and hit **OK**.
5. The command prompts you to: `Specify assembly baseline location:` Pick a point in the middle of the screen and a small symbol with a red line shows up. This indicates the assembly location to which you will attach subassemblies.

If you look in the **Prospector** and expand the tree for **Assemblies**, you see the assembly name. You now attach the subassemblies to both the left and right sides of the assembly location.

Exercise 7-3: Attach Subassemblies to an Assembly

1. From the **Imperial Basic** tool palette, click the **BasicLane** subassembly. The properties for the subassembly can be altered at that time. Look in the **Properties** palette under **Advanced Parameters** and ensure that the parameter for **Side** is set to **Right**. Set the **Depth** to **0.67**. Ensure that the **Width is** set to **12.0** and the **%Slope** is **−2.0**.
2. The command prompts to: `Select marker point within assembly:` Pick the assembly baseline location, which is the circular symbol in the middle of the assembly that you created. The lane should show up on the right side of the assembly marker.
3. From the **Imperial Basic** tool palette, click the **BasicCurbandGutter** subassembly. Look in the **Properties** palette under the **Advanced Parameters** and ensure that the parameter for **Side** is **Right**. Change the **Gutter width** to **2.0**. Change the **Gutter %Slope** to **−2.0**. Change the **Curb Height** to **0.5** and the **Curb Width** should be **0.5** as well. The **Curb Depth** should be **1.5**.
4. The command prompts to: `Select marker point within assembly:`

5. Click on the upper right circle of the lane you just placed and the **BasicCurbandGutter** subassembly appears connected to the lane.

6. From the **Imperial Basic** tool palette, click the **BasicSidewalk** subassembly. Look in the **Properties** palette and ensure that the parameter for **Side** is **Right**. Change the **Width** to **4.0**. Change the **Depth to −0.33**, and change both of the **Buffer widths** to **2.0**.

7. The command prompts to: `Select marker point within assembly:`

8. Click on the upper right circle of the **BasicCurbandGutter** subassembly, and the sidewalk appears connected to the back of the Curb.

9. Now repeat this process, but select the parameter for **Side** being **Left**, place the left side of the assembly, and connect it up as you did the right side.

10. When finished, the assembly should have a left and right travel lane, Curb and gutter, and a sidewalk on both sides.

11. Next, you handle the tie-out of the road to the existing conditions. Using the **Imperial-Daylight** tab in the **Tool** palette, click the **DaylightStandard** subassembly. Check the **Properties** dialog box and ensure that the side is **Right**, the **Ditch width** is **0.0**, the **Flat Cut Max Height** is **5.0**, the **Flat Cut Slope** is **6.0**, the **Flat Fill Max Height** is **5**, and the **Flat Fill Slope** is **6.0**.

12. Then click on the **DaylightStandard** icon and place it on the top right-most circle on the sidewalk. Repeat for the left side by setting the **Side** parameter to **Left**.

13. Press **<Enter>** twice to finish the command.

You have now created the assembly. Figure 7-9 shows how it should appear.

Figure 7-9   A completed assembly

Job skills benefit in that the designer can build the right side independently of the left if he or she chooses. The assembly does not have to be symmetrical.

The assembly for the Access Road is now complete. Save the file. You now move forward to create the roadway. In this task, the pieces to the corridor puzzle come together. You have all of the components for the corridor. Let us summarize. The primitive geometry for the alignment is developed and a completed alignment is defined. The Profile View is established with an existing ground profile. From there a finished ground profile (or **PGL**) was developed for the corridor. Then you set the settings for some subassemblies and created an assembly.

PGL—The profile grade line

Note that you can edit any particular subassembly by selecting it, right clicking, and choosing **Edit Subassembly** from the menu. Parameters can be altered as needed. You now create the corridor and the finished design surface.

Exercise 7-4: Create a Corridor

1. Open the drawing called Chapter-7b.dwg Notice the assembly.
2. Restore a presaved view to place the view around the centerline.
3. Type **V** for **View** at the command prompt to obtain the **View** dialog box. Double click the **Plan View** and hit **OK**. You are placed at the centerline of the Access Road. A view is set up for the profile as well in the event that you need it.
4. From the **Corridors** pull-down menu, select **Create Simple Corridor**....
5. Accept **Corridor 1** as the default name and hit **OK**.
6. The command prompt requests that you: Select the baseline alignment: Pick the red centerline in the view.
7. The command prompt then asks you to select a profile or hit **<Enter>** to obtain a list of available profiles.
8. Hit **<Enter>** and the **Select Profiles** dialog box appears. Choose **LAYOUT (5)** and hit **OK**. Note that the other options are the offsets created in a previous chapter.
9. The command prompt then asks you to select an assembly or hit **<Enter>** to select it from a list. Hit **<Enter>** and select **Assembly 1** from the list.
10. A large window of data displays for **Logical Name Mapping**.

If you recall from Chapter 2, logical name mapping is a technical way of assigning the particular alignments, profiles, and surfaces required by the subassemblies to create the complete assembly. For instance, the **DaylightStandard** subassembly needs to know which surface to compute to, and so the **Logical Name Map** for this would be the **Existing Ground** surface. See Figure 7-10.

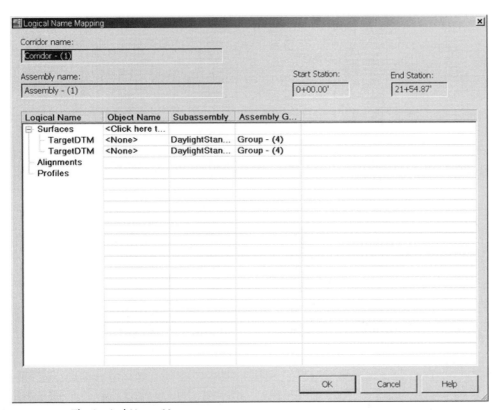

Figure 7-10   The Logical Name Map

11. In the second column and in the row for **Surfaces** is a cell that says **<Click here to set all>**.

12. When you click on the cell, a dialog box called **Pick a Surface** displays. Choose the **Existing Ground** and click **OK**. You see the **Target DTM** is now **Existing Ground**. Click **OK** and the corridor generates.
13. Close the **Panorama** if it appears and inspect the results of the processing. A Corridor model is evident in the drawing.

Let us learn how to control the appearance of the corridor data.

14. Click and select the roadway corridor. Then right click and choose **Corridor Properties**... shown in Figure 7-11.

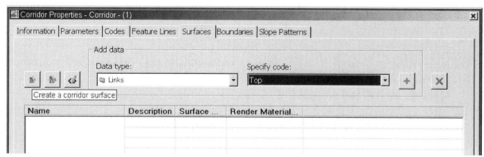

Figure 7-11   Corridor Properties

15. In the **Surfaces** tab, click the button on the left to **Create a corridor surface**.
16. After you choose this button, a **Corridor (1) Surface (1)** is created and appears in the window below.
17. In the **Data type** pop-down window, select **Links** if it is not already selected.
18. In the **Specify codes** pop-down window, choose **Top**. Then hit the + button to the right to add it to the matrix. The **Top** surface then appears under the **Name** column.
19. In the **Surface Style** column, click on the style that defaults, typically **Borders & Contours**.
20. A **Select Surface Style** dialog box displays with a pop-down menu. Select **Borders & Contours**. Click on the pop-down arrow to the right of **Borders & Contours** and go to the **Contours** tab. Set the **Contour interval** to **2** and **10** for the intermediate and index contours, respectively.
21. Now choose the **Boundaries** tab at the top.
22. Right click on **Corridor (1) Surface (1)**. Choose **Add Automatically**....
23. Select **Daylight** from the menu that displays.
24. The **Name** column now shows a boundary item. Hit **OK**.
25. You should see contours for the corridor in AutoCAD when it is finished processing.

This version of AutoCAD automatically creates a surface object upon creating a Corridor surface. You can see it in the prospector under surfaces.

Let us continue controlling the appearance of the corridor data and visualize it with faces and shading. To set the data up so they can be viewed in 3D, perform the following exercise.

Exercise 7-5: Set Up Corridor for 3D Shading

1. Click and select the roadway corridor. Then right click and choose **Corridor Properties**....
2. In the **Surface Style** column, click on the style that defaults, typically **Borders & Contours**.
3. A **Surface Style** dialog box displays with a pop-down menu. Select **Borders & Triangles and Directions**. Hit **OK**.

4. Use AutoCAD's **3DOrbit** command to view the corridor in a bird's eye perspective to help you see the road in 3D. While in the **3DOrbit** command, right click and choose **Visual Styles** and select **Conceptual** shading to shade the view. You see the road is 3D, complete with representation of the gutters and sidewalks.

5. If you have already left the **3DOrbit** command and the view is not shaded, use the **View** pull-down menu, and select **Visual Styles** and **Conceptual** shading.

6. Type **PLAN <Enter> <Enter>** to return to the plan view when you are finished viewing.

7. Save and close the file.

SUPERELEVATING A ROADWAY

The next step is to walk through a procedure for superelevating a roadway. Civil 3D has some excellent and very easy-to-use tools for computing superelevations. You will open another file and create a different template, which is for a more sophisticated highway condition. The specifications for this roadway are as follows: 12′ pavement lane, a curb and gutter, a sidewalk with a slope to drain into the roadway, and a multiple side slope configuration for tying out to daylight. These are multiple subgrade surfaces on this template for a base and subbase situation. You use AASHTO's superelevation criteria, and should the roadway design violate these criteria, the software provides a warning.

Exercise 7-6: Create an Assembly for Superelevations

1. Open the drawing called Chapter-7c.dwg.
2. Restore a presaved view to place your view around a location where you build a new corridor assembly.
3. Type **V** for **View** and restore the view called **Assembly**.
4. Under the **Corridors** pull-down menu, choose **Create Assembly**.... When the **Create Assembly** dialog box appears, accept the defaults by hitting **OK**.
5. When the prompt asks us to: `Specify assembly baseline location:` choose a point in the middle of the screen. Notice that a marker appears for **Assembly—1**.

Student Files

Using the **Tool Palettes** for **Civil 3D—Imperial** (which you have used before), select a variety of subassemblies from which to create the overall assembly.

6. From the **Imperial Roadway** tab, select the icon for **LaneOutsideSuper**. Before placing it, note that the **Properties** dialog box opens, allowing us to make changes to the parameters. Make the following changes: Select **Right** because you will do the right side first. The width is **12′**. The **Pave1** depth is **0.167** for a 2″ asphalt top surface. The **Pave2** depth will be **0.0**. The **Base** depth will be **0.33′** and the **Subgrade** depth will be **1′**.
7. The command prompt requests us to: `Select a marker point within the assembly`. Pick a point on the right side of the circle on the Assembly marker. A pavement lane with subgrades appears.
8. Look back in the **Properties** dialog box and change the side to **Left**.
9. Pick a point on the left side of the circle of the assembly marker. A pavement lane with subgrades appears.
10. Now pick the **Imperial—Structures** tab in the **Tool Palettes**. Notice many structures for assembly creation in this area.
11. Choose the **UrbanCurbGutterGeneral** structure.
12. Set the settings for this dialog box as shown in Figure 7-12.

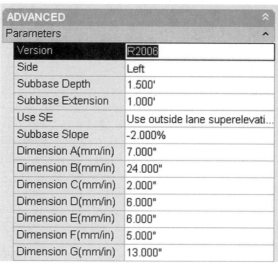

ADVANCED	⌃
Parameters	⌃
Version	R2006
Side	Left
Subbase Depth	1.500'
Subbase Extension	1.000'
Use SE	Use outside lane superelevati...
Subbase Slope	-2.000%
Dimension A(mm/in)	7.000"
Dimension B(mm/in)	24.000"
Dimension C(mm/in)	2.000"
Dimension D(mm/in)	6.000"
Dimension E(mm/in)	6.000"
Dimension F(mm/in)	5.000"
Dimension G(mm/in)	13.000"

Figure 7-12   UrbanCurbGutterGeneral settings

TIP   Note that if you click on any of the settings in the dialog box for **Dimension A—G**, the bottom of the dialog box provides an explanation of what these dimensions represent.

13. Do the left side of the template first.
14. Once these parameters are set, click a point on the upper left **Edge of Pavement** circle of the pavement.
15. Change the **Properties** dialog box to the **Right** side, and place the curb and gutter on the circle at the top right of the **Edge of Pavement**.
16. You now see curb and gutter on the left and right of the template.
17. Now, choose the **UrbanSidewalk** from the **Imperial—Structures** tab in the **Tool Palettes**.
18. Note back to the **Properties** dialog box that settings exist for the sidewalk now. Set the **Side** to **Right**. Set the **Inside boulevard width** to **2'**. Set the **Sidewalk width** to **4'**. Set the **Outside boulevard width** to **1'**. Set the **Slope** to **2%**. Set the **Depth** to **0.33'**.
19. Now pick a point on the circle at the upper right, top, back of curb. You see a sidewalk. Change the **Properties** setting for **Side** to **Left** and select a point on the circle at the upper left, top, **Back of Curb**. Sidewalks now exist on both sides of the template.
20. Now, choose the **DaylightStandard** from the **Imperial—Daylight** tab in the **Tool Palettes**. Set the settings as shown in Figure 7-13.
21. Then pick a point on the circle at the left edge of the sidewalk extension. A cut/fill subassembly appears.
22. Change the **Side** to **Right** in the **Properties** dialog box and place the subassembly on the right side as well.

This completes your assembly development. The assembly should appear as shown in Figure 7-14. Save your file.

Exercise 7-7: Establish Superelevation Criteria

You must now set the superelevation characteristics for the roadway. To do this you inspect the alignment properties. Let us run through this procedure and set alignment properties for superelevations.

1. Open the drawing called Chapter-7d.dwg.
2. Restore a presaved view to place your view around the centerline.

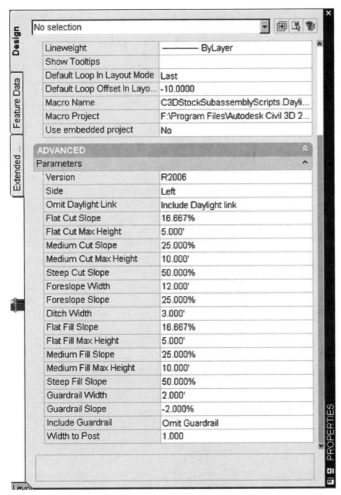

Figure 7-13   The DaylightStandard subassembly parameters

Figure 7-14   Completed assembly

3. Type **V** for **View** and restore the view called **Plan**.

4. Select the roadway and right click. Choose **Alignment Properties** from the menu.

5. Go to the **Design Speed** tab and click the button labeled **Add Design speed**....

6. Civil 3D places you in the file and wants to know what station to begin at. Type **0.0** and **<Enter>**.

7. When you are back in the dialog box, click in the **Design Speed** cell for that station number **1** and type **45** for the speed.

8. Then click on the **Superelevation** tab to the right. See Figure 7-15 and notice that there are three buttons on the upper left side of the box. Click on the third button, labeled **Set Superelevation Properties**.

9. The **Superelevation Specification** dialog box appears, as shown in Figure 7-16. Hit **OK**.

10. The superelevation rate table can be changed for the eMax rate, whether the roadway is two or four lane, whether the roadway is crowned or planar (such as in a divided highway). Take a moment to address the other parameters to see how they can be selected. Hit **OK**.

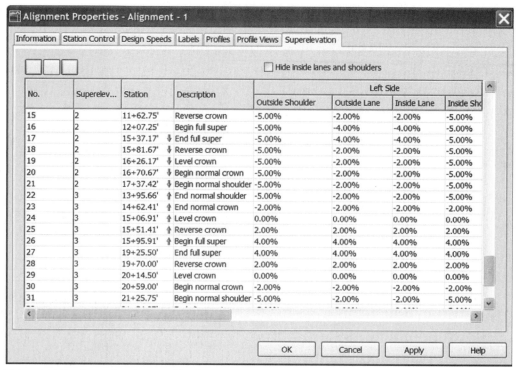

Figure 7-15   Superelevation specifications

11. Then a clean and well-laid-out dialog box appears showing parameters that dictate the superelevation of the roadway. See Figure 7-15 where each curve is broken into regions that can be addressed individually.
12. Any grayed-out items cannot be changed because they are set elsewhere, such as the design speed. Other parameters can be changed simply by selecting the item.
13. When finished inspecting, hit **OK**.

The computations occur for the superelevation data. Refer to Figure 7-16 showing the computations. All of the critical and most often needed stations are computed and shown in the table.

Civil 3D computes these data. If there are any errors or notifications that the user needs to be aware of, they appear in the **Panorama**. This might include warnings that curve radii are not up to par. Close the **Panorama** when you are through inspecting its contents.

Exercise 7-8: Build a Simple Corridor

1. Select **Create Simple Corridor** from the **Corridors** pull-down menu.
2. The **Create Corridor** dialog box appears and requests a name for the corridor. Hit **OK** for the default name and layer.
3. The prompt requests you to: Select a baseline alignment. **Click on the centerline for the road.**
4. Then it asks for you to select a profile. Zoom out and select the proposed profile in the profile view in the upper left part of the drawing.
5. It then asks you to select the **Assembly**. Pick the **Assembly** marker in the assembly you created.
6. The **Create Corridor** dialog box displays. Pick the button to **Set Logical Name Mapping** in the upper right corner of the dialog. The **Logical Name Mapping** dialog box shown in Figure 7-17 then appears so that Civil 3D can identify which surfaces are involved in the computations.

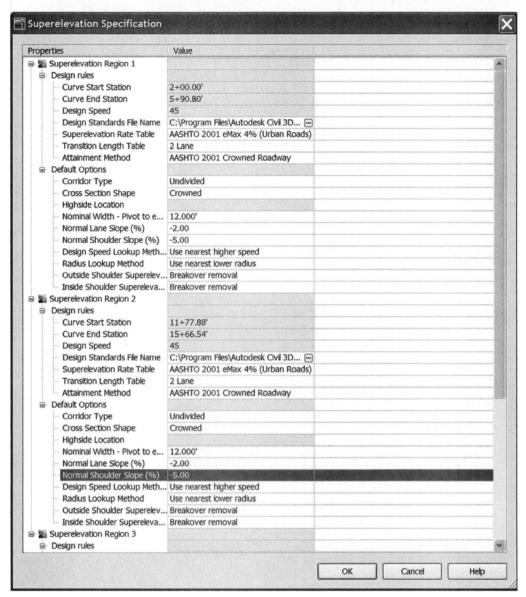

Figure 7-16   Superelevation computations

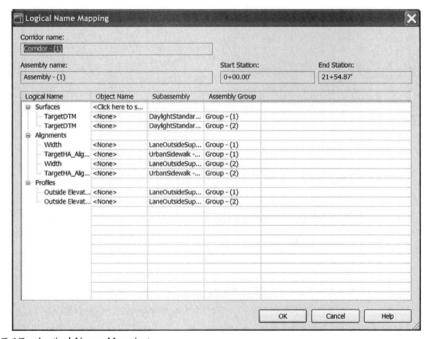

Figure 7-17   Logical Name Mapping

7. In the **Object Name** column, in the first row it says to **<Click here to set all>**. Click in that cell and choose the **Existing ground** surface. That will be the surface to which you tie out. Hit **OK** to return to the **Create Corridor** dialog box and hit **OK** to execute.

Exercise 7-9: Use View/Edit Corridor Sections

The corridor is now complete. Let us inspect it. If you zoom back to your centerline, it should appear as shown in Figure 7-18.

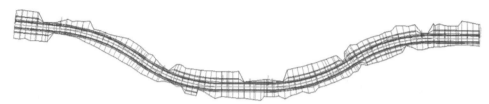

Figure 7-18   Completed Corridor

1. The daylight limits are shown around the perimeter of the site.
2. To inspect the cross section of the proposed road, choose **View/Edit Corridor Section** from the **Corridors** pull-down menu.
3. A table appears showing all of the pertinent data for the cross section at station 0+00. Refer to Figure 7-19.

View/Edit Corridor Section Tools - Corridor - (1)					
Alignment - 1 ⌄	0+00.00' ⌄				
Name	Design Value	Over...	Value	Comment	
⊟ 🏛 Assembly - (1)					
⊟ 🔧 Group - (1)					
⊟ 🔧 LaneOutsideSuper - (3)					
🔧 Side	Right		Right		
🔧 Base Depth	0.333'		0.333'		
🔧 Default Slope	-2.000%		-2.000%		
🔧 Lane Slope	-0.020		-0.020		
🔧 Pave1 Depth	0.167'		0.167'		
🔧 Pave2 Depth	0.000'		0.000'		
🔧 Sub-base Depth	1.000'		1.000'		
Section type: Begin alignment		Display Mode: All			

Figure 7-19   View/Edit Corridor Section tools

4. Take a moment to review this information. It should reflect a normal cross slope for the road.
5. A view of the road section also appears in AutoCAD. Zoom and pan freely to visually inspect it. See Figure 7-20.

Figure 7-20   A view of the road section

6. Use the drop-down menu for stations and select other stations, for instance, 2+50. Notice that the roadway is now superelevated.

7. A view of the superelevated roadway is also displayed in AutoCAD. (See Figure 7-21).

When through inspecting the sections, terminate the command by closing the toolbar.

Figure 7-21   A View of the superelevated roadway

CODES AND FEATURE LINES

Two additional items that should be mentioned with regard to corridor design are codes and feature lines.

Codes represent specific three-dimensional locations on the Corridor model that are identified with descriptions. They are defined with the subassemblies and are used to create and organize the assemblies. Codes can be exported for the corridor as point objects and uploaded to data collectors, or made available to 3D Machine Control for automated construction. Corridor section labeling is supported through the introduction of Code, Point, Link, and Shape label styles. Corridor assemblies and sections can be annotated with the new label styles.

Exercise 7-10: Review These Codes

1. In the same drawing, look in the **Prospector**. Expand **Corridors** and right click on **Corridor - (1)**. Choose **Properties** from the menu. When the dialog box appears, select the **Codes** tab. Refer to Figure 7-22.

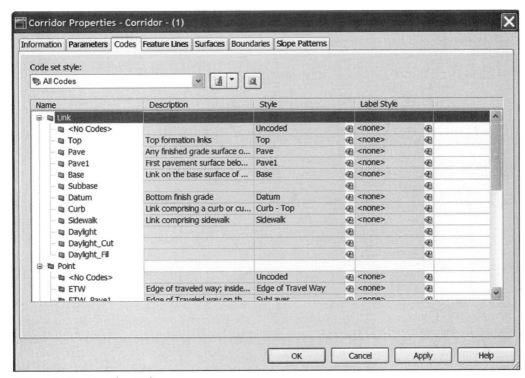

Figure 7-22   Corridor Codes

2. Change the **Code set style** to **All Codes** and click **OK**.
3. Look through the **Name** column and note that there are **Link** codes, **Point** codes, and **Shape** codes. Hit **OK** when finished observing.

A brief explanation of these items follows:

> **Link codes** are the codes assigned to each of the links that are part of the assembly definitions for the roadway. A link is a straight-line segment between endpoints on any cross section.

> **Shape codes** are closed cross sectional areas created by the subassembly. Shape codes can be used to visualize different materials within the assembly, in order to view the assembly more clearly.

Feature lines represent the linear elements of corridors such as centerlines, edge of pavement, ditch components, and daylight cut/fill lines. These can be used for visualizing the corridor but can also be exported as alignments or as grading feature lines.

Exercise 7-11: Change the Aesthetics of the Corridor

This next procedure walks you through how to assign a feature line style for an element of the corridor.

1. In the same drawing, look in the **Prospector**. Expand **Corridors** and right click on **Corridor - (1)**. Choose **Properties** from the menu. When the dialog box appears, select the **Feature Lines** tab.
2. Look through the **Code** column and note that there are descriptions for items such as **Top_Curb**, and so on.
3. Click on the **Top_Curb**, look in the **Feature Line Style** column, and you see it defaults to **Standard**. Click on the icon in the cell for that row. Refer to Figure 7-23.

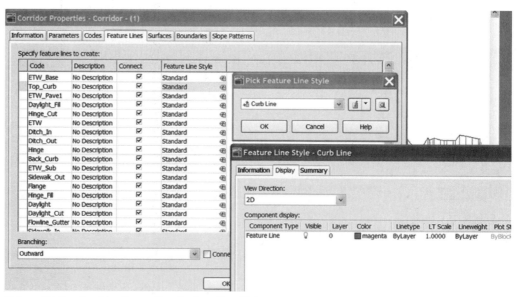

Figure 7-23   Corridor Properties, Feature Line Styles

4. A **Pick Feature Line Style** dialog box with a drop-down menu appears. Select the **Curb Line** style in the drop-down menu.
5. There is a button just to the right of the **Curb Line** item in the drop-down menu. Click on the triangle and select **Edit Current Style**. A **Feature Line Style – Curb Line** dialog box displays. Click on the **Display**

tab and note that you can control how this item appears in AutoCAD. In 2D it appears as a magenta color. Hit **OK** when finished observing. Hit **OK** and you are in the **Corridor Properties** dialog box again.

6. Click on **Daylight_Fill** and change the **Feature Line Style** for this to **Daylight Line – Fill**.
7. Click on **Daylight_Cut** and change the **Feature Line Style** for this to **Daylight Line – Cut**.
8. Hit **OK** to exit.
9. The corridor recomputes. When it is complete, zoom in and notice that the corridor shows a magenta line where the **Top of Curb** exists. Look just outside of the flowline for the curb.
10. Also observe that the **Daylight Lines** for **Cut** and **Fill** are red and green, respectively.

The next powerful feature is to export these items to AutoCAD for additional functionality. This linework can be sent to AutoCAD in 2D or 3D, which allows for customization.

11. Look in the **Corridors** pull-down menu and select the **Utilities** command.
12. You can export these objects as a Polylines, Feature Lines, Alignments, or Profiles. Select **Grading Feature Line From Corridor**.
13. You are asked to: `Select the corridor feature line.` Zoom in and select the magenta **Top of Curb** object.
14. You are asked again to: `Select the corridor feature line.` Zoom out and select the **Cut/Fill** linework. Hit <**Enter**> to complete.
15. When finished, notice that these objects are now in AutoCAD as separate entities and in three dimensions! Notice the grips in Figure 7-24.

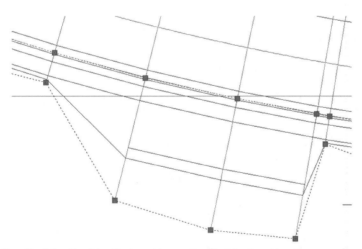

Figure 7-24   Result of the Corridor Feature Line as Grading Feature Line Command

One more concept to explore has to do with a major transportation corridor. This area of engineering warrants its own independent textbook; however, it needs to be touched on here.

The Civil 3D **Corridors** pull-down menu has another function for creating corridors called **Create Corridor….** You have used the **Create Simple Corridor** up to this point. The **Create Corridor…** command is for high-level roads that have multiple alignments, ramps, gores, multiple assemblies, and so on. It is in this function that multiple assemblies are assigned and act together in roadway design.

MULTIPLE BASELINES

The software makes more in-depth use of regions in this methodology. Although you have already seen regions used in the design of superelevations, these regions are used to expand that functionality.

Exercise 7-12: Multiple Baselines

Student Files

1. Open the file called Chapter-7-e.dwg. It has the same alignment and profile you have been working with but has two assemblies. One is a four lane, rural assembly with a shoulder, and the other is a two lane urban assembly with curb and gutter. A wide, white boundary is drawn for a surface to be created later. Type **V** for **View** and select **Assembly** and restore it. Zoom out a little and you see both assemblies.
2. Look in the **Corridors** pull-down menu and select the **Create Corridor...** command.
3. Select **Alignment—1** when asked by hitting **<Enter>** to access the library of alignments.
4. Select **Profile—Layout (5)** when asked by hitting **<Enter>** to access the library of profiles.
5. Select **Assembly—(1)** when asked by hitting **<Enter>** to access the library of assemblies.
6. A **Corridor Properties** dialog box appears, showing that the software already assumes a **Region 1**.
7. For **Baseline 1—Region 1**, change the ending station to **1000** for the use of **Assembly—(1)**.
8. Then right click on **Baseline 1** and a shortcut menu appears allowing you to **Insert a Region**.
9. First a **Select Assembly** dialog box appears, asking you to select the appropriate assembly for the second region. Select **Assembly—(2)**. See Figure 7-25.

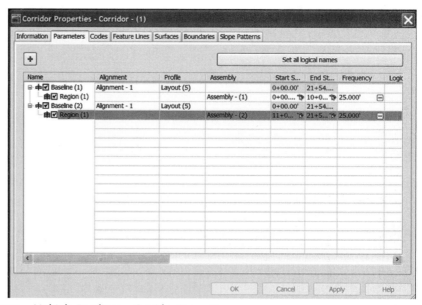

Figure 7-25   Multiple Baselines—Corridor Properties

10. Now change the starting station to **1000** for **Region 2**.
11. Click the button at the top right of the dialog box that says **Set all logical names** and select the **Existing ground**. Hit **OK**.

12. Hit **OK** and the system reprocesses the corridor.
13. When it is finished, click on the corridor and right click and select **Corridor Properties**.
14. In the **Surfaces** tab, click the button on the left to **Create a Corridor Surface**.
15. After you choose this button, a **Corridor—(1) Surface—(1)** is created and appears in the window below.
16. In the **Data Type** pop-down window, select **Links** if it is not already selected.
17. In the **Specify Code** pop-down window, choose **Top**. Then hit the **+** button to the right. The **Top** surface appears under the **Name** column.
18. In the **Surface Style** column, click on the style that defaults, typically **Borders & Contours**.
19. A **Surface Style** dialog box displays with a pop-down menu. Select **Borders & Contours** as the style. Click on the pop-down arrow to the right of **Borders & Contours**. Choose **Edit current style...** and go to the **Contours** tab. Set the **Contour interval** to **2** and **10** for the intermediate and index contours, respectively.
20. You should see contours for the corridor in AutoCAD when it is finished processing.
21. Zoom in and identify station 10+00 and notice that the software has built a corridor using the multiple assemblies. It has transitioned between the urban template and the rural template along the common edges, that is, the pavement. A 336′ contour runs through the transition between 10+00 and 10+25.

Note:
This exercise is no longer needed in the Civil 3D 2007 version; however, we will retain it for users of the 2006 version.

Exercise 7-13: Create a Corridor Surface

The next option to discuss is creating a corridor surface that will be evident in the **Surfaces** area of the **Prospector**.

1. In the same file, look under the **Corridors** pull-down menu. Select **Export > Corridor Surfaces**.
2. Pick your corridor when requested.
3. A **Create Surface** dialog box displays, confirming your surface naming and the styles to use. Use the defaults or change them here and hit **OK**.
4. Look under the **Prospector** and notice a new surface representing the corridor.
5. Expand **Corridor—(1), Surface—(1)**. Expand **Definitions**.
6. Right click on **Boundaries** and select **Add...**.
7. Give it a name of **Corridor Boundary** in the name field that displays in the **Add Boundaries** dialog box. Hit **OK**.
8. Pick the white polyline around the corridor as a boundary.
9. This can be used to perform surface-related functions such as pasting it into the Existing Ground to create a monolithic surface and the like.

Exercise 7-14: Earthworks on a Corridor Surface

Let us perform earthworks on the corridor surface and compare it to the existing ground surface.

1. From the **Surfaces** pull-down menu, choose **Create Surface...**.
2. In the **Create Surface** dialog box, in the upper left pop-down menu for **Type:** choose **TIN volume surface**, shown in Figure 7-26.
3. Under the Value column for the **Name** field, provide a name of **TIN Volume EG vs. Corridor**.
4. In the **Volume surfaces** field, there are fields for selecting the comparison surfaces.

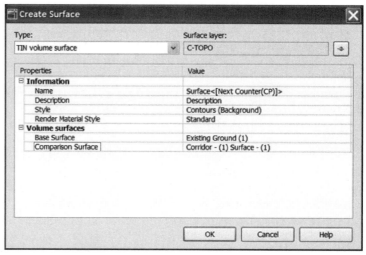

Figure 7-26   Create Surface dialog box

5. Select **Existing Ground** for the **Base Surface** and select **Corridor (1) – Surface (1)** for the **Comparison Surface**. Hit **OK**.
6. Look in the **Prospector** and you see a surface called **TIN Volume EG vs. Corridor**. Right click on it and select **Properties**....
7. Choose the **Statistics** tab in the dialog box for **Surface Properties** and expand the **Volume** item. Volumes are automatically computed for the two surfaces. See Figure 7-27.

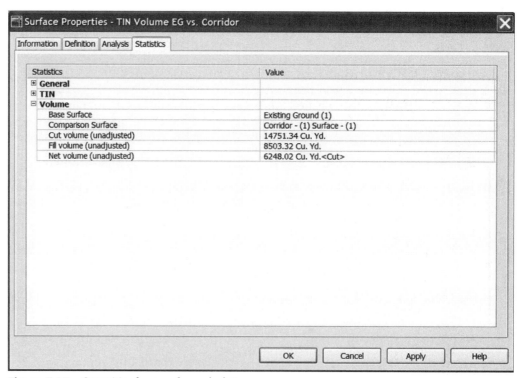

Figure 7-27   Quantities from surface calculations

8. If these two surfaces change, then the volumes recompute if the **Rebuild—Automatic** item is checked in the surface parameters.

Exercise 7-15: Corridor-Related Quantity Takeoffs

Now let us use the **Quantity Takeoff** commands under the **Sections** pull-down menu to compute some quantities from the corridor.

This task is often used due to the requirement to estimate the costs for providing certain quantities of material for a project. Contractors and engineers must be able to ascertain the cost of the project before it goes to construction.

1. Open Chapter 7-f.dwg; you will notice that we have already created the Sample Line Group as discussed in the earlier chapter on Sections in order to expedite this exercise.
2. Look under the **Sections** Pulldown and select **Define Materials**
3. A dialog will display called **Select a Sample Line Group**; select the SLG-1 group. The Setup Materials dialog then displays.
4. In the top left of the dialog is a pop-down window; select the option called Earthworks.
5. For the EG Criteria Object name select Existing groud (1).
6. For the Datum select Corridor (3) Corridor (3)—Surface (1).
7. Hit OK.
8. Look under the **Sections** pull-down and select **Generate Volume Report....**
9. The **Report Quantities** dialog box will display. Use Alignment - 1, SLG-1, Materials List - 1 and the Select Material Style sheet. See Figure 7-28. Hit **OK**.

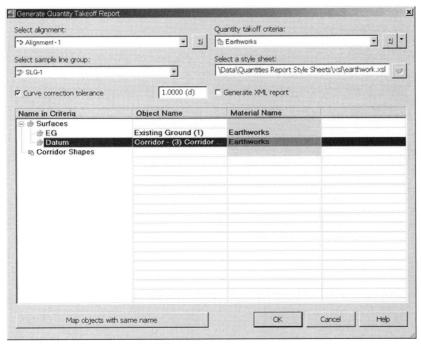

Figure 7-28   Generate Quantity Takeoff Report

10. For the Existing Ground and under Object Name select the Existing ground surface.
11. For the **Datum** and under **Object Name** select the **Corridor** surface.
12. The Quantity takeoff criteria we will use this time is the **Earthworks**, although another options exists for Materials if you need that later. The style sheet is **Earthworks** as well for our example. Hit **OK**.
13. A volume report is generated and should appear similar to that in Table 7-1.

Project: chapter 7-f.dwg   Alignment: Alignment—1   Sample Line Group: SLG-1
Start Sta: 0+00.000   End Sta: 21+54.874

Station	Cut Area (sq. ft.)	Cut Volume (cu. yd.)	Re-usable Volume (cu. yd.)	Fill Area (sq. ft.)	Fill Volume (cu. yd.)	Cum. Cut Vol. (cu. yd.)	Cum. Reusable Vol. (cu. yd.)	Cum. Fill Vol. (cu. yd.)	Cum. Net Vol. (cu. yd.)
0+00.000	230.93	0.00	0.00	12.77	0.00	0.00	0.00	0.00	0.00
0+50.000	249.93	445.24	445.24	19.05	29.47	445.24	445.24	29.47	415.77
1+00.000	237.50	451.33	451.33	24.24	40.08	896.57	896.57	69.55	827.02
1+50.000	149.15	358.01	358.01	29.75	49.99	1254.58	1254.58	119.53	1135.05
2+00.000	23.37	159.74	159.74	62.78	85.67	1414.32	1414.32	205.20	1209.11
2+25.000	12.87	15.88	15.88	68.78	61.40	1430.20	1430.20	266.61	1163.59
2+50.000	5.73	8.17	8.17	93.73	75.71	1438.37	1438.37	342.32	1096.05
2+75.000	4.94	4.69	4.69	88.94	85.13	1443.06	1443.06	427.45	1015.61
3+00.000	4.28	4.03	4.03	90.46	83.70	1447.09	1447.09	511.15	935.94
3+25.000	0.26	1.99	1.99	93.74	85.58	1449.08	1449.08	596.73	852.36
3+50.000	0.88	0.51	0.51	115.93	97.19	1449.59	1449.59	693.92	755.67
3+75.000	0.34	0.53	0.53	164.63	129.83	1450.12	1450.12	823.74	626.38
4+00.000	0.00	0.15	0.15	226.89	181.29	1450.27	1450.27	1005.03	445.24
4+25.000	0.00	0.00	0.00	313.84	249.79	1450.27	1450.27	1254.82	195.45
4+50.000	0.00	0.00	0.00	392.70	325.78	1450.27	1450.27	1580.60	−130.33
4+75.000	0.00	0.00	0.00	402.37	368.14	1450.27	1450.27	1948.74	−498.47
5+00.000	0.02	0.01	0.01	446.26	394.79	1450.28	1450.28	2343.53	−893.25
5+25.000	0.00	0.01	0.01	424.96	404.87	1450.29	1450.29	2748.39	−1298.10
5+50.000	0.00	0.00	0.00	372.11	369.68	1450.29	1450.29	3118.07	−1667.78
5+75.000	0.25	0.12	0.12	352.70	336.44	1450.42	1450.42	3454.51	−2004.10
5+90.799	0.01	0.08	0.08	327.22	199.01	1450.50	1450.50	3653.52	−2203.02
6+00.000	0.00	0.00	0.00	333.64	112.50	1450.50	1450.50	3766.02	−2315.52
6+25.000	0.00	0.00	0.00	256.94	272.76	1450.50	1450.50	4038.78	−2588.28
6+50.000	2.44	1.05	1.05	191.42	207.30	1451.55	1451.55	4246.08	−2794.53
6+75.000	7.60	4.33	4.33	136.95	152.02	1455.88	1455.88	4398.10	−2942.21
7+00.000	0.30	3.41	3.41	69.75	96.36	1459.30	1459.30	4494.46	−3035.16
7+25.000	9.11	4.55	4.55	14.39	39.40	1463.84	1463.84	4533.86	−3070.02
7+50.000	105.75	52.55	52.55	2.55	7.93	1516.39	1516.39	4541.79	−3025.39
7+75.000	119.49	102.62	102.62	1.34	1.86	1619.01	1619.01	4543.65	−2924.64
8+00.000	126.91	112.51	112.51	0.16	0.72	1731.52	1731.52	4544.37	−2812.85
8+25.000	205.70	152.86	152.86	0.00	0.08	1884.37	1884.37	4544.44	−2660.07
8+50.000	251.87	211.86	211.86	0.00	0.00	2096.24	2096.24	4544.44	−2448.21
8+75.000	212.69	215.08	215.08	0.00	0.00	2311.32	2311.32	4544.45	−2233.13
9+00.000	161.94	171.23	171.23	4.43	2.11	2482.55	2482.55	4546.56	−2064.01
9+25.000	141.41	136.40	136.40	28.53	15.66	2618.95	2618.95	4562.22	−1943.27
9+50.000	126.50	119.50	119.50	69.13	46.48	2738.45	2738.45	4608.70	−1870.26
9+75.000	114.29	107.19	107.19	81.12	71.59	2845.64	2845.64	4680.29	−1834.65
9+77.881	113.05	11.67	11.67	82.87	9.03	2857.31	2857.31	4689.32	−1832.01
10+00.000	105.83	89.66	89.66	89.62	70.65	2946.97	2946.97	4759.97	−1813.01
10+50.000	111.45	201.19	201.19	146.94	219.04	3148.15	3148.15	4979.01	−1830.86
11+00.000	80.40	177.64	177.64	218.61	338.47	3325.79	3325.79	5317.48	−1991.69

Station	Cut Area (sq. ft.)	Cut Volume (cu. yd.)	Re-usable Volume (cu. yd.)	Fill Area (sq. ft.)	Fill Volume (cu. yd.)	Cum. Cut Vol. (cu. yd.)	Cum. Reusable Vol. (cu. yd.)	Cum. Fill Vol. (cu. yd.)	Cum. Net Vol. (cu. yd.)
11+50.000	0.00	74.44	74.44	337.46	514.87	3400.23	3400.23	5832.36	−2432.13
11+77.881	0.03	0.01	0.01	411.42	386.66	3400.25	3400.25	6219.02	−2818.77
12+00.000	3.63	1.41	1.41	349.49	315.25	3401.65	3401.65	6534.27	−3132.62
12+25.000	68.19	31.24	31.24	297.95	302.94	3432.90	3432.90	6837.20	−3404.31
12+50.000	93.19	70.44	70.44	199.92	234.39	3503.34	3503.34	7071.59	−3568.25
12+75.000	124.56	95.33	95.33	103.30	144.29	3598.67	3598.67	7215.88	−3617.21
13+00.000	161.55	125.55	125.55	71.53	83.39	3724.22	3724.22	7299.27	−3575.05
13+25.000	165.57	144.57	144.57	42.22	54.33	3868.79	3868.79	7353.60	−3484.81
13+50.000	170.90	150.11	150.11	32.50	35.93	4018.89	4018.89	7389.53	−3370.63
13+75.000	164.95	150.21	150.21	39.50	34.70	4169.11	4169.11	7424.23	−3255.12
14+00.000	154.41	141.83	141.83	54.28	44.93	4310.93	4310.93	7469.16	−3158.22
14+25.000	146.96	132.16	132.16	64.48	56.67	4443.10	4443.10	7525.83	−3082.74
14+50.000	137.18	124.11	124.11	71.30	64.89	4567.20	4567.20	7590.72	−3023.52
14+75.000	131.24	117.92	117.92	77.86	71.27	4685.12	4685.12	7662.00	−2976.87
15+00.000	113.28	107.56	107.56	67.47	69.18	4792.69	4792.69	7731.18	−2938.49
15+25.000	85.00	87.11	87.11	149.85	103.33	4879.80	4879.80	7834.51	−2954.71
15+50.000	55.72	61.52	61.52	165.29	148.59	4941.31	4941.31	7983.09	−3041.78
15+66.536	83.55	40.22	40.22	91.67	79.73	4981.53	4981.53	8062.82	−3081.29
15+66.537	83.55	0.00	0.00	91.67	0.00	4981.53	4981.53	8062.82	−3081.29
15+75.000	103.49	30.84	30.84	64.06	23.89	5012.37	5012.37	8086.71	−3074.35
16+00.000	211.75	151.20	151.20	2.13	29.93	5163.56	5163.56	8116.64	−2953.08
16+25.000	390.31	281.89	281.89	0.00	0.94	5445.45	5445.45	8117.58	−2672.13
16+50.000	516.35	420.90	420.90	0.00	0.00	5866.36	5866.36	8117.58	−2251.23
16+75.000	630.68	530.89	530.89	0.00	0.00	6397.25	6397.25	8117.58	−1720.33
17+00.000	621.03	580.18	580.18	0.00	0.00	6977.42	6977.42	8117.58	−1140.16
17+25.000	588.40	563.78	563.78	0.00	0.00	7541.21	7541.21	8117.58	−576.38
17+50.000	533.25	524.78	524.78	0.00	0.00	8065.99	8065.99	8117.58	−51.59
17+75.000	470.20	469.74	469.74	0.00	0.00	8535.73	8535.73	8117.58	418.15
18+00.000	411.09	411.08	411.08	0.00	0.00	8946.81	8946.81	8117.58	829.22
18+25.000	317.45	341.69	341.69	0.34	0.15	9288.50	9288.50	8117.73	1170.76
18+50.000	263.18	275.74	275.74	18.36	8.29	9564.23	9564.23	8126.02	1438.21
18+75.000	204.76	223.36	223.36	49.05	29.65	9787.59	9787.59	8155.67	1631.92
19+00.000	160.84	174.42	174.42	45.20	41.30	9962.01	9962.01	8196.97	1765.04
19+25.000	153.34	149.21	149.21	34.24	34.92	10111.22	10111.22	8231.90	1879.32
19+50.000	162.22	149.78	149.78	25.34	26.34	10261.00	10261.00	8258.24	2002.76
19+54.873	163.99	30.23	30.23	22.97	4.16	10291.23	10291.23	8262.40	2028.83
20+00.000	133.33	248.46	248.46	22.59	38.08	10539.70	10539.70	8300.48	2239.22
20+50.000	275.76	378.79	378.79	10.66	30.79	10918.49	10918.49	8331.26	2587.22
21+00.000	218.93	458.05	458.05	8.29	17.55	11376.54	11376.54	8348.81	3027.73
21+50.000	153.64	344.97	344.97	19.36	25.61	11721.51	11721.51	8374.42	3347.09
21+54.874	0.00	13.87	13.87	0.00	1.75	11735.37	11735.37	8376.16	3359.21

Table 7-1 Volume Report

In computing earthworks, a section-based method is used where by the amount of cut and fill is computed between each section of the corridor. The volume of cut between station 0+00 and station 0+50 is 445.24 c.y. and the volume of fill between those same stations is 29.47 c.y. The difference between those two figures is 415.77 c.y., which is referred to as a Net Volume of 415.77 c.y. cut. Notice that the entire amount of cut and fill is assumed to be reusable, such that the earth cut from one area of the project can be used to fill another area.

The entire project's volumes are shown in the last row of Table 7-1. In the last row, the fourth column from the right shows the accumulated amount of cut removed from the existing ground. It is 11,735.37 c.y. of earth. The second column from the right shows the accumulated amount of fill needed to achieve the proposed conditions above existing ground. It is 8376.16 c.y. of fill. Therefore, the entire project must dispose of 3,347.09 c.y. of the earth that was cut; hence there is a net volume of 3,347.09 c.y. of cut. Typically it must be trucked off the site because it is excess earth. Clearly this is not a balanced site or all of the earth would have been used somewhere on the site.

ROADWAY DESIGN FOR REHABILITATION AND RECONSTRUCTION

The next exercise delves into roadway rehabilitation and reconstruction. This part of the software is new to most Autodesk users because the Land Desktop software could not perform this function as elegantly as Civil 3D can. The idea is that not all roads are new roads; in fact far more roads are maintained each year than built new. Civil 3D provides tools to assist engineers responsible for roadway maintenance, and these include subassemblies for asphalt overlays, median reconstruction, sidewalk rehabilitation, milling and overlaying, and curb and gutter replacements. There are parametric values that can even compute how much of a stripped material can be reused for the new overlay.

Exercise 7-16: Roadway Design for Rehabilitation and Reconstruction

1. Open Chapter 7-g.dwg; notice that you have already created an alignment and an existing ground profile. A new assembly has been created called **Assembly – Rehab**. This mill and overlay 2 subassembly was used to place a crowned surface over another, worn, existing crowned surface.
2. The method to perform these computations is to use the alignment in the drawing, which has been prepared to coincide roughly with the asphalt crown of the existing road. The profile used is the existing ground profile for the existing roadway because it already has reasonable grades on it. Then use the new assembly for rehabilitation of the new overlayed surface.
3. Select **Create Corridor** from the **Corridors** pull-down menu. Select the only **Alignment** available. Then select the existing ground profile for our proposed profile and select the **Assembly – Rehab** assembly for the new surface.
4. When the dialog box displays, choose the button for mapping logical names and then click in the cell for **<Click here to set all>**. Choose the **Existing Ground** surface and then select the **Assembly – Rehab** when prompted. Hit **OK**. Hit **OK** to compute.
5. Then zoom into the plan view of the road and notice a corridor exists.
6. From the **Corridors** pull-down menu, select **View/Edit Sections** and browse to stations 0+00, 4+00, and 17+00 and inspect the sections. Notice that milling and overlaying has occurred per your specifications. Refer to Figure 7-29, Figure 7-30, and Figure 7-31.

Student Files

Figure 7-29   Milling and Overlaying Result 1

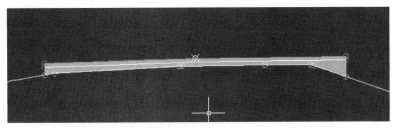

Figure 7-30   Milling and Overlaying Result 2

Figure 7-31   Milling and Overlaying Result 3

Exercise 7-17: Designing a Bifurcated Highway

This last roadway-related exercise explores a situation in which there are two baselines. This can occur in many projects, such as a mainlane with an offramp or a mainlane with a frontage or a potentially bifurcated highway.

This next design has an eastbound and westbound set of mainlanes. Although the alignments are parallel, the profiles are not the same. This is a *bifurcated highway*. This exercise uses multiple baselines to show how a DOT-level roadway might be designed.

Bifurcated highway—Independent PGLs

Student Files

1. Open Chapter 7-h.dwg; notice that you have already created two alignments and two profiles. Two new assemblies have been created called **Assembly – Eastbound** and **Assembly – Westbound**. There is also a third profile called **Profile – Middle** to guide the center of the median between the two mainlanes.

2. From the **Corridors** pull-down menu, select the **Create Corridor** command. You are asked to select an alignment, so pick the **Alignment – Eastbound**, either from the screen or by hitting <Enter> and choosing it from the library. Then select the profile called **Layout–Rt side** from the library as well. Last, select the **Assembly – Eastbound**.

3. This displays the **Create Corridor** dialog box. Click the **blue plus** sign in the top left of the dialog box to **Add a baseline** to your computations. Choose **Alignment – Westbound** when prompted.

4. Then right click on the **Baseline 2** and select **Add Region**.... Select the **Assembly – Westbound** when prompted. Set the **Profile for Baseline 2** to **Layout – Left**.

5. Click the button for **Set All Logical Names**. See Figure 7-32.
6. Click the cell where it says to <**Click here to set all**> and select the **Existing Ground (1)** surface for any tie-outs.

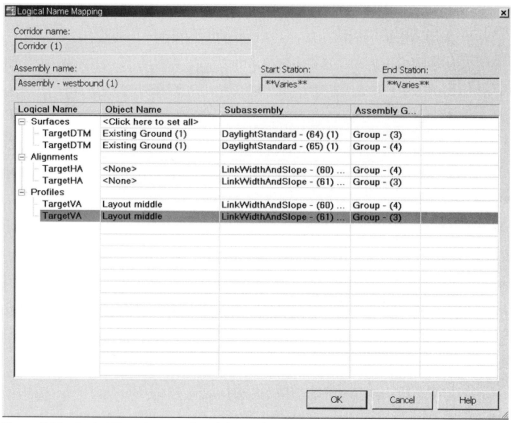

Logical Name	Object Name	Subassembly	Assembly G...
⊟ Surfaces	<Click here to set all>		
TargetDTM	Existing Ground (1)	DaylightStandard - (64) (1)	Group - (3)
TargetDTM	Existing Ground (1)	DaylightStandard - (65) (1)	Group - (4)
⊟ Alignments			
TargetHA	<None>	LinkWidthAndSlope - (60) ...	Group - (4)
TargetHA	<None>	LinkWidthAndSlope - (61) ...	Group - (3)
⊟ Profiles			
TargetVA	Layout middle	LinkWidthAndSlope - (60) ...	Group - (4)
TargetVA	Layout middle	LinkWidthAndSlope - (61) ...	Group - (3)

Corridor name:
Corridor (1)

Assembly name:
Assembly - westbound (1)

Start Station:
Varies

End Station:
Varies

Figure 7-32   Logical Name Mapping for a bifurcated roadway

7. Then under **Profiles, Object Name**, select **Layout middle** for both of the **Target Profile** items. This guides the center of the median between the two baselines. Click **OK**. Click **OK** to compute.
8. When the processing is completed, select **View/Edit Corridor** sections from the **Corridors** pull-down menu. Notice how the center of the median now follows the **Layout middle** profile established. Refer to Figure 7-33.

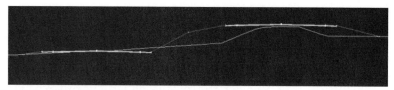

Figure 7-33   Cross section of resultant bifurcated roadway

9. Then feel free to establish the surface and view it using the **3DOrbit** command. You can see the two baselines interconnected, each with its own alignment and profile activity. Refer to Figure 7-34.

Figure 7-34   Isometric view of bifurcated roadway

At this point you have developed basic roads as well as a more sophisticated roadway with subgrades and superelevations. There are many more structures for building assemblies within the software, and now you have a strong foundation to explore them.

Begin to combine the concepts you have learned. For instance, a powerful feature of the software is that after a corridor is developed, very often changes occur to the design. You have already seen the ripple-through effect that occurs when modifications are made to the various object components. If the existing ground surface is modified, that change ripples through the corridor and redevelops the corridor. If the profile is changed, it ripples through the design and affects the sections. Take some time to try out these features similar to what has been accomplished in previous chapters.

CHAPTER TEST QUESTIONS

Multiple Choice

1. According to the laws of mechanics, when a vehicle travels on a horizontal curve:

    a. It is forced outward by centrifugal force
    b. It is forced inward by centrifugal force
    c. It is unaffected
    d. None of the above

2. Engineers use which method(s) to counteract centrifugal forces that are trying to send the car off of the road on curves?

    a. Superelevation where the pavement lanes are tilted
    b. Vertical curves where the pavement is curved in crests and sags
    c. Tapers and widenings of pavement
    d. None of the above

3. The rate of superelevation varies:

    a. Based on the vertical curves in the profile
    b. Based on the radius of the horizontal curve
    c. Based on the tangent lengths of the alignment
    d. Based on the centrifugal forces involved

4. Variables and terms for superelevations include:

    a. e, e_{max}, R, V
    b. R, K, L

    c. PVC, PRC, PI
    d. All of the above

5. A Maximum superelevation (SE) is required:

    a. When the radius (R) is a minimum
    b. When the radius (R) is a maximum
    c. In all cases
    d. Never

6. At a reverse curve, the roadway section is often:

    a. At e_{max} because the roadways are fluctuating the most at that point
    b. Flat at the PRC as the superelevation in one direction transitions to the other direction
    c. In a tangent; therefore, there is no need to address it
    d. None of the above

7. The Tangent runout defines:

    a. The length of highway needed to bring a normal crown section to a section where the outside lane has a zero percent slope instead of normal crown
    b. The distance needed to bring a section from flat to fully superelevated, or vice versa
    c. The length of vertical curve needed to negotiate the arc
    d. All of the above

8. Using the Virginia DOT superelevation chart in Figure 7-6, what superelevation rate would be appropriate for a 2500′ radius curve and a design speed of 60 mph?
   a. 5%
   b. 6%
   c. 6.5%
   d. 5.5%

9. Civil 3D provides subassemblies for which of the following?
   a. Tunnels
   b. Lanes, curbs, guardrails, and shoulders

c. Bridge trusses and cable stayed bridges
d. All of the above

10. Corridors can be checked for quality assurance using the following methods:
   a. 3DOrbit
   b. View/Edit Sections
   c. Surface Styles
   d. All of the above

True or False

1. True or False: Horizontal alignments, profiles, and subassemblies are used to build corridors.

2. True or False: Assemblies are created by drawing polylines to represent the proposed roadway conditions.

3. True or False: Centrifugal forces act on the vehicle as it negotiates through horizontal curves.

4. True or False: Superelevated roadway sections are always used when the roadway goes through turns.

5. True or False: Superelevations are used to counter the centrifugal forces acting on vehicles as they go in and out of curves on roadways.

6. True or False: The *AASHTO Green Book* is a good reference for highway design.

7. True or False: e_{max} is the maximum a road can be superelevated to before the car will leave the pavement on a arc.

8. True or False: Two components used in superelevating a roadway are runout and runoff.

9. True or False: Subassemblies are the primitive components of assemblies, and they instruct the software in how to handle lanes, curbs, guardrails or shoulders, related side sloping and tie-out.

10. True or False: Civil 3D provides many subassemblies to begin with and the user accesses them from **Tool Palettes**.

STUDENT EXERCISES, ACTIVITIES, AND REVIEW QUESTIONS

1. Describe the process that a superelevated roadway goes through as it moves from normal crown to a fully superelevated section.

2. If a roadway has a −2.08% cross slope, how many feet will it take to superelevate to flat or 0% if it has superelevation rate of 6%?

3. If a roadway has a −2.08% cross slope, how many feet will it take to superelevate to a fully superelevated roadway if it has superelevation rate of 5%?

4. If a roadway is fully superelevated to 4% and needs to go back to a normal crown of −2.08%, how many feet will it take to reduce its superelevation to a crowned section roadway if it has superelevation rate of 4%?

5. Using the Virginia DOT superelevation chart shown in Figure 7-6, what superelevation rate would be needed to counteract a curve radius of 5000′ and a speed of 65 mph?

6. Using the Virginia DOT superelevation chart shown in Figure 7-6, what curve radius would be appropriate for a superelevation rate 3.0% and a design speed of 40 mph?

7. Using the Virginia DOT superelevation chart shown in Figure 7-6, what superelevation rate would be appropriate for a 3500′ radius curve and a design speed of 55 mph?

8. Using the Virginia DOT superelevation chart shown in Figure 7-6, what design speed would be allowable for a superelevation rate of 5% and a curve radius of 2800′?

view of a house template. With the road and houses laid out in 3D, it was fairly easy to design. See Figure 8-9.

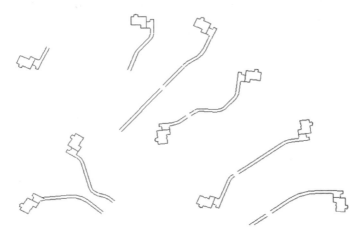

Figure 8-9   Subdivision driveways

Figure 8-10 shows a section of the design completed. The houses float in 3D and the point data represent the tie-out to the existing ground. It should be noted here that the tie-out can be seen immediately when performed. The benefit here is that if the tie-out projects into the adjacent property, the designer can adjust the side slopes at that point to rectify it.

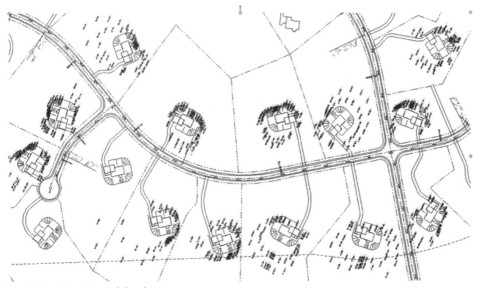

Figure 8-10   A section of the design

The graphics in Figures 8-11 through 8-15 show other sites that are readily developed as 3D models. This methodology plays very strongly into the rapidly growing 3D/GPS Machine Control market. This is a GPS based technology by which the earthmoving equipment is located on the existing ground based on GPS. An onboard computer stores the Digital Terrain Model for the proposed surface(s), which may include subgrade surfaces. Robotics handle the hydraulic movement of the earthmoving blade. This technology requires an accurate 3D model from which to operate. If engineering firms do not learn to work in 3D, they will lose more and more work to those that do work in 3D. Civil 3D provides add-on software that can convert the Civil 3D data into a formatted data conducive for use by Topcon, Trimble, and Leica equipment. They are Carlson Connect, which addresses Topcon; Trimble Link from Trimble; and Leica X-Change.

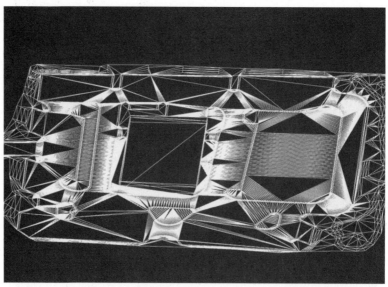

Figure 8-11   A TIN model of a commercial site

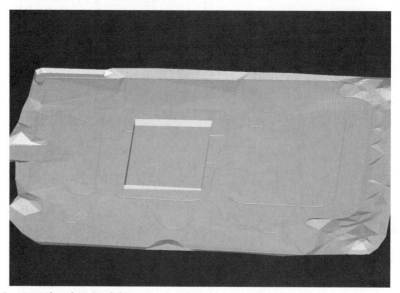

Figure 8-12   A rendered view of the commercial site

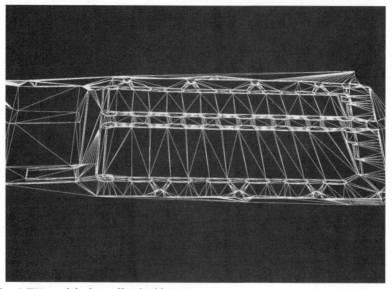

Figure 8-13   A TIN model of an office building site

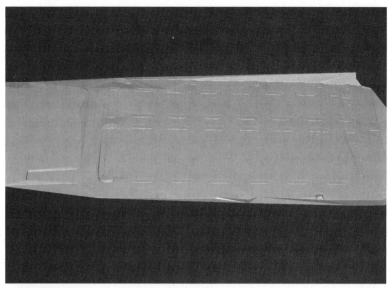

Figure 8-14   A rendered view of an office building site

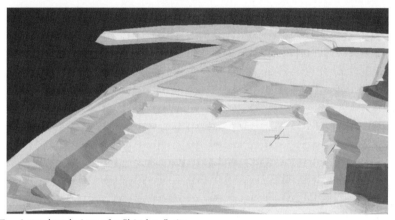

Figure 8-15   A rendered view of a "big box" site

The concept of engineers designing their sites in three dimensions is called Virtual Site Design, which is designing a site in the computer using slopes, grades, and elevations. The contouring, labeling, and analyses should be computer generated from that point on. With the site in 3D, some very visual viewpoints can be established to verify quality control, and all editing of the surface would be done to the 3D elements, which consist of breaklines, points, and grading objects.

Once a designer has a surface, one of the quality assurance functions to be performed is earthworks. You need to know how much earth you are moving or cutting or filling. There are several methods for earthworks computations, some of which are explored next.

EARTHWORKS ALGORITHMS

In order to discuss the validity of results from an earthworks computation, one should know which algorithms were used in the processing because the results can vary based on the method selected. Some of the common methods are:

- **Section method**
- **Grid method**
- **Composite method**

Each of these methods has positive and negative aspects. Please review the next set of graphics. The Section method entails computing a cross section through the existing ground and the proposed surface. The area of cut is computed as well as the area of fill for that section. Then, at some interval away, another section is computed and the area of cut and fill in that section is computed. Then the two areas of cut are averaged together to obtain an average end area of cut. Similarly, the two areas of fill are averaged together to obtain an average end area of fill. These numbers yield a square footage of material. Then the average cut is multiplied over the interval distance yielding the cubic footage of material between the two sections. Divide that by a factor of 27 to obtain the number of cubic yards. Repeat this for each section in the project.

TIP   Divide by 27 to obtain cubic yards from cubic feet because 1 cubic yard is 3′ × 3′ × 3′.

Example 8-1: The Section Method Done by Hand

In this example problem computed by hand, refer to Figure 8-16. Here is a simple situation in which you have a road section with two 12′ lanes and 6′ of 3:1 embankment.

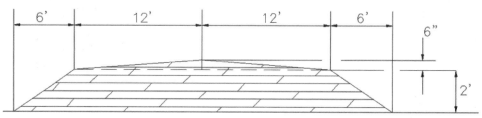

Figure 8-16   A roadway section in fill

In order to compute this earthwork, you do not need to compute the amount of cut because there is none; this entire section is in fill. Therefore, the fill area is computed using a series of triangle and rectangular area formulas as follows:

Fill in embankment:

(6′ × 2′)/2 = 6 s.f.; there are two embankment areas, so the fill in the embankment is 12 s.f.

Fill under the pavement is computed using a rectangular area formula for the 2′ of fill and then a triangular area formula for the fill above the 2′ but under the grade of the road.

12′ × 2′ = 24 s.f.; there are two lanes, so 24 s.f. × 2 = 48 s.f.
Then: 12′ × .5′ = 6 s.f.; there are two lanes, so 6 s.f. × 2 = 12 s.f.
Summing, we have 48 s.f. + 12 s.f. + 12 s.f. = 72 s.f. of material.

For the second section, which was cut 50′ away from the first, please refer to Figure 8-17. This section has both cut and fill. Let us compute the fill first.

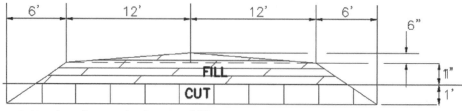

Figure 8-17   A roadway section in cut and fill

Fill in embankment:

$(3' \times 1')/2 = 1.5$ s.f.; there are two embankment areas, so the fill in the embankment is 3 s.f.

Fill under the pavement is computed using a rectangular area formula for the 1' of fill and then a triangular area formula for the fill above the 1' but under the grade of the road.

$12' \times 1' = 12$ s.f.; there are two lanes, so 12 s.f. $\times$ 2 = 24 s.f.
Then: $12' \times .5' = 6$ s.f.; there are two lanes, so 6 s.f. $\times$ 2 = 12 s.f.
Summing, we have 24 s.f. + 12 s.f. + 3 s.f. = 39 s.f. of material.

Now compute the amount of cut.
Cut in embankment:

$(3' \times 1')/2 = 1.5$ s.f.; there are two embankment areas, so the fill in the embankment is 3 s.f

Cut under the pavement is computed using a rectangular area formula as well.

$12' \times 1' = 12$ s.f.; there are two lanes, so 12 s.f. $\times$ 2 = 24 s.f.
Summing, we have 24 s.f. + 3 s.f. = 27 s.f. of cut for the second section.
Computing the average fill between the two sections we have
39 s.f. + 72 s.f. = 111/2 = 55.5 s.f.
Computing the average cut between the two sections we have
0 s.f. + 27 s.f. = 27/2 = 13.5 s.f.
The volume of fill between section 1 and section 2 is 72 s.f. $\times$ 50'
= 3600 s.f./27 = 133 c.y.
The volume of cut between section 1 and section 2 is 27 s.f. $\times$ 50'
= 1350 s.f./27 = 50 c.y.
Therefore, a net fill of 83 c.y. exists between the two sections.

Example 8-2: The Grid Method Done by Hand

The next type of earthworks is the Grid method. In this case a grid is identified and laid across the site. The grids are uniform and the elevations of the proposed surface and the existing surface are computed for each corner of each grid. The difference between the proposed and existing elevations are computed for each corner. The four corners of each grid are averaged and multiplied by the area of the grid to obtain the cubic feet of material. This is then divided by 27 to obtain c.y. This example contains 4 grids, each 50' $\times$ 50', as shown in Figure 8-18.
Beginning with the grid in the upper left, compute as follows:

$-1 + -5 + -2 + -2 = -10'/4$ corners $= -2.5'$ of average cut.
With a grid area of 2500 s.f., that grid contains 2500 $\times$ 2.5'
= 6250 c.f. of cut. Divided by 27 yields 231.5 c.y. of cut.

The grid in the upper right corner computes as follows:

$-5 + 0 + -2 + 0 = -7'/4 = -1.75'$ of cut. There is also some
fill in this grid of 0 + 1 + 0 + 1 = 2'/4 = .5' fill.
Therefore, the cut for the grid is 2500 $\times$ 1.75 = 4375 s.f., divided
by 27 = 162.0 c.y. of cut. The fill for the grid is 2500 $\times$ .5
= 1250 s.f. divided by 27 = 46.3 c.y. of fill. The net is 162 − 46.3
= 115.7 net cut.

The lower left grid computes as follows:

$-2 + -2 + -1.5 + -3 = -8.5/4 = -2.125'$ net cut. With a grid area of
2500 s.f., that grid contains 2500 $\times$ 2.125' = 5312.5 c.f. of cut.
Divided by 27 yields 196.8 c.y. of cut.

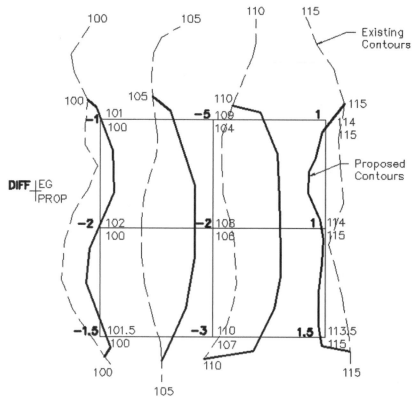

Figure 8-18   The grid method

The grid in the lower right corner computes as follows:

$-2 + 0 + -3 + 0 = -5'/4 = -1.25'$ of cut. There is also some fill in this grid of $0 + 1 + 0 + 1.5 = 2.5'/4 = .625'$ fill.
Therefore, the cut for the grid is 2500×1.25
$= 3125$ s.f., divided by $27 = 115.7$ c.y. of cut. The fill for the grid is $2500 \times .625 = 1562.5$ s.f. divided by $27 = 57.8$ c.y. of fill.
The net is $115.7 - 57.8 = 57.9$ net cut.

The earthworks for the site then is:

Cut = $231.5 + 162 + 196.8 + 115.7 = 706.0$ c.y.
Fill = $0 + 46.3 + 0 + 57.8 = 104.1$ c.y.
The net is 601.9 c.y. of cut.

The last method is the Composite (in Land Desktop terminology) or Prismoidal method. Civil 3D uses this method for its site-based earthworks, and it is a TIN comparison-based method. It is more math intensive than the others but Civil 3D can rapidly perform it. It produces a more accurate volume than the Average End Area or the Grid method.

Pros and Cons of Earthworks Algorithms

Section Method

- Pro: It is a traditional method; is understood by all; and produces good results.
- Con: Accuracy varies according to interval and can be enhanced by running the section computations twice, where the second run is perpendicular to the first run.

Grid Method

- Pro: It is also a traditional method; is understood by all; and produces good results.

- Con: Interpolation occurs. Hence, some loss of accuracy exists; the perimeter can be lost if the grid falls off the surface; highs and lows inside of each grid can be lost; and accuracy varies according to grid size.

Composite Method

- Pro: It is the best theoretical method; uses all surface data; and no interpolation is involved.
- Con: It seems to be sensitive to vertical faces (which we should not have anyway).

Figure 8-19   The Grading pull-down menu

Now that grading capabilities have been discussed, let us explore some of Civil 3D's functions. The commands for the **Grading** pull-down menu (Figure 8-19) are described next.

Grading Creation Tools... invokes the toolbar for performing grading functions.

Draw Feature Line allows for the creation of feature lines used in many grading functions. The options for creating them include slope, grade, and elevational data combined with horizontal distances.

Create Feature Lines from Objects allows selecting existing AutoCAD objects and converting them to feature lines.

Edit Grading allows for a submenu that contains the following routines:

Grading Editor allows for editing grading objects in a **Panorama**.

Edit Grading allows for editing the grading parameters for objects such as the **Stepped Offset** command.

Change Group allows for changing the grading group that an object resides in.

Edit Feature Lines pull-down menu has a large number of new commands that are critical to 3D site design. These commands include:

Elevation Editor... dialog-based command allows for making changes to the elevations of objects.

Quick Elevation Edit allows for quick, on-the-fly alterations to the slopes/grades of feature lines.

Edit Elevations command prompt–based command allows for making changes to the elevations of objects.

Set Grade/Slope Between Points is a command prompt system for altering the grades and elevations of feature lines at user-selected locations.

Set Elevation by Reference sets the elevation based on a referenced 3D elevation from another location.

Insert/Delete Elevation Point inserts/deletes vertices. These points do not alter the horizontal geometry of the object; they affect only the third dimension.

Insert High/Low Elevation Point inserts a point elevation based on the incoming and outgoing user-selected grades.

Elevations from Surface sets the elevations of feature line vertices based on surface sampling.

Insert/Delete PI inserts/deletes vertices. These do affect the geometry of the feature line.

Join joins feature lines similar to AutoCAD's PEDIT.

Reverse reverses the direction of the feature line.

Fillet places a curve at the PI of an feature line.

Fit Curve fits the PI with a curve similar to PEDIT's **Fit** command.

Smooth smooths the PI with a curve similar to PEDIT's **Smooth** command.

Weed deletes unneeded vertices.

Stepped Offset offsets the feature line and raises or drops it. Think curb faces or retaining walls here!

Grading Utilities

Grading Volume Tools... invokes the toolbar to adjust cut/fill volumes for grading groups.

Create Detached Surface detaches the surface from the grading group so it no longer updates.

Polyline Utilities

Convert 2D to 3D Polylines converts polylines to allow them to be used to set different elevations on each vertex.

Convert 3D to 2D Polylines converts polylines to flatten them.

Edit Polyline Elevations edits elevations at vertices.

Add Feature Line Labels... labels slopes on feature lines.

Quick Profile... draws a temporary profile based on linework selected and surfaces chosen. These profiles are deleted on saving and are used for quality assurance/control.

The Grading Layout Tools

The **Grading Layout Tools** are described next (see Figure 8-20).

Figure 8-20   The Grading Layout Tools

Sets the **Grading Group**.

Sets the **Target Surface**.

Sets the **Grading Layers**.

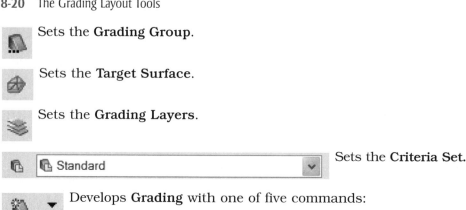

Sets the **Criteria Set**.

Develops **Grading** with one of five commands:

- **Create Grading** creates a grading object based on the criteria.
- **Copy Create Grading** copies the criteria and style from an existing grading and applies it to a grading you wish to create.
- **Create Transition** allows for transitions around objects.
- **Create Infill** connects the interior of grading objects such as on the top of a berm of a detention pond where one grading develops the inside of the pond and the other grading develops the outside of the pond.
- **Create Feature Lines** creates linework that can be used with the grading commands.

 Edit Grading allows for absorbing elevations from the surface and some management commands.

 Grading Volume Tools.

 Grading Editor and **Grading Elevation Editor**.

 Grading Group Properties and **Grading Properties**.

Exercise 8-1: Laying Out a Basic Parking Lot in 3D

Now explore Civil 3D and run through some procedures using the grading objects.

1. Open Chapter-8-a.dwg.
2. Type **V** for **View**, restore the **Plan** view, and you see some existing ground contours.
3. There is a white object on the **C-Site-Grad-3D** layer. Use this as the construction limits for this site.
4. Let us set it to the existing ground elevation and drape it on the existing ground in order to guarantee that no grading occurs outside these limits.
5. From the **Grading** pull-down menu, select the command **Create Feature Lines from Objects**. Choose the white polyline object on the **C-Site-Grad-3D** layer, accept the defaults in the **Create Feature Lines** dialog box, hit **OK**, and one feature line is added to **Site 1**. Turn on the **C-Topo-Grad-Flin** layer to see it.
6. From the **Grading** pull-down menu, choose the **Edit Feature Lines** command and then choose **Elevations from Surface**.
7. Select the **Existing ground surface**.
8. Turn on the toggle to insert intermediate breaks. Hit **OK**.
9. Select the feature line and hit **<Enter>**. Notice it processes and places small markers where it has inserted intermediate breaks. Hit **<Enter>** to terminate.
10. This object is now "draped" onto the existing ground data with a 3D vertex at each TIN crossing. Click on the feature line, and notice all of the vertices that now have the elevations of the existing ground embedded into them.
11. Turn on the layer called **C-Site-Grad-Text** for the design parameters for this site.
12. They indicate a Finished and Basement floor elevation for the building in the upper right of the site and a 4' sidewalk with a 2% cross slope draining toward the parking lot. It has a 6" high curb on it. The parking lot drains to the Southeast at −2%.

Next you will create the 3D objects that meet these specifications.

13. Zoom into the cyan building polylines and notice two lines: one representing the outside building wall and the inside linework representing the building's basement. From the **Grading** pull-down menu, select **Create Feature Lines from Objects**. The software will respond with `Select lines, arcs, polylines or 3d polylines to convert to feature lines or [Xref]:` Select the two cyan building polylines and hit **<Enter>**.

14. Accept the defaults in the **Create Feature Lines** dialog box, and hit **OK**.

15. Choose **Edit Feature Lines >> Elevation Editor** from the **Grading** pull-down menu.

16. When asked, select the outside cyan polyline representing the building pad. This is the outermost cyan linework.

17. In the **Panorama** that displays, set the elevations to **362.0** for all vertices.

18. Repeat the command and select the inside cyan polyline representing the building basement.

19. Set the elevations to **353.0** for all vertices.

20. Draw a polyline along the west and southern edge of the building by snapping to the edge of the finished floor of the building. This inherits the elevations of the finished floor.

21. From the **Grading** pull-down menu, select **Edit Feature Lines >> Stepped Offset**. The following prompts and responses should occur:

    `Specify offset distance or [Through/Layer]:` **4**

    `Select a feature line, survey figure, 3d polyline or polyline to offset:` Select the polyline you just drew.

    `Specify side to offset or [Multiple]:` Pick a point outside the building.

    `Specify elevation difference or [Grade/Slope] <0.000>:` **g**

    `Specify grade or [Slope/Difference] <0.000>:` **−2**

22. Hit **<Enter>** to terminate. This object represents to the top, face of curb.

NEW to AutoCAD **2007**

TIP   You can use the **<Shift><Space>** keys to select objects on top of other objects if needed.

Repeat this command and the following prompts and responses should occur.

    `Specify offset distance or [Through/Layer] <4.0000>:` **.1**

    `Select a feature line, survey figure, 3d polyline or polyline to offset:` Select the Top of curb line you just created.

    `Specify side to offset or [Multiple]:` Pick a point outside the building.

    `Specify grade or [Slope/Difference] <−2.000>:` **d**

    `Specify elevation difference or [Grade/Slope] <−0.080>:` **−.5**

23. Select a feature line, survey figure, 3d polyline, or polyline to offset: Hit **<Enter>** to terminate. This represents the flowline of the curb.

The building and the sidewalk are established. Let us work on the parking lot now.

24. Draw a pline that represents the drainage divide in the parking lot from the edge of sidewalk at the northwest corner of the building diagonally to the southwest corner of the parking lot.

25. Then use **Grading >> Draw Feature Line** to draw a feature line from the flowline of the curb at a **−2%** grade to the lower left corner of the parking lot designated by the red polyline. Accept the defaults in the dialog box when it displays. Refer to Figure 8-21 for the location of the swale. Then the following prompts and responses should occur: `Specify start point:` Snap to the flowline of the curb near the northwest corner of the building. This is shown in Figure 8-22. It should have an elevation of 361.40.

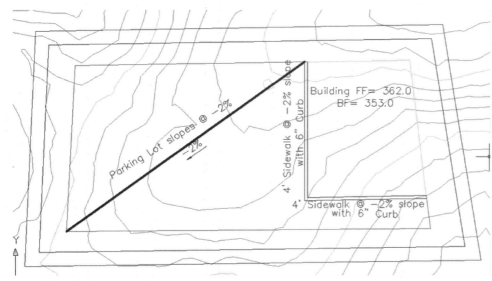

Figure 8-21   Parking lot swale

```
Specify elevation or [Surface] <361.420>: <Enter>
Specify the next point or [Arc]: Distance 426.778', Grade -84.686,
Slope -1.181:1, Elevation 0.000'.
```

Snap to the lower left corner of the red polyline in the southwest corner of the parking lot.

```
Specify grade or [SLope/Elevation/Difference/SUrface] <0.000>: -2
Specify the next point or [Arc/Length/Undo]: <Enter>.
```

This elevation should be 352.88 when you are finished.

26. Next draw a pline by snapping to the flowline of the edge of sidewalk, which is the same spot that Figure 8-22 indicates. Then snap to the upper right corner of the red polyline depicting the parking lot. Snap to the end of the swale drawn at −2% to the lower left corner of the parking lot. This will draw a polyline from the curb flowline near the northwest corner of the building around the north and west perimeter of the parking lot to the southwest corner of the parking lot.

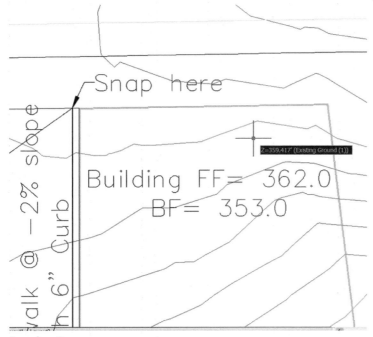

Figure 8-22   Snap location

Figure 8-23   Polyline on the north and west perimeter

27. From the **Grading** pull-down menu, select the command **Create Feature Lines from Objects**. Choose the white polyline object that you just drew around the north and west side of the parking lot, accept the defaults in the **Create Feature Lines** dialog box, hit **OK**, and one feature line is added to **Site 1**.

28. Choose **Edit Feature Lines >> Elevation Editor** from the **Grading** pull-down menu. Select the new feature line you just converted. When the **Panorama** displays, set the elevation for the second vertex to **356.387**. The three elevations for the vertices should then be **361.42**, **356.387**, and **352.88**, respectively. Close the **Panorama**. This will be the polyline shown in Figure 8-23.

   Now set the perimeter of the east and southern edge of the parking lot.

   Use the same command sequence to establish both grade and elevation along this object. From the flowline of the sidewalk beginning at the southeastern edge of the building, design a −1% grade drop to the southeastern corner of the parking lot, and then tie into the drainage divide in the southwest corner of the parking lot at an elevation of 352.8861.

29. Draw a pline from the flowline of the sidewalk beginning at the southeastern edge of the building to the southeastern corner of the parking

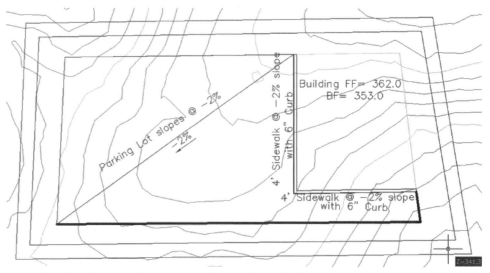

Figure 8-24   Location of south and east perimeter of parking lot

lot, and then end it at the drainage divide at the southwestern corner of the parking lot. Refer to Figure 8-24 for the location of this polyline.

30. From the **Grading** pull-down menu, select the command **Create Feature Lines from Objects**. Choose the polyline object that you just drew around the east and south side of the parking lot, accept the defaults in the **Create Feature Lines** dialog box, hit **OK**, and one feature line is added to **Site 1**.

31. Choose **Edit Feature Lines >> Elevation Editor** from the **Grading** pull-down menu. Select the new feature line you just converted. When the **Panorama** displays, set the grade for the first vertex to −1.0%. The three elevations for the vertices should then be, **361.42, 360.968**, and **352.88**, respectively. Close the **Panorama**. Save your file.

This object now represents the southern edge of the parking lot and is designed to achieve a –1% drop from the sidewalk to the southeastern corner and tie into the diagonal divide at the southwest corner. The next task is tie the parking lot out to the natural ground and make a surface. In the file you will be working with, the feature lines were exploded to turn them into 3D polylines. There is no longer any reason to do this in the 2007 version. You can send these directly into the surface generation.

32. Open Chapter-8-b.dwg.
33. Select **Grading Creation Tools** from the **Grading** pull-down menu.

Student Files

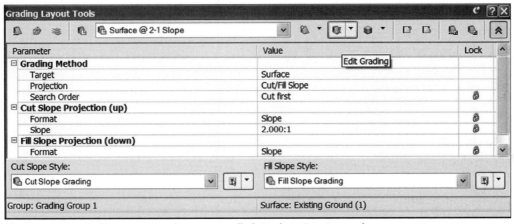

Figure 8-25   Grading Layout tools are now called Grading Creation tools.

34. The **Grading Creation Tools** toolbar appears as in Figure 8-25.
35. Move from left to right across the toolbar, setting or verifying the settings as you go.
36. Choose the first button on the left to set the **Grading Group** for the next step. A **Create Grading Group** dialog box displays.
37. Accept the default name of **Grading Group <Next Counter>**. Turn **on** the **Automatic Surface Creation**. The **Use the Group Name toggle** should be **on**. The **Surface Style** should be set to **Borders & Contours**. Accept **10.0** for tessellation spacing and 3 degrees for angle. This sets intermediate computations. Turn **off** the **Volume base surface**. Hit **OK**.
38. When the **Create Surface** dialog box appears, hit **OK** for the defaults.
39. Next click on the second button, set the **Target Surface**, and choose the **Existing ground**.
40. Click on the third button to set the **Grading Layers** and accept the default layers.
41. Click on the fourth button to set the **Criteria Set**. Select **Basic** as the criteria set to use.
42. In the pop-down window, choose **Surface @ 2-1 Slope**. With the **Details** button enabled (the chevron icon on the right of the toolbar), notice that

details for the **Grading Method** and **Slope Projection** are shown as in the figure.

43. Next choose the first button to the right of the pop-down window that says **Surface @ 2-1 Slope**. It should default to **Create Grading**.

44. The prompt requests you to: Select the feature: Pick the feature line at the north edge of the parking lot. It then asks to: Select the Grading side: Pick a point to the outside of the parking lot. It then asks: Apply to entire length? Hit **<Enter>** for **Yes**. Grading appears along the north side of the parking lot.

45. It then requests you to: Select the feature: again, assuming that you will repeat this sequence. Pick the feature line along the southern edge of the parking lot. It then asks to: Select the Grading side: Pick a point to the outside of the parking lot. It then asks: Apply to entire length? Hit **<Enter>** for **Yes**. Grading appears along the north side of the parking lot. Hit **<Enter>** to exit.

46. Pick on a contour and right click. Select **Edit Surface Style**. In the **Contour** tab, expand the **Contour intervals** item and change the minor and major intervals to **2** and **10**, respectively.

Notice that the grading ties out to existing ground along the north, west, and southern sides. See Figure 8-26.

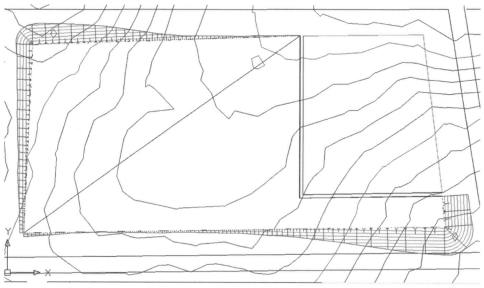

Figure 8-26   The grading tie-outs

Now tie the north and eastern sides of the building out to the north-eastern corner of the site.

47. Explode the outside cyan building feature line.

48. Use the **BREAK AT POINT** AutoCAD command to place a break on the outside cyan polyline at the northwest corner of the building. Use the **BREAK AT POINT** AutoCAD command to place a break on the outside cyan polyline at the southeast corner of the building. By doing this you can project the top and right sides of the building toward the existing ground.

49. Select **Grading Creation Tools** from the **Grading** pull-down menu.

50. The **Grading Creation Tools** toolbar displays.

51. Click on the fourth button to set the **Criteria Set**. Select **Basic** as the criteria set to use.

52. In the pop-down window, choose **Surface @ 2-1 Slope**.

53. Once again, choose the first button to the right of the pop-down window that says **Surface @ 2-1 Slope**. It should default to **Create Grading**.

54. The prompt requests you to: Select the feature: Zoom in and pick the outer edge of the building that you just broke. Hit **OK** to the **Create Feature Lines** dialog box that displays. It then asks to: Select the Grading side: Pick a point well outside of the building. It then asks: Apply to entire length? Type **Y** for **Yes**. Hit **<Enter>** to accept and terminate.

55. Grading appears as shown in Figure 8-27, in which the entire perimeter is now tied to existing ground.

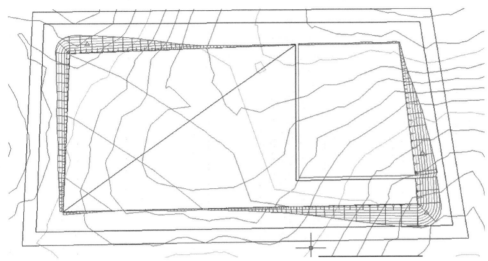

Figure 8-27   Resultant grading

56. Let us add in the sidewalk and drainage divide in the parking lot.

57. Look in the **Prospector** under **Surfaces**. Expand the surface called **Grading Group 1**. Expand **Definition**. Right click on the **Breaklines** item and select **Add....**

58. Hit **OK** to accept all of the defaults that come up in the **Add Breaklines** dialog box.

59. When asked to select objects, place a crossing window to include the sidewalk edges, basement, the left and southern edge of the building, and the drainage breakline in the parking lot.

60. If the **Panorama** appears with messages, close it.

61. You see now that the contours reflect the sidewalk and drainage divide. See Figure 8-28.

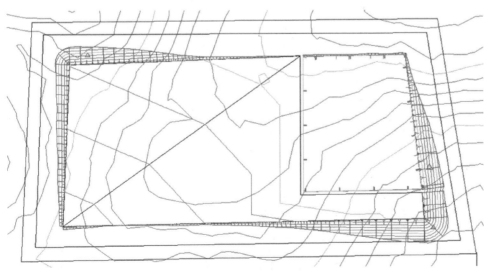

Figure 8-28   The contours reflect the sidewalk and drainage divide

62. Using **3DOrbit**, rotate the view into a 3D bird's eye perspective. The Surface style for 3D kicks in and a 3D triangulation appears automatically.
63. Turn off the layer for the existing ground, **C-Topo-Existing Ground**, and you isolate the proposed parking lot. As discussed in Chapter 7 on corridors, you can select **View>> View Style>> Conceptual** shading and then observe something like Figure 8-29. It shows the basement, the divide, and the sidewalk clearly.

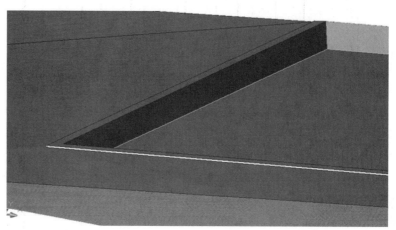

Figure 8-29   Gouraud shading

In the next exercise, you will compute earthwork volumes for the site.

Earthwork volumes, sometimes called takeoffs, are performed on projects in order to estimate the amount of earth being moved. This is important to the engineer in that the design goal is often to produce a "balanced site." This means that whatever earth is cut or dug out, is then used to lift or fill the site in areas where that needs to be done. Balancing means that we have no excess cut or fill material. Consider that a dump truck might have a capacity of 10 c.y. and you can estimate how many truckloads it would take to move excess earth on the site.

Exercise 8-2: A Quick Earthworks Volume

Let us compute the earthworks for the site.

1. Open Chapter-8-c.dwg.
2. Select **Utilities** from the **Surfaces** pull-down menu. Then choose **Volumes**....
3. When the **Panorama** appears, click the first button from the left to create a new volume entry. A row of data displays in the window.
4. Click in the cell for **Base surface** where it says to <Select surface> and choose **Existing Ground**.
5. Then click in the cell for **Comparison surface** where it says to <Select surface> and choose **Grading Group 1**. Click somewhere in the **Panorama** to execute the processing. It should yield almost 12,000 c.y. of material in a net fill condition.

Exercise 8-3: "Popping Up" a Curbed Parking Lot Island

In order to provide another example of the power of the **Grading** tools in Civil 3D, let us see how to pop a curbed island out of the asphalt paving area. The curb will be 6″ high and have a slight kilter to its face so as to avoid a vertical edge on the face of curb.

1. Open Chapter-8-c.dwg or stay in the drawing from the last exercise.
2. Zoom into the parking area to the west of the building. You see some plines representing the flowline of some parking islands.
3. From the **Grading** pull-down menu, select **Create Feature Lines from Objects**. Select the four parking lot islands in the parking lot. Hit **OK** to the **Create Feature Lines** dialog box.
4. From the **Grading** pull-down menu, select **Edit Feature Lines** and then **Elevations From Surface**. When the **Select Surface** dialog box displays, select **Grading Group 1** as the **Surface to compute to**. Hit **OK**.
5. Turn on the toggle to insert intermediate grade breaks. Hit **OK**, and the software computes additional elevations from the asphalt pavement over and beyond the vertices in the plines.
6. Then you are requested to: `Select the feature line:` Pick all four islands one by one. The pline islands are now "draped" onto the parking lot surface.
7. You are now ready to pop the islands up from the surface. From the **Grading** pull-down menu, select **Edit Feature Lines >> Stepped Offset**. The following prompts and responses should occur:

    `Specify offset distance or [Through/Layer] <0.1000>:` **.1**
    `Select a feature line, survey figure, 3d polyline or polyline to offset:` Pick an island.
    `Specify side to offset or [Multiple]:` Pick a point inside the island.
    `Specify elevation difference or [Grade/Slope] <-0.500>:` **.5**

8. Repeat this for each island. Inspect the results when complete and notice that the interior object is 0.5′ higher than the outside object, hence achieving the top of curb.
9. Let us add the islands into the surface of the parking lot.
10. Look in the **Prospector** under **Surfaces**. Expand the surface called **Grading Group 1**. Expand **Definition**. Right click on the **Breaklines** item and select **Add**....
11. Hit **OK** to accept all of the defaults that come up in the **Add Breaklines** dialog box.
12. When asked to select objects, select the eight objects representing the islands.
13. If the **Panorama** appears with messages, close it.
14. You see now the islands and the contours "jump" the curb as expected.
15. See Figure 8-30 for an example of the 3DOrbit that you can perform.

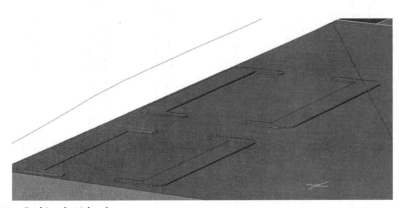

Figure 8-30 Parking lot islands

Exercise 8-4: A Brief Example of Intersection Design

One more example that might introduce the gradings to readers would be the age-old intersection design.

Intersections are one of the most often asked about design tasks that people want to automate. You open a file that has some edge of pavement in it. The linework is in 3D and can be achieved from the corridor design using the **Corridor** >> **Export** >> **Corridor Feature Line** as **Grading Feature Line** command. The linework for each component of the corridor can be imported into AutoCAD. Then, using the **EXPLODE** command, explode the feature lines so that they can be trimmed, extended, or broken as needed to begin developing the desired site features. That is where the exercise begins. You add the intersection returns and illustrate an example of how intelligent feature lines can be. Note to the reader: Generalize on this example!

TIP   Note that intersections and cul-de-sacs can be graded using the **Corridor** tools as well, but these might be considered high-end skills.

Student Files

1. Open Chapter-8-d.dwg and note that a basic intersection has been developed as described previously. The edge of pavement and centerlines are in 3D. The red arcs denote where the returns are supposed to be and they are in 2D.
2. Under the **Grading** pull-down menu, select the **Create Feature Lines from Objects** command.
3. Select the four edges of pavement, identified with numbers 1, 4, 3, and 6, plus the one in the intersection between 3 and 6. Also select the two centerlines on **Layer 0**, and hit **<Enter>**. Hit **OK** for the dialog box that displays.
4. Now draw a polyline by snapping to the edge of pavement at the point marked by a number **1**. Make sure you Endpoint snap to the black edge of pavement and not the red arc. The black edge has a 3D elevation. Then type **A** for **arc** and then type **S** for **second point** (of the arc) and Near snap to the point on the red arc marked with a number **2**. Then Endpoint snap to the edge of pavement marked by a number **3** and terminate.
5. Repeat the **Polyline** command and perform this on the right side of the intersection. Snap to the edge of pavement at the point marked by **4**. Make sure you Endpoint snap to the black edge of pavement and not the red arc. Then type **A** for **arc** and then type **S** for **second point** (of the arc) and Near snap to the point on the red arc marked with a **5**. Then Endpoint snap to the edge of pavement marked by **6** and terminate.
6. Under the **Grading** pull-down menu, select the **Create Feature Lines from Objects** command. Select the two polylines that you just drew. Now use the **Grading** pull-down menu, and select **Edit** and then **Elevations** to check the elevations of the two arcs that you converted to feature lines. Notice that Civil 3D recognized that the plines were being

This is a revolutionary function in case you did not realize it. This is the first time in AutoCAD history that there can be a 3D arc! Notice that there are no chords to simulate the arc as has been the case in the past using various routines.

connected to feature lines and automatically inherited the elevations at the endpoints.

7. From the **Surfaces** pull-down menu, select **Create Surface...** Provide a **Name** for the surface, call it **Intersection**, and use the **Borders & Contours** style.

8. In the **Prospector**, under **Surface**, expand the intersection surface that was just created. Expand **Definition** and right click on **Breaklines**. Type **Intersection** in the description for the breaklines when the dialog box appears.

9. Select all of the feature lines, including the edge of pavement, the arcs that you created at the intersection, and the centerlines. Do not forget the edge of pavement breakline in the intersection between the arcs.

10. A **Surface** generates showing contours inside the intersection. Use **3DOrbit** to inspect the intersection in 3D.

11. Now, generalize on this concept because if the intersection had curb and gutter it could be created just as easily and within a minute or two. Refer to Figure 8-31 to see the contours created in the intersection.

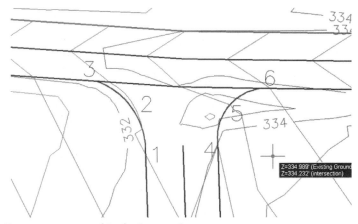

Figure 8-31   Contours are created in the intersection

You could take a lot more time performing grading features but would likely need a separate textbook to cover all the situations. In the meantime, explore these commands and find that almost any type of site can be designed in 3D.

> **TIP** Be aware that you can create your own grading criteria. Look in the **Toolspace Settings** under the **Grading** item. Expand **Criteria Sets** and notice the **Basic** grading. Right click on **Basic** grading and select **New...** to create your own criteria. When the dialog box appears, click on the **Criteria** tab and explore how you can customize these tools.

Congratulations! You have graded a parking lot, a building with a basement, and its related tie-outs and sidewalks. You have also added parking lot islands and computed earthworks. You have even had a taste of modeling an intersection using **Grading** commands! Enormous powers exist in the grading objects and only experimentation can yield their true powers.

CHAPTER TEST QUESTIONS

Multiple Choice

1. When grading a site, which of the following information items are useful?

   a. Elevations
   b. Slopes
   c. Grades
   d. All of the above

2. The **Elevation Editor** can be used to perform which design tasks?

   a. Establish profiles and vertical tangents grades
   b. Set grades and elevations
   c. Establish the elevations for superelevations
   d. All of the above

3. Methods of creating proposed topographic grading data include which of the following?

   a. Developing Zones of Slope Influence
   b. Contour digitizing
   c. Developing 3D breaklines and Developing 3D spot shots
   d. All of the above

4. You would consider using "thinking points" for which of the following design tasks?

   a. Annotating parcel lines
   b. Balancing traverses
   c. Establishing pipe inverts
   d. None of the above

5. The sites developed in Civil 3D can be made ready for use by 3D/GPS Machine Control equipment by using which of the following add-ons?

   a. Carlson Connect, which addresses Topcon
   b. Trimble Link
   c. Leica X-change
   d. All of the above

6. Understanding that the concept of engineers designing their sites in three dimensions is a "must achieve" goal, then what do they rely on the computer to do for them?

   a. Assisting in designing the site in the computer using slopes, grades, and elevations
   b. Creating the contouring objects and analyses

   c. Annotating the contour, spot shot, and grade labeling
   d. All of the above

7. What are some of the methods for performing earthworks takeoffs?

   a. Section method
   b. Grid method
   c. Composite method
   d. All of the above

8. In the Section earthworks method, there are the following computations. Section 1 has a cut area of 100 s.f. Section 2 has a cut area of 200 s.f. Section 1 has a fill area of 50 s.f. and Section 2 has a fill area of 150 s.f. The distance between Section 1 and Section 2 is 50 feet. What is the net value for cut or fill for the volume between Section 1 and Section 2 in cubic yards?

   a. 278 c.y. cut
   b. 185 c.y. fill
   c. 93 c.y. cut
   d. 463 c.y.

9. Which of the following earthworks algorithm properties are true?

   a. The Grid method is a traditional method; however, some loss of accuracy can occur for a variety of reasons.
   b. The Section method is a traditional method; however, some loss of accuracy can occur for a variety of reasons.
   c. The Grid method produces good results when used properly, but highs and lows that may occur inside individual grids may not be accounted for in the computations.
   d. All of the above.

10. When working with some of the **Grading** tools, what results can be achieved?

   a. Grading objects can be created according to our design criteria.
   b. Surfaces can be automatically created as we create grading objects.
   c. Contours for the surfaces can automatically be displayed.
   d. All of the above.

True or False

1. True or False: The civil engineering field should prepare its roadway and site designs using contours because they represent the proposed site conditions better than any other data type.

2. True or False: A Zone of Influence is a method of design, not a command in Civil 3D.

3. True or False: Once a proper 3D site is created, sections can be pulled to observe extraordinary

detail such as loading docks and parking lot islands.

4. True or False: Civil 3D can use 3D house templates for grading purposes.

5. True or False: There are several types of earthworks algorithms are available in Civil 3D.

6. True or False: The Grid method can be performed only manually by the user.

7. True or False: To convert cubic feet of earthworks into cubic yards, you multiply by 27.

8. True or False: The Grid method gets less accurate as the size of the Grid becomes smaller.

9. True or False: The Section method becomes more accurate as the size of the interval between sections becomes smaller.

10. True or False: A "balanced site" means that whatever earth is cut or dug out, is then used to lift or fill the site in areas where that needs to be done.

STUDENT EXERCISES, ACTIVITIES, AND REVIEW QUESTIONS

1. What is the difference between slope and grade in the software's viewpoint?

2. What is a "tie-out"?

3. What is daylight?

4. What types of quality control can be used to check the proposed design?

5. What type of CAD data can be used for rendering the site?

6. What is a "thinking point"?

7. How can someone perform quality control on his or her grading?

8. What do the commands **Distance @ Grade**, **Distance @ Slope**, **Relative Elevation @ Slope**, **Surface @ 2:1**, **Surface @ 3:1**, **Surface @ 6:1**, and **Surface @ Slope** do?

9. Could you use the **Distance @ Grade** tool in the **Grading Layout** toolbar to create an island with a curb?

10. Compute the earthworks for the site in Figure 8-32 based on a 100′ × 100′ grid method. Show your work.

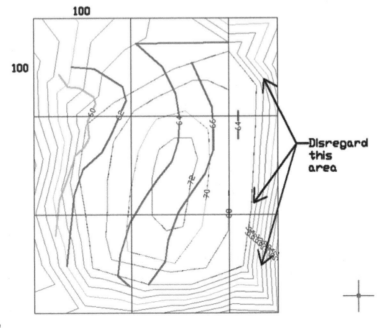

Figure 8-32

Piping for Storm Sewers and Drainage

9

Chapter Objectives

After reading this chapter, the student should be able to:
- Understand basic hydrology and hydraulic concepts.
- Perform piping layout and drafting for utilities and storm sewers.

INTRODUCTION

This chapter discusses developing storm sewer plan and profile data. This is part of a larger engineering task related to hydrology and hydraulics. Civil 3D performs some limited hydrologic computations. The hydraulics is also limited but does use object-oriented piping for storm sewers on your project. It extracts the rim elevations and develops inverts based on criteria established. This feature, although not a totally complete hydraulics design tool, is a welcomed addition to the software's capabilities.

The purpose of this chapter is to introduce the reader to some basic hydrology and hydraulic concepts and provide some basic procedures for using Civil 3D toward these purposes.

REVIEW OF FUNCTIONS AND FEATURES TO BE USED

This chapter provides instruction in the functions for producing hydrologic watersheds from the surface object as well as producing pipes in plan and profile views. These will carry into sections should they be imported as well. The features for piping are objects and react to the environment around them. These functions can be used for typical drafting purposes as well as to react to design changes and conflict resolution. The piping of multiple networks can be imported into any number of profiles or sections to test for conflicts.

HYDROLOGY

Runoff

When rain or snow falls onto the Earth, it follows the laws of gravity and conveys to the lowest point. Part of the precipitation is absorbed by the ground and replenishes the groundwater reservoirs. Most of it, though, flows downhill and is called runoff. Runoff is extremely important in site design because it changes the landscape through erosion. Engineers typically must design retention facilities to detain the amount of additional runoff their design creates when they replace natural grass-covered ground with asphalt-paved parking lots and the like. If the additional runoff were not captured and held, then it would likely flood or erode downstream properties.

A Definition of Runoff

Runoff is a portion of precipitation, snowmelt, or irrigation water that flows over ground in uncontrolled, above surface streams, rivers, or sewers. Computation factors include the speed of its arrival after a rainfall or a snowmelt; the sum of total discharges during a specific period of time; and the depth to which a drainage area would be covered if all of the runoff for a given period of time were uniformly distributed over it.

Meteorological Factors Affecting Runoff

The following meteorological factors affect runoff:

- Type of precipitation such as rain, snow, sleet, and so on
- Rainfall intensity, amount, and duration
- Distribution of rainfall over the watershed
- Other meteorological and climatic conditions that affect evapotranspiration, such as temperature, wind, relative humidity, and the season

Physical Characteristics Affecting Runoff

The computations of hydrologic study often involve the type of land use, the vegetation on the land, the soil type evident in the area, the total drainage area involved, the shape of the watershed basin, elevations, slopes, drainage patterns, topography, bodies of water, and biogeomorphism, which accounts for the impedances that may exist that could alter the flow of water in the basin.

The watershed for Front Royal, Virginia, is shown in Figure 9-1.

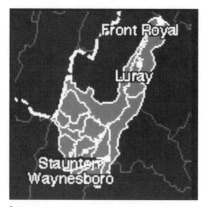

Figure 9-1   A simple watershed map

- Rivers and streams in this watershed: 15 (provided by EPA's River Reach File, version 1)
- Lakes in the watershed: 68; total number of acres: 679.8
- River and stream miles:
    1358.7 total river miles
    988.9 perennial river miles
    % of total rivers and streams that have been surveyed

Software for Hydrologic Estimations and Computations

Software for hydrologic estimations and computations include TR-20—rainfall runoff calculations; HEC-1—water surface profiles; and TR-55—Technical Release 55 (TR-55). These present simplified procedures for small urbanizing watersheds to calculate storm runoff volume, peak rate of discharge, hydrographs, and storage volumes required for floodwater reservoirs.

In these computations there are some of the following items

IDF (Intensity-Duration-Frequency) Curves.
These indicate the relationship between the intensity (I) and duration (D) of a rainfall event with a given return period (T).

The Rational Method for Predicting Runoff.
This may be used to predict the peak runoff with the equation

$q = CIA$, where:
q = Peak runoff (CFS) C = Runoff coefficient
I = Rainfall intensity (in./hr) A = Area (acres)

The method indicates that the units have been "rationalized" where 1 cubic foot per second (CFS) = 1.01 in.-acre/hr.

The rational method is dependent on the selection of C and I, where C is based on the soil, ground cover, and other factors, and I is obtained from the local IDF curve for a given return period and duration.

One of the challenges of the rational method is that of choosing the correct duration. The rational method is intended only to determine peak runoff.

SCS Storm Distributions

By studying the Weather Bureau's Rainfall Frequency Atlases, the Soil Conservation Service (SCS) determined that the continental United States could be represented by four dimensionless rainfall distributions each of a twenty-four-hour duration. These distribution curves were developed from the same depth-duration-frequency data used for IDF curves.

The major advantage of the SCS storm distributions is that each curve contains depth information for storm events of all durations up to twenty-four hours. These distributions also provide the cumulative rainfall at any point in time which makes them useful for volume-dependent routing calculations.

Determining the Time-of-Concentration

For overland flow computations related to storm activity a time-dependent factor must be introduced in order to determine how the runoff is distributed over time. The time-of-concentration (T_c) is used for SCS methods. The T_c is usually defined as the time required for a theoretical drop of water to travel from the most hydrologically remote point in the watershed to the point of collection.

Some Important Equivalents

1 cfs = 448 gallons/minute
1 cfs-day/mi^2 = 0.03719″ of runoff
1″ of runoff/mi^2 = 26.89 cfs-day = 2,323,200 ft^3 = 53.33 acre-ft
1 acre = 43,560 sq. ft = 4047 sq. m = .4047 ha
1 sq. mi = 640 acres = 2.59 sq. km
Acceleration of gravity = 32.17 ft/s^2 = 9.806 m/s^2

Example 9-1: Watersheds

In Figure 9-2, delineate the watershed boundaries for the site.

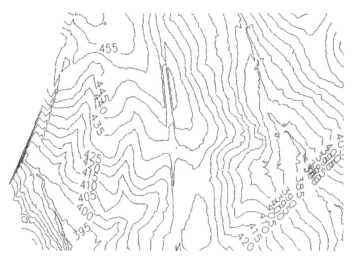

Figure 9-2   Delineate the watershed boundaries

The graphic in Figure 9-3 shows an example of watershed boundaries.

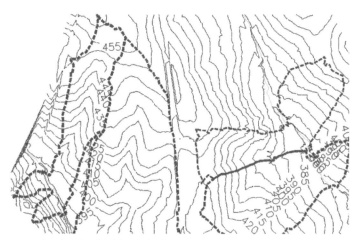

Figure 9-3   An example of watershed boundaries

Exercise 9-1: Civil 3D Watersheds

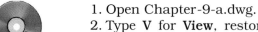

You now explore Civil 3D and its ability to develop watersheds from the surface model.

1. Open Chapter-9-a.dwg.
2. Type **V** for **View**, restore the **Plan** view, and you see existing ground contours.
3. These contours belong to a surface model. Watershed generation is built into every surface model. Click on and select the **Surface** in AutoCAD; right click and select **Edit Surface Style**.
4. When the **Surface Style** dialog box comes up (Figure 9-4), click on the **Display** tab and deselect or turn off the contours and borders. Select the **Watersheds** for display and turn it on.
5. Then click on the **Watersheds** tab. Several types of watersheds are automatically computed.
6. Hit **OK** in the **Surface Style** dialog box and the surface immediately represents itself as a watershed analysis, as shown in Figure 9-5.

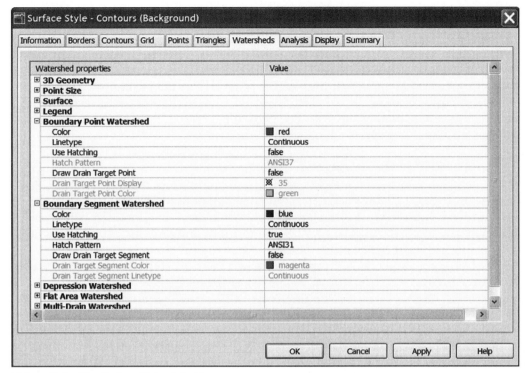

Figure 9-4   The Surface Style dialog box

TIP

The type of watersheds automatically computed include the following.

Boundary Point, a watershed in which water leaves the shed at a single location.

Boundary Segment, a watershed in which water leaves the shed over a levee or wall type object (technically it is likely to be the side of a TIN triangle).

Depression, a watershed in which water is in a basin and has no way out.

Flat Area, a watershed that is flat (this is interesting to observe in proposed roadways because flat areas are not desirable in the paved lanes due to safety reasons as they can cause hydroplaning).

Multi-Drain, a watershed in which water has multiple ways out of the shed.

Multi-Drain Notch, a watershed in which there is a flat area between two points on the surface, reminiscent of a trapezoidal channel.

Each of these watersheds can be colored or shaded differently so that it stands apart for analysis.

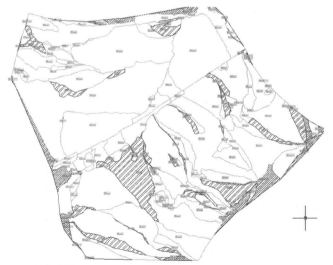

Figure 9-5   A watershed analysis

HYDRAULICS

Hydraulics, like hydrology, is a huge field within civil engineering. The word *hydraulics* comes from the Greek word *hydraulikos*, meaning water. This term comes from another term meaning water pipe. Hydraulics systems, whether retaining or conveying water, are designed to deal with water at rest or in motion. This discussion is concerned with the principles and methods of planning, controlling, transporting, conserving, and using water.

Basic Concepts Related to Flowing Water

Flow is classified into two conveyances, open channel and closed conduit flow. Open channel flow occurs when the flowing stream has a free or unconstrained surface that is open to the air. Flows in channels or in pipelines with partial flow and venting are examples. In open channels, the only force that can cause flow is the force of gravity on the fluid.

Hydraulically speaking, a pipe is a closed conduit that carries water under pressure. It can be of any shape but is often round for friction purposes.

Flow occurs in a pipeline when a pressure or "head" difference exists between the ends. The discharge Q that occurs depends mainly on the amount of pressure or head difference that exists from the inlet to the outlet; the friction or resistance to flow caused by pipe length, pipe material, bends, constrictions, changes in shape and size, and type of fluid; and the cross sectional area of the pipe.

Discharge–Area–Velocity Relationships

Flow rate or discharge, Q, is the volume of water in cubic feet passing a flow section per unit time, usually measured in cubic feet per second (ft^3/s). To get the time rate of flow or discharge, Q, in cubic feet per second, use the equation: $Q = AV$, or $V = Q/A$.

Hydraulic Mean Depth and Hydraulic Radius

An irregular flow cross section often uses a hydraulic radius and depth for computing area. The hydraulic radius, R_h, is defined as the area of the flow section divided by the wetted perimeter, P_w: $R_h = A/P_w$. The wetted perimeter times the hydraulic radius is equal to the area of irregular section flow.

Velocity Head Concept

It is known that a dropped object gains speed as it falls due to acceleration caused by gravity (g) which is equal to 32.2′ per second per second (ft/s^2). Measurements show that an object dropping 1′ reaches a velocity of 8.02′ per second (ft/s). An object dropping 4′ reaches a velocity of 16.04 ft/s. After an 8′ drop, the velocity attained is 22.70 ft/s.

If water is stored in a barrel and an opening is made in the barrel wall 1′ below the water surface, the water shoots from the opening with a velocity of 8.02 ft/s, which is the same velocity that a freely falling object reaches after falling 1′. Similarly, at openings 4′ and 8′ below the water surface, the velocity of the shooting water is 16.04 ft/s and 22.68 ft/s, respectively. Therefore, the velocity of water shooting from an opening under a given head (h) is the same as the velocity that would be attained by an object falling that same distance. The equation that shows how velocity changes with head defines velocity head: $V = \sqrt{2gh}$, which is also written as $h = V^2/2g$.

CIVIL 3D STORM SEWERS

This chapter explores Civil 3D and its ability to draft storm sewer piping using the piping objects. The main menu for piping is the **Pipes** pull-down, shown in Figure 9-6. The first thing a user must do is **Set Pipe Network catalog** and select either **Imperial** or **Metric** piping. Then one would select the **Create by Layout** tool, which, as you know by now, invokes the **Layout** toolbar for placing pipes and structures. Editing can occur after the system is in place. **Draw Parts in Profile View** allows the network to be automatically drawn in the profile of your choice. **Add Labels** can be done afterward if the labeling style chosen did not perform labeling.

Several new operations were added to the 2007 software. They include the following:

- **Create from Object**, by which the user can create pipe systems from preexisting AutoCAD objects.

- **Parts List**, by which new pipe parts can be developed. This includes both structures and pipe types. They can be created or edited.

- **Parts Builder**, by which new graphic parts can be developed with a version of the new dynamic blocks editor.

- **Create Interference Check**, by which the user can visualize where conflicts exist between utilities. This allows you to solve for physical conflicts as well as tolerance conflicts. In other words, you can look for conflicts that appear within a user definable distance from the objects. This is important for jurisdictions that have tolerance where, say, the water lines cannot be within 5′ of a sanitary line.

- **Create Alignment from Network Parts**, by which once the network is established, you may need to create alignments and corresponding profiles for its parts.

- **Change Flow Direction** which allows for changing the downhill flow of water.

Your job here is to place a storm structure network beginning at station 9+00 using a manhole. It connects with a 12″ concrete pipe to another manhole at station 12+00. Then a 12″ pipe is used to connect to a manhole at 15+00. This connects to a headwall just off the paving perpendicular to station 15+00. Use the **Network Layout Tools** toolbar to create the network. The tools (Figure 9-8) are described here:

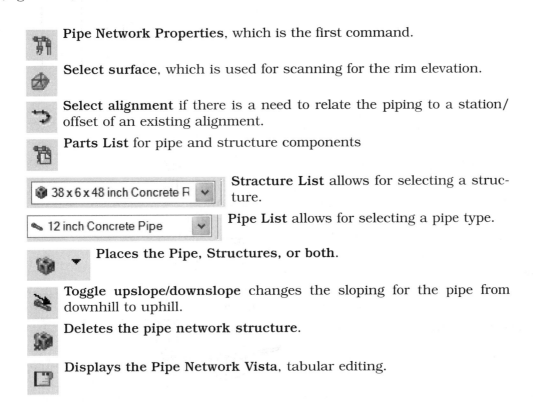

Pipe Network Properties, which is the first command.

Select surface, which is used for scanning for the rim elevation.

Select alignment if there is a need to relate the piping to a station/offset of an existing alignment.

Parts List for pipe and structure components

38 x 6 x 48 inch Concrete F   **Stracture List** allows for selecting a structure.

12 inch Concrete Pipe   **Pipe List** allows for selecting a pipe type.

Places the Pipe, Structures, or both.

Toggle upslope/downslope changes the sloping for the pipe from downhill to uphill.

Deletes the pipe network structure.

Displays the Pipe Network Vista, tabular editing.

Exercise 9-2: Create Pipes by Layout

1. Open Chapter-9-b.dwg.
2. Type **V** for **View**, restore the **Plan** pipe view, and you see a corridor between station 9+00, a high point, and 15+00, a low point.
3. From the **Pipes** pull-down menu, Figure 9-6, select **Set Pipe Network Catalog**.

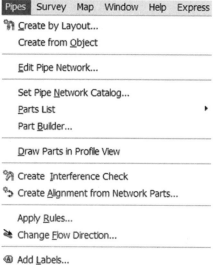

Figure 9-6   The Pipes pull-down menu

4. Choose **Imperial** for each selection item.
5. Then choose **Create by Layout** from the **Pipes** pull-down menu.
6. A **Create by Layout** dialog box displays. Set the defaults as shown in Figure 9-7, where the **Surface name** is **Corridor—(1), Surface—(1)** and the **Alignment name** is **Alignment—1**.

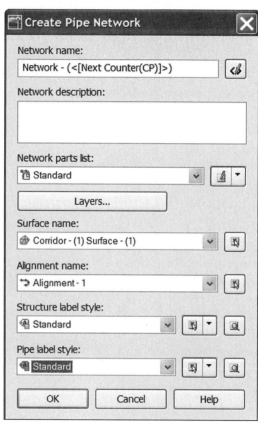

Figure 9-7   Create by Layout

7. The **Network Layout Tools** toolbar displays.
8. Choose the first button on the left in Figure 9-8, **Pipe Network Properties**. It invokes a dialog box for **Pipe Network Properties**. Set the parameters for the **Layout Settings**, as shown in Figure 9-9.

Figure 9-8   Pipe Network Properties

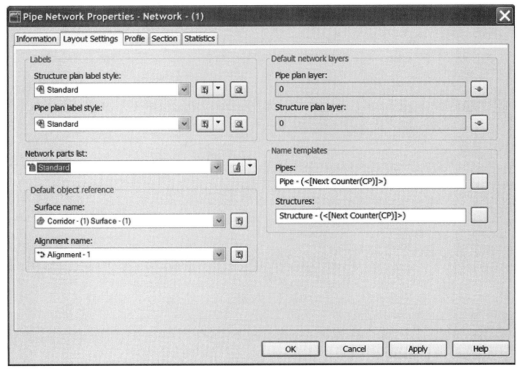

Figure 9-9   The Layout Settings

9. Then pick the tab for **Profile**, and set the parameters for **Structure profile label style** to **Standard** and the **Pipe profile label style** to **Standard**. Hit **OK**.
10. In the pop-down **Structure List**, select **Concentric Cylindrical Structure 48 dia. 18 frame 24 cone 5 wall 6 floor**, from the **Concentric Cylinder Structure** item.
11. Pick **12 inch concrete pipe** from the **Pipe List**.
12. Pick on the command in the toolbar for **Draw Pipes and Structures**. The prompt asks you to: `Specify the structure insertion point:` Pick a point on the corridor near station 9+00.
13. Pick another point at station 12+00 while staying in the command.
14. Pick another point at station 15+00 while staying in the command.
15. Now in the pop-down **Structure List**, change the Concentric Structure 48 dia. 18 frame 24 cone 5 wall 6 floor, to a **38×6×48 inch Concrete Headwall VH** from the **Concrete Rectangular Headwall Variable Height** item.
16. Place the last point near the end of the 3:1 side sloping (approximately 40′ right of the centerline for the corridor at station 15+00). Then hit **<Enter>** to complete the network.

On inspection, note that the piping is drawn in the plan view. Next you lay it into the profile and see how it appears. After that you move the piping around to refine it for your needs and observe how it recomputes itself.

Exercise 9-3: Create Pipes in the Profile

1. Type **V** for **View**, restore the **Prof** pipe view, and you see a profile between station 9+00, a high point, and 15+00, a low point.
2. From the **Pipes** pull-down menu, select the command to **Draw Parts in Profile View**.
3. It then says to: `Select network(s) to add to profile view (or Selected parts only):` Zoom out and select a pipe in the **Pipe Network** you just created and hit **<Enter>**.
4. The prompt says to: `Select the profile view:` Zoom back to the profile and select any grid line in your profile.
5. Observe your profile because the pipe system is then drawn into the profile for you. It appears something like Figure 9-10.

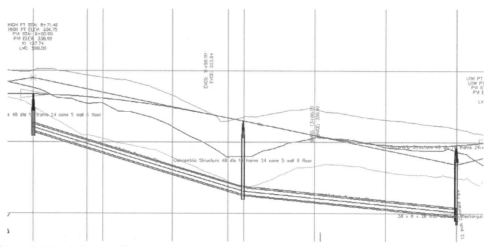

Figure 9-10   A Pipe Profile

Notice that the rim elevations are all automatically taken from the roadway's finished ground surface. The labeling appears as directed by the Standard labels selected in the styles chosen. You need to modify the label styles to reflect the way you want your plans to appear.

Exercise 9-4: Add a Structure

Take a few moments to edit and examine the system. You will add another manhole structure at station 11+00.

1. Choose **Edit Pipe Network** from the **Pipes** pull-down menu. It asks you to: `Select part from the pipe network:` Pick a pipe in the plan view.
2. The **Network Layout Tools** toolbar displays.
3. In the pop-down **Structure List**, change the current structure to **Concentric Cylindrical Structure 48 dia. 18 frame 24 cone 5 wall 6 floor**.
4. Under **Draw Pipes and Structures**, click the **down arrow** to choose **Structures Only**. It asks you to: `Specify the structure insertion point.`
5. Use a Nearest snap to pick a point on the pipe near station 11+00 and notice a new structure is placed. Hit **<Enter>** to terminate.
6. From the **Pipes** pull-down menu, select the command to **Draw Parts in Profile View**.
7. It then says to: `Select network(s) to add to profile view (or Selected parts only):` Zoom out and select a pipe in the **Pipe Network** you just created and hit **<Enter>**.

8. The prompt says to: `Select the profile view:` Zoom back to the profile and select any grid line in your profile. Notice that the system added the new structure at station 11+00.

9. Now let us adjust the labeling for the profile to add more information. Choose the **Add Labels** (Figure 9-11) command from the **Pipes** pull-down menu.

Figure 9-11 Add Labels

10. Change the **Standard Structure** label style by clicking on the **down arrow** next to the **Structure label style** pop-down window and selecting **Edit** label style.

11. This brings up the **Label Style Composer** (Figure 9-12).

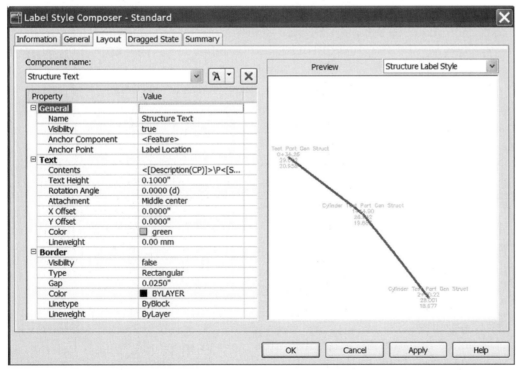

Figure 9-12 The Label Style Composer

12. In the **Layout** tab look in the **Value** column for the **Text Property Contents**. In that cell is a button with ellipses to enter the dialog. Click that button.
13. This causes the **Text Component Editor – Label Text** dialog box to display.
14. Place your cursor to the right of the data in the **Preview** window on the right, and using the **Properties** pop-down window add in the following data items for labeling: **Structure Station**, **Structure Offset**, and **Insert Elevation**. Hit the **blue arrow** to the right of the **Properties** pop-down window after each selection to populate the window to the right. When complete, hit **OK**, **OK**, and **OK** to exit. Notice that the profile labeling has already been altered. See Figure 9-13 for results when complete.

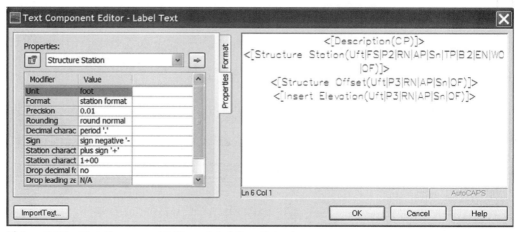

Figure 9-13   Text Component Editor—Label Text dialog box

Exercise 9-5: Move Structures

Now you move the beginning structure from station 9+00 to station 8+50.

1. Zoom back to the plan view or type **V**, select **Plan** view, and restore it.
2. Click on the manhole structure and notice two cyan grips appear. See Figure 9-14.

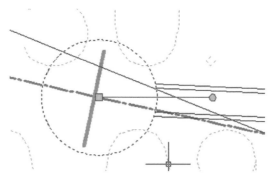

Figure 9-14   Grips on a structure

3. Click on the square grip and stretch the grip to approximately station 8+50.
4. Notice that the structure and pipe move at the same time.
5. Now zoom back to the profile, or type **V**, select **Prof** view, and restore it.
6. Notice that the pipe and structure are already relocated in the profile view.

Exercise 9-6: Create Pipes by Layout

You now perform a brief exercise in identifying pipe conflicts.

1. Open Chapter-9-c.dwg.
2. Type **V** for **View**, restore the **Plan** pipe view, and you see two piping systems in the corridor between station 10+00, a high point, and 13+00. Let us examine whether a conflict occurs here.
3. Select **Create Interference Check** from the **Pipes** pull-down menu.
4. The software responds with `Select a part from the first network:` Select the pipe that crosses the corridor centerline.
5. Then `Select another part from the same network or different network:` Select the other pipe on the north side of the corridor centerline.
6. The **Create Interference Check** dialog box displays. See Figure 9-15. Click on the button at the bottom of the dialog box that says **3D proximity check criteria....**

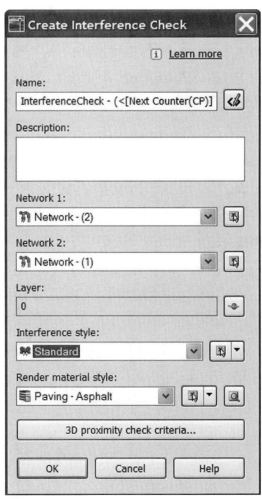

Figure 9-15   The Create Interference Check dialog box

7. When the **Criteria** dialog box appears, turn on the **Apply 3D proximity** check button.
8. Then type in 5′ for the **Use Distance** field and hit **OK**. Hit **OK**.
9. You get a response that says 1 interference was found, which is correct.
10. To see the interference, type **V** for **View**, select **Pipe Conflict**, and set it current.

11. You see a sphere indicating the conflict. Figure 9-16 shows a 2′ sphere for a 5′ conflict. This is just one way to show conflicts. Try experimenting with the settings and identify other methods on your own.

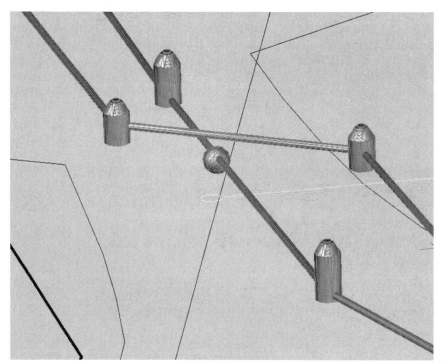

Figure 9-16 An interference of two pipes

With experimentation, you will notice that very powerful drafting features exist in the piping abilities of the Civil 3D product. The 2007 version has added huge capabilities to the product, and its modeling and drafting features are unparalleled.

CHAPTER TEST QUESTIONS

Multiple Choice

1. Meteorological factors affecting runoff include which of the following?

   a. Type of precipitation such as rain, snow, sleet, etc.
   b. Rainfall intensity, amount, and duration
   c. Distribution of rainfall over the watershed
   d. All of the above

2. Physical characteristics affecting runoff include which of the following?

   a. Location such as the tropics, piedmonts, mountains
   b. Land use, vegetation, soil type, and total drainage area involved
   c. Shading of the watershed basin
   d. Sunlight strength

3. Various software is usually used to compute which of the following hydrological characteristics:

   a. Storm runoff volume
   b. Peak rate of discharge
   c. Hydrographs and storage volumes
   d. All of the above

4. Intensity-Duration-Frequency curves indicate which relationships?

   a. Between duration and area
   b. Between storage volumes and flows
   c. Between the intensity and duration of a rainfall event with a given return period
   d. Between erosion and siltation factors

5. The time-of-concentration is:
   a. Used for SCS methods
   b. Defined as the time required for a theoretical drop of water of water to travel from the most hydrologically remote point in the watershed to the point of collection
   c. Also called a T_c
   d. All of the above

6. Several types of watersheds are automatically computed in Civil 3D. They include:
   a. Boundary Profiles
   b. Boundary Points and Segments
   c. Weirs
   d. Low Flow Areas

7. The discharge of water depends mainly on:
   a. The amount of pressure that exists from the inlet to the outlet
   b. The friction or resistance to flow caused by pipe length
   c. Pipe material, bends, constrictions, the type of fluid, and the cross sectional area of the pipe
   d. All of the above

8. The hydraulic radius, R_h, is defined as:
   a. The area of the flow section divided by the wetted perimeter
   b. The wetted perimeter divided by the area of the flow section
   c. $Q = AV$
   d. None of the above

9. There are several terms associated with fluid flow and they include which of the following?
   a. Velocity head
   b. Hydraulic radius
   c. Discharge–area–velocity relationships
   d. All of the above

10. The word *hydraulics* comes from:
   a. The Greek word *hydraulikos*
   b. The English term *water*
   c. Hydrology
   d. Water at rest or in motion

True or False

1. True or False: Runoff is that part of precipitation that is absorbed by the ground, thereby replenishing our reservoirs.

2. True or False: Runoff can cause erosion.

3. True or False: Hydrologic study often involves such items as land use, vegetation, soil type, and total drainage area.

4. True or False: Meteorological factors affecting runoff include rainfall quality and contaminants.

5. True or False: As evidenced by the watershed for Front Royal, Virginia, watersheds tend to be small and usually less than one acre.

6. True or False: IDF curves can be found on the Internet and stand for Internet-Definable-Frequency curves.

7. True or False: The rational method for predicting runoff may be used to predict the peak runoff.

8. True or False: The rational method formula is $q = CIA$, where q is the peak runoff being computed.

9. True or False: Although the rational method appears straightforward, it is totally dependent on the "correct" selection of the area involved.

10. True or False: The time-of-concentration is usually defined as the time required for a drop of water to be absorbed into the ground.

STUDENT EXERCISES, ACTIVITIES, AND REVIEW QUESTIONS

1. What is the volume of water in a full 24″ pipe if the velocity of the flow is 5 ft/s?

2. What is the velocity of water in a 48″ pipe, flowing full with a volume of 60 cfs?

3. What size pipe is needed to convey 50 cfs at a velocity of 4 ft/s?

4. Develop a storm network in Civil 3D that uses different pipe sizes and structures as an exercise in placing the objects.

5. Edit these objects to increase, decrease, or change slopes of the pipes.

Civil 3D's Visualization Capabilities

10

Chapter Objectives

After reading this chapter, the student should be able to:
- Understand the art of visualizing his or her designs for display to the public.
- Have an awareness of a variety of examples in which renderings helped projects succeed through public hearings.

INTRODUCTION

Civil 3D 2007 includes a new rendering engine that replaces the VIZ Render in Civil 3D 2006. This new engine may appear very similar to the popular Autodesk VIZ software, which was created expressly for the architectural/engineering/construction industry.

NEW to AutoCAD 2007

This chapter provides a brief introduction to students on the art of visualizing their designs for display to the public. Visualizations can develop stunning and photorealistic computer renderings of projects. The objective here is simple: to introduce visualization to Civil 3D users. This feature should be explored by the user whenever the opportunity presents itself and may actually warrant its own textbook. This chapter commences with some examples of where these renderings might be used in the civil engineering business. Indeed, the examples provided were extracted from some of my actual projects.

In today's very visual society, communicating designs is a paramount necessity for engineers and designers. Civil engineers seem to lag behind other disciplines when it comes to visualizations, and the reason probably goes back to the idea that most engineers do not design in 3D. That is why Civil 3D could cause a paradigm shift in the way engineers work if it assists them in moving into three dimensions.

SAMPLES OF RENDERINGS

Some examples of renderings of projects that I have performed include the following.

Figure 10-1 shows a 3D model and rendering of a Virginia high school to depict it at night with lighting. It was developed for public hearings for new construction at the school.

The site pictured in Figure 10-2 was a 1.5 square mile model developed for public hearings on the new road and bridge. The graphic is a computer rendering of the site.

Figure 10-1   A rendering of a Virginia high school sports field

Figure 10-2   A 1.5 square mile model of a road and bridge for public hearings

The roadway and bridge were cut from this model and overlayed geographically into a photograph of the site to create the photomontage in Figure 10-3. The existing conditions can be viewed in the upper right of the photomontage.

The rendering in Figure 10-4 was done for a local parks organization for public hearings to show that a quality structure and related grounds were to be built in its neighborhood.

The site in Figure 10-5 was developed as a 3D model and rendered for use in a litigation case. It had to do with zoning issues and included several scenarios of waivers that might be applicable.

Figure 10-3   Photorealistic Photomontage

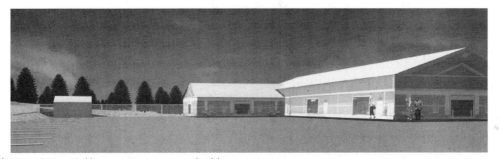

Figure 10-4   Golf course maintenance building

Figure 10-5   Zoning issues and shadow studies

The model and photomontage in Figure 10-6 were done for a DOT project to show what a lane widening would appear like. The lane on the left with its respective shoulder did not exist when this rendering was done.

Figure 10-6   Photorealistic Photomontage

The rendering and 3D model in Figure 10-7 were done for a DOT project and include a new bridge and entrance monument with decorative lighting. The existing conditions are shown in the upper right of the photomontage.

Figure 10-7   A new gateway to Annapolis

The 3D model in Figure 10-8 was built for an international airport and displayed within the state legislature to show the legislators what the airport would appear like after construction in the year 2010. New interstate highways, railway spurs, and overhead taxiways were all part of the project. New parking and concession facilities were in the model as well as expanded service areas.

Figure 10-8   An international airport

In the model pictured in Figures 10-9 and 10-10, lighting was a factor, and the office building and parking are depicted with both daytime and nighttime lighting. Everything is done to engineering specifications including parking lot drainage, cross slopes on sidewalks, and so forth.

Figure 10-9   A commercial office building—daylight

Figure 10-10   A commercial office building—night lights

The example in Figure 10-11 is of another airport but shows a 3D model with an orthophoto draped onto it. Some relatively simple structures were placed on the model depicting planned expansion.

Figure 10-11   Orthophoto draping

The project shown in Figure 10-12 included improvements to both the site as well as the building for this hospital. It was shown to the directors of the hospital prior to construction so they could visualize what the plans indicated.

Figure 10-12   V.A. Hospital improvements

The last 3D model and rendering in Figure 10-13 was performed for public hearings to allow for a roadway extension to be built. The model was projected into photographs of the site to show a photorealistic view of the project.

Figure 10-13   Public hearing presentation for a roadway extension

Exercise 10-1: Civil 3D Example

Let us take a short trip into the visualization world by opening a file you have worked on and viewing it in a 3D perspective.

Civil 3D has visualization built in. In order to expedite the lesson, once you open the file, also notice that some closed polylines have been drawn on the **C-Mask** layer. The surface modeling in Civil 3D allows for masks to be placed on the data, which can affect several things, among them, rendering. The green masks represent grass-covered areas, the blue masks represent sand areas, the red areas indicate gravel drives, and the black mask indicates the paved road. There are five grass areas, three sand areas, two gravel areas, and one paved road.

1. Let us begin by opening Chapter-10-a.dwg. You see the surface you have been dealing with in various parts of this text as well as the masks.
2. In the **Toolspace**, under the **Prospector** tab, expand **Surfaces** to see the **Existing ground**.
3. Expand **Existing ground** and you see an item for **Masks**. Right click on **Masks** and select **New**....
4. The **Create Mask** dialog box appears, as shown in Figure 10-14.

Student Files

Properties	Value
⊟ **Information**	
Name	Mask 1
Description	Grass
⊟ **Masking**	
Mask type	Render Only
Mid-ordinate dista...	3.000'
Render Material S...	Grass - Thick

OK   Cancel   Help

Figure 10-14   The Mask Properties dialog box

5. Accept the name as the default; enter **Grass** for the description. Click on the Value for **Mask type** and select **Render Only**. Change the **Mid-ordinate distance** to 3′, and select **Grass-Thick** for the **Render Material Surface**. Hit **OK**.

6. When prompted, select a **green** colored mask to apply the grass material to.

7. Repeat steps 3 through 6 for the additional four grass areas.

8. Right click on **Masks** and select **Add**....

9. The **Mask Properties** dialog box appears.

10. Accept the name as the default; enter **Sand** for the description. Click on the Value for **Mask type** and select **Render Only**. Change the **Mid-ordinate distance** to 3′, and select **Sand** for the **Render Material Surface**. Hit **OK**.

11. When prompted, select a **blue** colored mask to apply the sand material to.

12. Repeat steps 8 through 11 for the additional two sand areas.

13. Right click on **Masks** and select **Add**....

14. The **Mask Properties** dialog box appears.

15. Accept the name as the default; enter **Gravel-Mixed** for the description. Click on the Value for **Mask type** and select **Render Only**. Change the **Mid-ordinate distance** to 3′, and select **Gravel-Mixed** for the **Render Material Surface**. Hit **OK**.

16. When prompted, select a **red** colored mask to apply the gravel material to.

17. Repeat steps 13 through 16 for the additional gravel area.

18. Right click on **Masks** and select **Add**....

19. The **Mask Properties** dialog box appears.

20. Accept the name as the default; enter **Asphalt** for the description. Click on the Value for **Mask type** and select **Render Only**. Change the **Mid-ordinate distance** to 3′, and select **Asphalt** for the **Render Material Surface**. Hit **OK**.

21. When prompted, select a **black** colored mask to apply the asphalt material to. Save the file.

Because the file is developed, you now view it in 3D.

22. Type **V** for **View**, select the view called **Corridor ISO**, set it **Current**, and hit **OK**.

23. Turn on all layers and thaw them out.

24. Then select **View Style >> Realistic** from the **View** pull-down menu.

25. Use the **Zoom** command to zoom out and observe the view, and you should see grass, sand, gravel, and asphalt, as shown in Figure 10-15.

We are now in a new world and that is 3D modeling and visualizations!

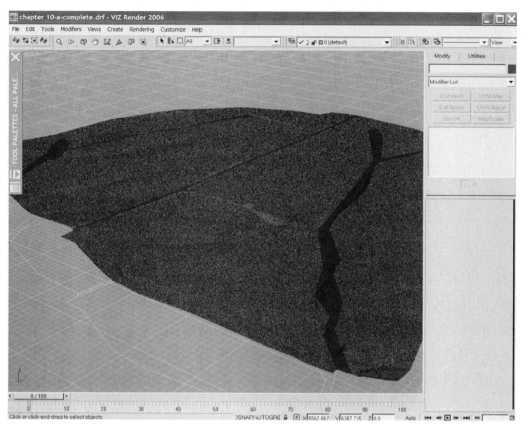

Figure 10-15   Materials are applied

Surveying

Chapter Objectives

After reading this chapter, the student should be able to:
- Understand how the Surveying interface operates within Civil 3D.
- Understand the application of Surveying Objects.
- Perform traverse balancing tasks.
- Understand how to reduce field data collection files.

NEW
to AutoCAD
2007

INTRODUCTION

Chapter 11 introduces surveying to the Civil 3D user. This technology was added to the 2007 release and largely completes the offering for use throughout the enterprise. With this technology, the surveying department can now develop the entire base data required for a project; the planning department can perform initial, preliminary designs; and the engineering department will take the project to final design. On completion, the project's data are likely to return to the surveying department for quality assurance/control and stakeout, or Machine Control data preparation.

The Civil 3D 2007 surveying functionality includes the following functions and abilities.

- The Prospector's Survey interface. Surveyors can view the numeric data along with previews; locate data visually, such as traverse or figures, in AutoCAD; and even change properties and values of the data using simple, succinct pull-down menus.

- Import, manage, and edit data collector field data.

- Enter, edit, balance, and report on open or closed traverses.

- Manage and edit survey equipment libraries.

- Manage and edit figure prefix libraries for automatic linework generation. This concept is similar to what has been offered in Autodesk's Land Desktop software, in which linework is collected in the field and given a prefix. For example, tops of curbs can be given prefixes such as TC1, TC2, and so on. The figure prefix library contains an entry called TC, so the software treats anything beginning with TC as a TC item, thereby correctly establishing the layering for the linework. This allows multiple edges of pavement to be open in the data collector simultaneously. The user should build this carefully so as not to have other linework items mistakenly fall into the prefix. Consider something like a Tree and a Traverse station. If a Tree is prefixed with TR*, TRV descriptions would also be captured.

- The Prospector's inclusion of an interface collection for traverse networks, figures and survey points, and the ability to manage and oversee these collections visually.
- Create survey figures directly from AutoCAD linework. This is even new to Land Desktop users.
- Develop terrain model breaklines from figures that have a specific prefix. In other words, perhaps linework coded with a BRKL prefix is intended to be breaklines for the Digital Terrain Model (DTM) and so many include tops and toes of channels, ridge lines, and so on. Users can now send those figures that have a BRKL prefix, for example, directly into the surface.
- The use of the Vault to protect the data for which surveyors are ultimately responsible.
- Output design data and surfaces in formats compatible with Topcon (via Carlson Connect), Trimble (via Trimble Link), and Leica (via Leica X-change) GPS Machine Control equipment.

The purpose of this chapter is to introduce the reader to some fundamental surveying concepts and provide basic procedures developing survey data on projects.

REVIEW OF FUNCTIONS AND FEATURES TO BE USED

The text provides instruction in the functions for traverse entry, balancing traverse angles, how to compute the errors that are inherent in traverse computations, and the mathematics behind balancing the error of a traverse. Once the theories and algorithms are covered in the chapter, the discussion moves into automating these calculations in Civil 3D.

SURVEYING FUNCTIONS

Surveyors perform many functions in the process of developing engineering projects. They are responsible for collecting the base information consisting of existing conditions prior to design. This often includes identifying the property's boundary; the environmental conditions affecting the site such as wetlands and water bodies; the planimetric features, which are the buildings and other infrastructure on the site; and the topography, which consists of the three-dimensional conditions.

The method for collecting these data begins at the courthouse to identify the legal bounds of the property. These bounds are drawn up by the surveyor in a Civil 3D file using the Parcels and Rights of Way discussed in Chapter 4. These might be augmented by existing centerlines, easements, and so on.

Following or simultaneously, field collection techniques are employed to find physical evidence for the information on the site. GPS or total station equipment may be used to locate these data accurately. Precise 2D and 3D Control must be established for the project before any significant work is performed. In other words, monuments are established on the properties that are tied into state-certified control. These monuments can be occupied time and time again, and allow specific relations to be performed, as surveying and construction occur so that all aspects of the site are perfectly related.

The developed control has errors in its initial calculations due to: either equipment or human error. The error is then eliminated from the control, creating a perfectly closed polygon, the vertices of which become the control monuments.

Other surveying techniques include aerial photogrammetry, GIS research, and GPS surveying. An example of an aerial photogrammetry project in Chapter 5 consisted of about 200 acres of natural ground. Photogrammetry is the surveying task of flying above a project and photographing it using a zigzag pattern in which similar images at slightly different angles are photographed. Photographs are placed in the stereodigitizer and calibrated using Visually Identifiable Objects (VIOs) that are set using GPS or total station equipment. These now have accurate 3D coordinates that are set to the VIOs in the photos, thus calibrating everything in the photos to a coordinate. Digitizing then captures features and breaklines. This excellent method for collecting large amounts of data quickly has accuracies of about 0.01' for every thousand feet the aircraft is flying above the site. These data are typically verified by localized equipment such as total stations to ensure that the accuracies are met. The data produced from the stereodigitizing consist of CAD data usable by software such as Civil 3D.

A rapidly increasing methodology for collecting data is through the use of GPS equipment. Base stations are set on the project, again tied into state monumentation. The surveyor uses a rover with a GPS antenna that communicates with the base station to compute relative coordinates of the distance the rover is from the base station. Because the base station is tied down, the rover can be tied down relative to it, thus producing very accurate results. A new technology under development all over the world is the use of Imaginary Reference Stations. Using this technology, organizations collect a variety of certified base stations, tie them into a central server, amass errors, and correct for those errors. When a rover ties into the network, its location is computed and an imaginary base station is created virtually at the location of the rover. Interpolated error corrections occur on the server and send postcorrected data (i.e., coordinates) directly to the rover for immediate, and direct use in surveying or construction. See Figures 11-1 and 11-2 for graphics on how this works.

These systems are operational in many parts of the world. In the United States they tend to be accessible in certain regions where this technology is in development. Many state DOTs are behind this development, as are private interests.

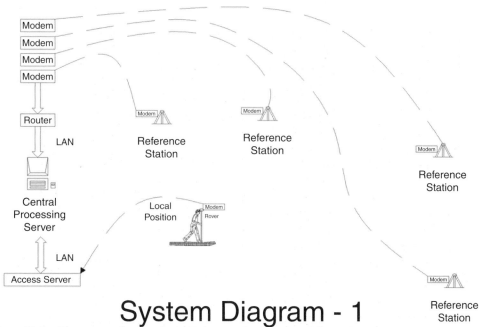

System Diagram - 1

Figure 11-1   The process for Imaginary Reference Stations—initializing the rover

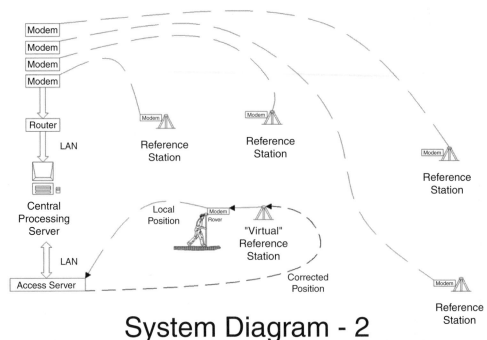

System Diagram - 2

Figure 11-2   The process for Imaginary Reference Stations—error corrections obtained

Developing Traverse Data

When a surveyor creates the traverse, he or she occupies a known location and typically backsights another known location. The angle between the two locations is now established. Then foresights are shot to new locations that may be accessible to the project. The idea is to locate positions that will be needed as the site is surveyed and built. These positions should be in as open an area as possible so they can be seen from as many points of view as possible in order to establish control for that portion of the project. A position would be useless if the surveyor could not see it. This then determines how many traverse points are required.

Once the setup station is identified and a backsight is created, turned angles and distances are captured to establish the next location (turned angles were discussed in Chapter 4).

Exercise 11-1: Creating a Traverse

You now create a traverse using the following data. The procedure here is simple in order to convey the ideas involved, but the same techniques can be used on larger, more sophisticated traverses.

The traverse setup is on Point *A* located at 1000,1000. Backsight *B* is located 500′ at a bearing of South 45-00-00 West of Point *A*. A turned angle of 90-00-00 degrees and a distance of 500′ produces Point *C*. Then *C* is occupied with Point *A* being the backsight. A turned angle of 269-30-00 and a distance of 499.90′ produces Point *D*. Point *D* is then occupied with *C* being the backsight to locate Point *E*. A turned angle of 270-00-00 and a distance of 500′ are used to locate Point *E*. Then *E* is occupied with a backsight to *D* to locate Point *A* again at a turned angle of 269-30-00 and a distance of 500. Refer to Figure 11-3. There is not a closed traverse because error has been placed into the computations. Clearly, surveyors do not intentionally add error, yet the equipment might, or perhaps a prism is not held perfectly plumb to the shot. Maybe ambient conditions such as heat or fog have affected the line of sight. To replicate this error, the bust of 0.1′ and 1 degree has been entered, and a method known as the Compass rule will be used to correct this error. This method has been used for many

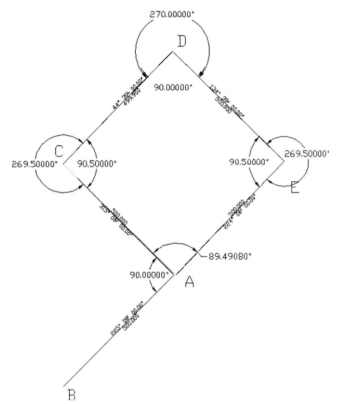

Figure 11-3   Traverse data

years and still holds some favor because it is easily hand calculated for verification purposes. Computer automation has improved traverse balancing by allowing surveyors to use a much more sophisticated method called least squares, but this exercise concentrates on the compass rule.

Using the Compass rule, first correct for the bad angles and then for the bad distances. To correct for the angles, this figure should be closed. The intention was to start at Point *A* and return to Point *A*, but with the error it was impossible to close back precisely. You need to determine how much angular error and distance error there is. It is necessary to know how many degrees should be in the figure, and certain simple methods can obtain that information. Closed figures with *n* sides are *n*-gons and the angles sum up to

$$(n - 2) \times 180$$

TIP

So a figure with

3 sides is a triangle and has 180°.

4 sides is a rectangle and has 360°.

5 sides is a pentagon and has 540°.

6 sides is a hexagon and has 720°.

7 sides is a heptagon and has 900°.

8 sides is an octagon and has 1080°.

9 sides is a nonagon and has 1260°.

10 sides is a decagon and has 1440°.

12 sides is a dodecagon and has 1800°.

15 sides is a 15-gon and has 2340°.

So, based on this information there should be 360° in Figure 11-3. Let us find out what we have and correct it in the following manner (see Table 11-1).

Point	Measured Interior Angle	Average Correction (Degrees)	Adjusted Angle (Degrees)
A	89.4908	−0.1227	89.3681
C	90.5	−0.1227	90.3773
D	90.0	−0.1227	89.8773
E	90.5	−0.1227	90.3773
Result	360.4908	.4908	360.00

Table 11-1   Adjustment of Angles

Note that the sum of the interior angles is 360.4908 (it should be 360-00-00), therefore, the average correction is .4908/4 vertices = 0.1227". This is applied to the angles, and because it is an average, it is subtracted. By the time the error is applied to Point E, the entire error of .4908 degrees has been distributed.

Computation of Azimuths for the Traverse

Now compute the relative azimuths of each traverse segment. The azimuth of AB = 225.0 degrees and with the turned angle of BAC = 90.0 degrees, the azimuth of AC = 315.0 degrees. Next you occupy each setup, then backsight the previous setup, and turn an interior angle to obtain the next setup. Then apply the corrected turned angle from Table 11-1 to each vertex.

315.0 = AC
−180, to flip the line of sight from C to A
_____
135.0 = CA
+90.3773 = C, subtract adjusted interior angle of C from azimuth
_____
225.3773 = CD
−180, to flip the line of sight from D to C
_____
45.3773 = DC
+89.8773 = D
_____
135.2546 = DE
−180
_____
315.2546 = ED
+90.3773 = E
_____
405.6319 = EA
−180
_____
225.6319 = AE
+89.3681 = A
_____
315.0 = AC

This check proves it is now balanced because this result is the same angle where you began.

Exercise 11-2: Computing a Traverse Misclosure and Relative Precisions

Here you compute the nonclosure or linear misclosure based on the incorrect distance(s). You compute how far off the error is in the x direction and the y direction. These values are the departure and latitude (refer to Figure 11-4).

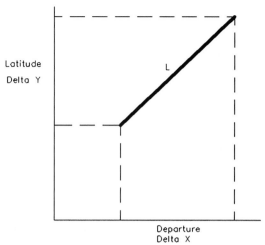

Figure 11-4   Departure and latitude diagram

Departures are eastings/westings, where as latitudes are northings/southings.

Departures = L sin(alpha)

Latitudes = L cos(alpha)

where the angle alpha is from north (straight up) to line *L*.

Linear misclosure is the square root of the departure of misclosure squared plus the latitude of misclosure squared or

$$\sqrt{(\text{dep. misclosure})^2 + (\text{lat. misclosure})^2}$$

See Table 11-2.

Station	Preliminary Azimuths	Length	Departures	Latitudes
A	315.0	500	−353.554	353.554
C	45.3773	499.9	355.801	351.147
D	135.2546	500	351.979	−355.121
E	225.6319	500.0	−357.431	−349.633
A				
		Sum = 1999.9	−3.205	0.053

Table 11-2   Table Computation of Departures and Latitudes

Linear misclosure = $\sqrt{(-3.203)^2 + (.053)^2}$

= $\sqrt{(10.272 + .0028)}$ = 3.205′

Relative precision = Linear misclosure/Traverse length

= 3.203/1999.9 = 0.0016 = 1/624

The compass rule adjusts the departures and latitudes of traverse courses in proportion to their lengths. See Table 11-3.

Correction in departure for *AC* =

$$-\frac{(\text{Total departure misclosure})}{\text{Traverse perimeter}} \times \text{Len } AC$$

Correction in latitude for *AC* =

$$-\frac{(\text{Total latitude misclosure})}{\text{Traverse perimeter}} \times \text{Len } AC$$

| Station | Preliminary Azimuths | Unadjusted | | | Balanced | | Coordinates | |
		Length	Dep.	Lat.	Dep.	Lat.	X east	Y north
A			(.80)	(.013)			1000	1000
	359-59-59.5	500	−353.554	353.554	−352.754	353.567		
C			(.80)	(.013)			647.246	1353.554
	270-00-01	499.9	355.803	351.147	356.603	351.16		
D			(.80)	(.013)			1003.849	1704.714
	180-00-00.5	500	351.979	-355.121	352.779	−355.108		
E			(.80)	(.013)			1356.628	1349.606
	90-00-00	500	−357.431	−349.633	−356.631	−349.62		
A							999.997	999.986
Summary =		1999.9	−3.203	.053	−.003	−.001		

Table 11-3   Table of Compass Rule

$$-3.203/1999.9 \times 500.0 = .8'$$
$$0.053/1999.98 \times 500.0 = .013'$$

For C to D:

$$0 - 3.203/1999.9 \times 499.9 = 0.8'$$
$$0.053/1999.9 \times 499.9 = .013'$$

ELECTRONIC FIELD RECORDERS OR DATA COLLECTORS

Using field records, surveyors also collect data. When the rod person places the prism on the shot to be located, a shot code and a possible linecode will be determined if the shot also represents linework such as a crown or edge of pavement. Point codes discussed in Table 3-1. These codes describe the points, and the libraries place the shot on the correct layers, with the correct symbology and annotation. In the case of linework, there are also linecode libraries.

SURVEY LANGUAGE

This is also known as the field book format.

Point Creation Commands

AD = ANGLE DISTANCE, locates points/linework by a turned angle, distance
AD (point) [angle] [distance] (descript)
AD [VA] (point) [angle] [distance] [vert angle] (descript)
AD [VD] (point) [angle] [distance] [vert distance] (descript)

AUTO PT NUMBERING, for automatic point numbering
AP ON (point)
AP OFF

BEARING DISTANCE, locates points/linework by bearing, distance
BD (point) [bearing] [quadrant] [distance] (descript)
BD [VA] (point) [bearing] [quadrant] [distance] [vert angle] (descript)
BD [VA] (point) [bearing] [quadrant] [distance] [vert distance] (descript)

DEFLECTION DISTANCE, locates points/linework by deflection angle, distance

DD (point) [angle] [distance] (descript)

DD [VA] (point) [angle] [distance] [vert angle] (descript)

DD [VD] (point) [angle] [distance] [vert distance] (descript)

FACE 1, locates point from a traverse station. Using horizontal angle measured on face 1, distance

FACE 2, locates point from a traverse station. Using horizontal angle measured on face 2, distance

F1 (DIRECT), F2 (INVERTED)

F1/F2 (point) [angle] [distance] (descript)

F1/F2 [VA] (point) [angle] [distance] [vert angle] (descript)

F1/F2 [VD] (point) [angle] [distance] [vert distance] (descript)

LAT LONG (point) [latitude] [longitude] (descript)

NE, sets a 2D point.

Is a Control Point and is not moved in traverse balancing

NEZ, sets a 3D point.

Is a Control Point and is not moved in traverse balancing

NE SS, 3D pt side shot (not fixed in observation database, can be over-written)

NE (point) [North] [East] (descript)

NEZ (point) [North] [East] (elev) (descript)

NE SS (point) [North] [East] (elev) (descript)

PRISM [height]

PT OFFSET (point) [offset] (ahead)

SKIP

STADIA, reduces a distance measured using stadia rod intercept

STADIA (point) [angle] [distance] [rod] (vert angle) (descript)

AZ=AZIMUTH DISTANCE, locates points/linework by azimuth, distance

ZD (point) [azimuth] [distance] (descript)

ZD [VA] (point) [azimuth] [distance] [vert angle] (descript)

ZD [VD] (point) [azimuth] [distance] [vert distance] (descript)

Point Location Commands

AZ [point 1] [point 2] [azimuth]

B [point1] [point 2] [bearing] [quadrant]

BS [point] (orientation), **backsight a known point**

STN [point] (inst. height) (descript), **occupy a station**

Point Information Commands

A [point 1] [point 2] [point 3]

AZ [point 1] [point 2]

B [point 1] [point 2]

D [point 1] [point 2]

DISP PTS (point 1) (point 2)

GRADE [point 1] [point 2]

INV PTS [point 1] [point 2]

INVERSE RADIAL [backsight] [station]

SLOPE [point 1] [point 2]

SD [point 1] [point 2]

VD [point 1] [point 2]

Point Editing Commands

DEL PTS [point 1] (point 2)

MOD DESC [point 1] [descript]   MOD EL [point 1] [elev] MOD ELS [point 1] [point 2] [elev]

MOD ELS BY [point 1] [point 2] [amount]

Figure Commands

AREA [figure]

BEG [figure], **begins a figure called {FIGURE}**

C3, next three shots are on a curve. The first shot should be on the beginning of the curve, the next shot about midway around the arc, and the last shot should be on the end of the curve.

CLOSE

CLOSE BLD, closes to connect the last point to the first point by creating two right-angled segments to close a building

CLOSE RECT [offset], closes objects such as sheds, storm structures

CONT [figure], continues a figure called {FIGURE}. This command assumes that the last point in the existing figure is the point to continue from. It continues a figure from the back end.

CRV = TANGENT CURVE

CRV [DELTA, LENGTH, DEFL, MID, TAN, CHORD] [radius] [value]

END, ends the current figure

These Figure commands draw figures using a data collector or manual input; they do not set points.

FIG AD [angle]   [distance] FIG DD [deflection] [distance] FIG ZD [azimuth] [distance]

FIG BD [bearing]   [quadrant] [distance]   FIG NE [Northing] [Easting]

ID FIG

INVERSE FIG [figure]

MAPCHECK [figure]

OFFSET [figure] [distance]

PC, next shot is on PC. The shot after that must be a PT.

POINT [point]

RT [distance], is a great command for drawing a building that has 90-degree corners and measured wall lengths

SET (point)

START [figure], restarts a figure called {FIGURE} from the beginning of the figure

These XC commands are for drawing nontangential curves.

XC = EXTEND CURVE

XC ZD (BULB) [radius] [chord-az] [chord-dis]

XC BD (BULB) [radius] [chord-brg] [quad] [chord-dis]

XC AD (BULB) [radius] [chord-angle] [chord-dis]

XC DD (BULB) [radius] [chord-defl] [chord-dis]

XC C3 [pt on curve] [endpt]

XC PTS [radius] [radpt] [endpt]

Intersection Commands

AZAZ [pnt 1] [az 1] [off 1] [pt 2] [az 2] [off 2]

BB [pnt 1] [brg 1] [quad 1] [off 1] [pnt 2] [brg 2] [quad 2] [off 2]

LNLN [pnt 1] [pnt 2] [off 1] [pnt 3] [pnt 4] [off 2]

RKAZ [point] [radius] [pnt 1] [azimuth] [offset]

RKBRG [point] [radius] [pnt 1] [brg] [quad] [offset]

RKLN [point] [radius] [pnt 1] [pnt 2] [offset]

RKRK [pnt 1] [radius 1] [pnt 2] [radius 2]

SAVE [NORTH, SOUTH, EAST, WEST, 1, 2, ALL] (point) (descript)

SAVE [NEAR, FAR] (point) [ref pt#] (descript)

SAVE PICK (point) (descript)

SQ [ref 1] [ref 2] [ref]

Figure Correction and Manipulation Commands

DEL FIG [fig]   DISP FIGS

Equipment Correction Commands

ANGLES [RIGHT, LEFT] [ZENITH, NADIR, HORIZ]

ATMOS [ON, OFF]   COLL [ON, OFF]   CR [ON, OFF]

EDM OFFSET [offset]

HORIZ ANGLE [RIGHT, LEFT]

PRESS [pressure] [INCH, MBAR, MM]

SF [factor]

PRISM CONSTANT [constant]   PRISM OFFSET [offset]

TEMP [temperature] [F, C, K]

UNITS [METER, FOOT] [DMS, DECDEG, GRAD, MILS, RADIANS]

VERT ANGLE [ZENITH, NADIR, HORIZ]

Baseline Commands

BL IS [point 1] [point 2] [station]   BL PT (point 1) [station] (offset)
(descript)

BL INV [point 1] (point 2)

Centerline Commands

CL IS [figure] (station) (point)

CL INV [point 1] (point 2)

CL PT (point) [station] (offset) (skew angle) (descript)
CL PT BY [point] [station 1] [offset] [distance] (station 2) (descript)
CL ELEV (point) [station] [offset] [elev] (descript)
CL ROD (point) [station] [offset] [rod] (descript)
CL VD (point) [station] [offset] [vert distance] (descript)
HI [elev]
XS [station]
XS ELEV (point) [offset] [elev] (descript)
XS ROD (point) [offset] [rod] (descript)
XS VD (point) [offset] [vert distance] (descript)

AutoCAD Related Commands

PAN   SHELL   REDRAW   ZOOM PT [point]
ZOOM [WINDOW, EXTENTS, PREVIOUS, W, P, E, A]

Miscellaneous Commands

CALC [formula]   DITTO [ON, OFF]   HELP (command)   HISTORY
OUTPUT [ON, OFF]   TRAV [ON, OFF]

Unit Conversions for Distance

Unit	Symbol	Equation
Inches	"	
Feet	'	1 ft = 12 inches
Yards	Y	1 yd = 3 ft
Rods	R	1 rod = 16.5 ft
Chains	C	1 chain = 66 ft
Links	L	1 link = 1/100 chain = 0.66 ft
Meters	M	1 ft = 0.3048 m
Millimeters	MM	1 mm = 1/1000 m
Centimeters	CM	1 cm = 1/100 m
Kilometers	KM	1 km = 1000 m

Mathematical Operations

Sometimes a surveyor needs to compute distances, angles, bearings, or azimuths, and these commands also perform mathematical calculations. Occasionally, an angle, bearing, or distance is needed, but the surveyor may be unsure of what the value is. However, it might be possible to define this information from existing points.

The following commands—Angle, Azimuth, Bearing, Slope, Grade, Vertical Distance, Slope Distance, and Distance commands—display angle, azimuth, or distance values. However, when they are placed inside parentheses, they can be used to compute these values and send them into other commands. For instance:

ZD 5000 (AZ 500 501) (D 503 504) uses the azimuth from point 500 to point 501, and the distance from point 503 to point 504 to set point 5000.

The built-in functions can also be used in conjunction with the [Calc] command. For example:

CALC (A 10 20 30)/2 returns half the angle from point 10 to point 20 to point 30.

The Angle and Azimuth commands always return an angle. This value can be used in any command that requires an angle or an azimuth. For example:

ZD 5000 (SLOPE 10 20) 1000 uses the slope angle between point 10 and point 20 to set point 1000.

AD VA 5000 250.3000 600.00 (SLOPE 10 20).

Files Produced When Balancing

When an analysis is run and the network is adjusted, the following files are created and stored:

- **an<traversename>.trv**: Displays the horizontal closure and angular error
- **fv<traversename>.trv**: Displays a report of raw and adjusted elevations from the vertical adjustment methods
- **ba<traversename>.trv**: Displays the adjusted station coordinates derived from balancing the angular error and horizontal closure with no angular error
- **<traversename>.lso**: Displays the adjusted station coordinates based on Horizontal Adjustment Type setting (Compass rule)

Secrets of Survey Functions

A file exists in the folder C:\Documents and Settings\All Users\Application Data\Autodesk\C3D 2007\enu\Survey\ called **translat.ref** File. This file can be customized to control the order in which the user would prefer to enter the data parameters.

Translator File for SURVEY

Line

—

1 Description

2 Command Number, # of Command Phrases, # of data elements,

Auto Point Data #, Check for last literal,

Variable command element, Variable data element,

Batchable command

3 Command Phrase1 #, Command Phrase2 #, ...

4 Data Mnemonic1, Data Mnemonic2, ...

5 Data Order1, Data Order2, ...

6 NULL

7 Blank Line

Example

BD

32,1,4,1,1,0,0,1

12

P,B,Q,D

1,2,3,4

NULL

Another file that exists in the same folder is called **language.ref file**. It contains the potential synonyms for each of the commands in the field book language. For instance, if the user would prefer to call the AD command "LOC" for *locate*, then LOC can be added to command #1.

A listing of the default synonyms is shown here; however, this file can be customized for use in Civil 3D.

```
# SYNONYM FILE FOR SURVEY
# Phrase Number, SYNONYM1, SYNONYM2, SYNONYM3, ....
1,AD,ANG-DIST
2,ADJ,ADJUST
3,A,ANG,ANGLE,ANGLES,ANGS
4,AUTO,AP
5,ATMOSPHERE,ATMOS
6,AZ,AZM,AZIMUTH
7,AZAZ
8,AZ-DIST,ZD
9,AZSP,AZSPI,AZ-SPI,AZ-SPIRAL
10,B,BEARING,BRG,BRGS,BEARINGS
11,BB,BRG-BRG
12,BD,BRG-DIST
13,BEG,BEGIN,BEGINS
14,BL,BASELINE
15,BOWDITCH,COMPASS
16,BRK,BRG-ARC,BARC,BRG-RK
17,BS,BACKSIGHT,BACKSITE
18,BSP,BRG-SPI,BSPI
19,BY,INC,INCRE,INCREMENT
20,CHORD,CHRD
21,CL,CENTERLINE,CLINE
22,COLL,COLLIMATION
23,CONSTANT
24,CONT,CONTINUE
25,CR,CURVATURE
26,CRANDALL,CRAND
27,CV3,C3
28,CURVE,CRV,CV
29,DD,DEFL-DIST
30,DESC,DESCR,DESCRIPT,DESCRIPTION
31,DEL,DELETE
32,DEPTH
33,DISPLAY,DISP,DSP,LIST
34,D,DIST,DISTANCE,DIS,HD,HDIST,LENGTH
35,DELTA
36,EDM
37,EL,ELEV,ELEVATION
38,ELEVATIONS,ELS,ELEVS
39,END,ENDS,E
40,ENTITY,ENT
```

41,EXIT,QUIT,STOP

42,FC1,F1,FACE1

43,FC2,F2,FACE2

44,FACTOR,FACT

45,DEFL,DEFLECTION

47,FILE

49,FOR

50,FROM

51,FS,FORESIGHT,FORESITE

53,HI

56,INDEX

57,INV,INVERSE

58,IS

59,JOB

61,LEVEL,LV

62,LINE,LN

63,LINE-LINE,LNLN

64,LINE-SPI,LNSP

65,LOC,LOCATE,LOCATION

66,LOW,LO

72,MOD,MODIFY

74,NE,NORTH-EAST

75,NEW

76,NEZ

77,NULL

78,OBJ,OBJECT

79,OFF

80,OFFSETS,OFFSET

81,ON

82,PT,POINT,PNT

83,PNTS,PTS,POINTS

84,PAN

85,PC

86,PD,PAR-DIST,PAR-DIS

87,PRC

88,PRESS,PRESSURE

89,PRISM

90,RAD,RADIAL,RADIUS

91,RECALL

92,REDRAW,R

94,RK,ARC

95,RKB,ARCB,ARC-BRG,RKBRG

96,RKAZ,ARCAZ,ARC-AZ,RKZ

97,RKLN,ARCLN,ARC-LINE

98,RKRK,ARCARC,ARC-ARC

99,RKSP,ARCSP,ARCSPI,ARC-SPIRAL,ARC-SPI

```
100,ROD
101,ROTATE,ROT
102,SAVE,KEEP
103,SCALE
104,SCS
105,SIDESHOTS,SIDESHOT,SS
106,SF,SCALEFACTOR
107,SL,SEA-LEVEL
108,SYMBOLS,SYM,SYMBOL
109,SPIRAL,SPI,SP
110,SPSP,SPISPI,SPI-SPI,SPIRAL-SPIRAL
111,SQ,SQ-OFF,SQOFF,SQUARE-OFF,SQUARE
112,STADIA
113,START
114,STN,STATION,STA
115,TAN-OFF,TANOFF
116,TAN,TANGENT
117,TEMP,TEMPERATURE
118,TIME,DATE
119,TO
120,TRANSIT
121,TRANSLATE,TRANS
122,TRAVERSE,TRAV,TRV
123,UNIT,UNITS
125,VA
126,VD,VDIST
127,VERTICAL,VERT
128,VIEW
129,W,WINDOW
130,XSECT,XS,XSECTION
131,SD,SDIST
132,ZOOM,Z
137,COMMAND,CMD
138,INFO,INFORMATION
139,EXTENTS,E
140,PREV,P,PREVIOUS
141,DITTO
142,SQRT
144,WAIT
145,WHILE
146,SIGNATURE
147,OUTPUT,OUT
148,MID,MIDORD,MID-ORDINATE
149,PICK
150,NEAR
151,FAR
```

152,ALL
153,XC,NC
154,BULB
155,RUN
156,BATCH
157,FBK
200,RIGHT,RT,RT-TURN
201,ZENITH
202,HORIZ,HORIZONTAL,HOR
203,NADIR
204,LEFT,LT
205,USFOOT,USFEET
206,METER,METERS,METRIC
207,FOOT,FEET
208,DMS
209,GRAD,GRADS
210,DECDEG
211,MIL,MILS
212,FAHREN,F,FAHRENHEIT
213,K,KELVIN
214,CELSIUS,CENTIGRADE,C
215,MM
216,INCH,INCHES
217,MBAR
218,GRADE,G
219,SLOPE
220,NORTH
221,SOUTH
222,EAST
223,WEST
224,MAPCHECK,MAPCHK,CHECK,CHK
225,AREA
226,RECT,RECTANGLE,RECTANG
227,CIRC,CIRCLE
228,CLOSE
229,RENAME
230,NAME
231,SSS
232,SAS
233,SAA
234,SSA
235,ASS
236,ASA
237,AAS
238,TRIANGLE,TRI,TRIANG
239,SOLVE

241,BLDG,BUILDING,BLD,BUILD
242,FIGURE,FIG
243,FIGURES,FIGS
244,SKEW
245,SET
246,LAT,LATITUDE
247,LONG,LON,LONGITUDE
248,ID,IDENT,IDENTIFY
249,SKIP
250,BACKUP
251,HISTORY,HIST,H
252,CALCULATE,CALC,CA
253,HELP,?
254,EDIT
255,MCS
256,MCE
257,RADIANS,RADS

Exercise 11-3: Computing a Traverse Adjustment in Civil 3D

The following procedure takes the reader through an entire survey project beginning with a closed traverse taken from field notes to entering the field shot locations and then checking the traverse with a cross loop. This is usually done for quality assurance/control and accuracy purposes.

Field Notes

BS	STA @	FS	Angle RT.	DISTANCE
4	1	2	278-05-06	513.43
		(6)Mon 101	61-14-49	214.84
		(7)IPF A	311-23-25	23.85
1	2	3	291-45-07	440.52
		(8)Mon 102	163-14-58	221.69
2	3	4	260-24-37	308.20
		(9)IPF B	49-04-04	54.45
3	4	1	249-45-10	347.68
		(10)IPF C	54-31-01	49.62

Cross Loop Field Notes

BS	STA @	FS	Angle RT.	DISTANCE
2	1	1A	53-28-13	204.12
1	1A	1B	150-14-00	198.54
1A	1B	3	230-27-34	171.85
1B	3	4	297-59-58	308.20

Coordinates for Mon 101: $N = 10177.65460149$ $E = 10446.77951350$

Mon 102: $N = 10094.69083254$ $E = 9532.42708331$

The survey task is as follows:

1. Enter and adjust traverse loop 1–4.
2. Rotate to control monuments.
3. Enter and rotate boundary to pipe found and complete ties.
4. Enter and adjust cross loop.

   The following procedure should allow the user to complete this task.

5. Set up drawing, set up points settings, description keys.
6. Set point for Traverse #1 by N/E (occupied station).
7. Set point for Traverse #4 by N/E (BS).
8. Traverse settings are no vertical, angles right.
9. Traverse entry (occupy #1, BS 4, Locate 2,3,4,1).
10. Traverse loops, define loops, should be OK.
11. Traverse loops, check adjust loop.
12. Traverse loops, adjust loop.
13. Command line, traverse/sideshots, sideshot settings
14. Command line, traverse/sideshots, sideshot entry

    Now the enter monuments.

15. Set point for Mon 101, 102 by N/E (using real coordinates).
16. Set layer for boundary.
17. Draw boundary using COGO, Lines, By direction.
18. Set label preferences.
19. Label property lines.
20. Check closure of property lines.
21. Fillet, R = 0, to close.
22. Align to fit into place.
23. Checkpoints, modify project.
24. Update labels if needed.
25. Enter cross loop.
26. Edit Points, Erase 101-102.
27. Edit field book file (ensure that N/E exists for points 1–4).
28. Define traverse loop.
29. Traverse analysis.

Civil 3D File for Adjusting a Closed Traverse

The following FBK file can be built to automate the traverse development in Exercise 11-3. This can be conducted in Windows Notepad and run by using the **Import Field Book** option, which is available when right clicking on the traverse created in the **Survey Toolspace.**

NE 1 2000 2000 "TRV 1"

NE 4 2347.68 2000 "TRV 4"

STN 1

BS 4

AD 2 278.050600 513.43 "TRV 2"

STN 2

BS 1

AD 3 291.450700 440.52 "TRV 3"

STN 3

BS 2

AD 4 260.243700 308.20 "TRV 4"

STN 4

BS 3

AD 1 249.451000 347.68 "TRV 1"

Civil 3D File for Adjusting a Traverse Cross Loop

This FBK file for Exercise 11-3 can be edited in Notepad as well and run by right clicking on a new traverse created in the **Survey Toolspace**.

NE 1 10070.5376 10260.5491 "TRV 1"

NE 2 10132.4980 9750.8695 "TRV 2"

NE 3 10518.9622 9962.3056 "TRV 3"

NE 4 10418.1471 10253.5486 "TRV 4"

STN 1

AUTO OFF

BS 2

AD 101 53.281300 204.12 "TRV 1A"

STN 101

BS 1

AD 102 150.140000 198.54 "TRV 1B"

STN 102

BS 101

AD 3 230.273400 171.85 "TRV 3"

STN 3

BS 102

AD 4 297.595800 0.00 "TRV 4"

Refer to Figure 11-5 for an image of the completed traverse following import.

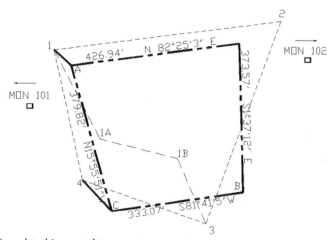

Figure 11-5   Completed imported traverse

Exercise 11-4: Computing a Traverse Reduction in Civil 3D—1

1. Open the drawing Chapter 11-1a.dwg.
2. In the **Toolspace**, click on the **Survey** tab.
3. Right click on the **Survey Databases** item; select **New local survey database...**
4. Name it **Chapter 11**, when the dialog box appears to name the database.
5. Right click on **Networks** and choose **New...**; when the dialog box for naming appears, type in **Chapter 11**.

6. Expand **Networks**, right click on **Chapter11**, and select **Edit Field Book...**; select **Survey-Trav.fbk** from where your files are stored for the textbook. It contains the following data, all in the field book language:

UNIT FOOT DMS

HORIZ ANGLE RIGHT

PRISM CONSTANT 0

PRISM OFFSET 0

EDM OFFSET 0

CR OFF

ATMOS OFF

COLLIMATION OFF

JOB CHAP-11

!NOTE TRAVERSE DEMO

TRAVERSE DEMO

SCALE FACTOR 1.000000

VERT ANGLE ZENITH

END

NE 1 1000 1000 "STA1"

NE 2 646.44660941 646.44660941 "STA2"

END

STN 1

!AZ 1 2 225.000000

!PRISM 0.000000

BS 2

AD 3 90.0 500.000 "STA3"

STN 3

BS 1

AD 4 269.5 499.9 "STA4"

STA 4

BS3

AD 5 270 500 "STA5"

STA 5

BS 4

AD 1 269.5 500

STA 1

BS 5

7. This data set stores Points 1 and 2 using a **Control Point** command (NE). It then occupies Point 1, backsights 2, and locates 3. It then occupies 3, backsights 1, and locates 4. Then it occupies 4, backsights 3, and locates 5. Then it occupies 5, backsights 4, and locates 1 again.
8. Close the file.
9. Expand **Networks**, right click on **Chapter11**, and select **Import Field Book...**; select **Survey-Trav.fbk** from where your files are stored for the textbook. The **Import Field Book** dialog box displays, as shown in Figure 11-6. Accept the data as shown and hit **OK**.
10. When the import is complete, the traverse shown in Figure 11-7 is computed.

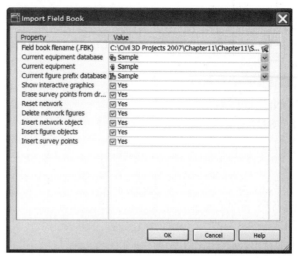

Figure 11-6   Import Field Book dialog box

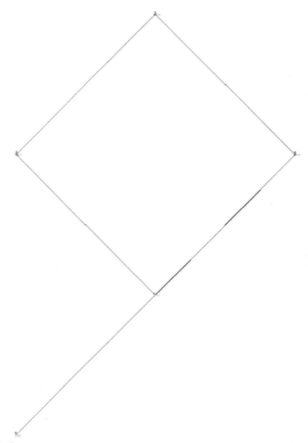

Figure 11-7   Imported traverse

11. Note in the **Survey** tab that two control points, 1 and 2, have been created. These will not be moved during traverse balancing. Expand **Setups** and notice that Stations 1, 3, 4, 5, and 1 are described.
12. Right click on **Traverses**, select **New...**, and type in **chapter 11** in the naming dialog box.
13. Then click on **Traverses, chapter 11**. Fill in the Initial Station (1), Initial Backsight (2), Stations (3-5,1), and Final Foresight (1) fields, as shown in Figure 11-8.

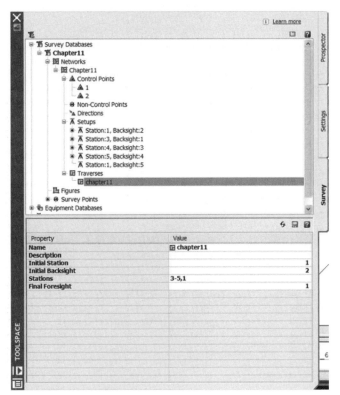

Figure 11-8   Defining the traverse stations

14. Click **chapter 11** and right click. From the choices, select **Traverse Analysis...**
15. The dialog box shown in Figure 11-9 is displayed.

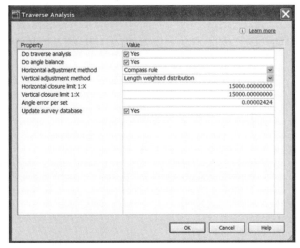

Figure 11-9   Traverse Analysis dialog box

16. Accept the settings and hit **OK**.
17. You may get a dialog box that reads, **First Occupied Point does not have known coordinates.** Hit **OK**.
18. Then a dialog box, shown in Figure 11-10, may appear that indicates you have made changes and not applied them. Hit **Yes** to apply the changes now.

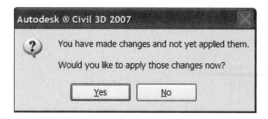

Figure 11-10

19. Then click **chapter 11** and right click again. From the choices, select **Traverse Analysis....**
20. The Notepad files shown in Figures 11-11, 11-12, and 11-13 display. They are a traverse analysis, a balanced angle analysis, and the Compass rule.

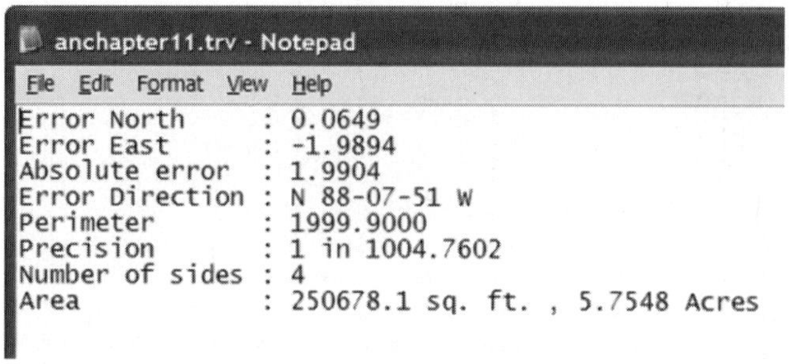

```
anchapter11.trv - Notepad
File  Edit  Format  View  Help
Error North    : 0.0649
Error East     : -1.9894
Absolute error : 1.9904
Error Direction : N 88-07-51 W
Perimeter      : 1999.9000
Precision      : 1 in 1004.7602
Number of sides : 4
Area           : 250678.1 sq. ft. , 5.7548 Acres
```

Figure 11-11   Traverse analysis report

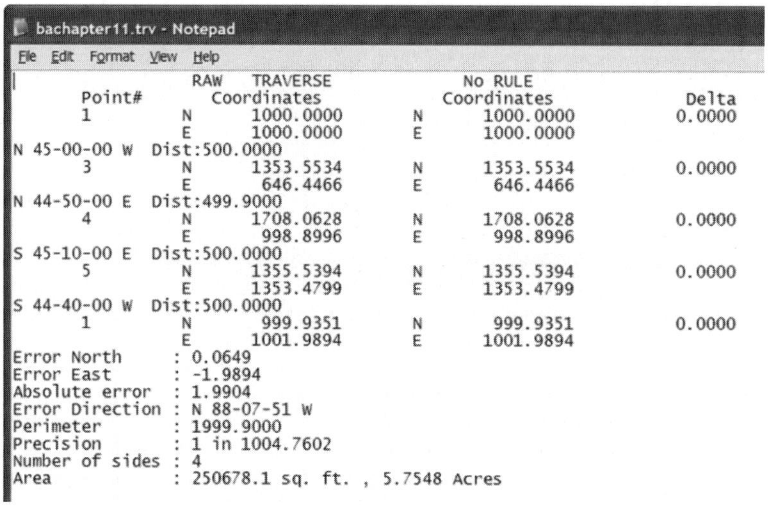

```
bachapter11.trv - Notepad
File  Edit  Format  View  Help
                  RAW    TRAVERSE               NO RULE
        Point#           Coordinates            Coordinates         Delta
          1       N      1000.0000       N      1000.0000         0.0000
                  E      1000.0000       E      1000.0000
N 45-00-00 W  Dist:500.0000
          3       N      1353.5534       N      1353.5534         0.0000
                  E       646.4466       E       646.4466
N 44-50-00 E  Dist:499.9000
          4       N      1708.0628       N      1708.0628         0.0000
                  E       998.8996       E       998.8996
S 45-10-00 E  Dist:500.0000
          5       N      1355.5394       N      1355.5394         0.0000
                  E      1353.4799       E      1353.4799
S 44-40-00 W  Dist:500.0000
          1       N       999.9351       N       999.9351         0.0000
                  E      1001.9894       E      1001.9894
Error North    : 0.0649
Error East     : -1.9894
Absolute error : 1.9904
Error Direction : N 88-07-51 W
Perimeter      : 1999.9000
Precision      : 1 in 1004.7602
Number of sides : 4
Area           : 250678.1 sq. ft. , 5.7548 Acres
```

Figure 11-12   Traverse analysis, balanced angles report

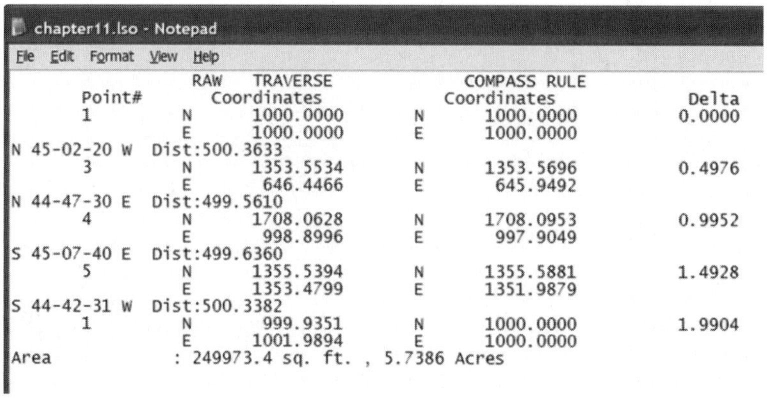

```
chapter11.lso - Notepad
File  Edit  Format  View  Help
                  RAW    TRAVERSE               COMPASS RULE
        Point#           Coordinates            Coordinates         Delta
          1       N      1000.0000       N      1000.0000         0.0000
                  E      1000.0000       E      1000.0000
N 45-02-20 W  Dist:500.3633
          3       N      1353.5534       N      1353.5696         0.4976
                  E       646.4466       E       645.9492
N 44-47-30 E  Dist:499.5610
          4       N      1708.0628       N      1708.0953         0.9952
                  E       998.8996       E       997.9049
S 45-07-40 E  Dist:499.6360
          5       N      1355.5394       N      1355.5881         1.4928
                  E      1353.4799       E      1351.9879
S 44-42-31 W  Dist:500.3382
          1       N       999.9351       N      1000.0000         1.9904
                  E      1001.9894       E      1000.0000
Area             : 249973.4 sq. ft. , 5.7386 Acres
```

Figure 11-13   Traverse analysis—compass rule applied report

21. This shows the compass rule being applied. Note that Point 1 starts at 1000,1000 and returns to 1000,1000 at the end of the processing.

Exercise 11-5: Computing a Traverse Reduction in Civil 3D—2

Student Files

In this exercise you execute the field books described earlier in the traverse loop and cross check example. Use the field books constructed based on the field notes and run them through the Civil 3D software.

1. Open the drawing Chapter 11-1b.dwg.
2. In the **Toolspace**, click on the **Survey** tab.
3. Right click on the **Survey Databases** item; select **New local survey database....**
4. Name it **Chapter 11-b,** when the dialog box appears to name the database.
5. Right click on the **Chapter 11-b database** and select **Open database**.
6. Right click on **Networks** and choose **New...**; when the dialog box for naming appears, type in **Chapter 11-b**.
7. Expand **Networks**, right click on **Chapter 11-b**, and select **Import Field Book...**; select **Chapter 11-b.fbk** from where your files are stored for the textbook. Accept any defaults that appear on executing.
8. Right click on the **Networks** item under Chapter 11-b, and select **New...**; name it **Chapter 11-b**.
9. Expand **Networks** and expand **Chapter 11-b**.
10. Right click on **Chapter 11-b** and choose **Import Field Book.**
11. Select the field book called **Chapter 11-b.fbk** from where your files are stored for the textbook.
12. The **Import Field Book** dialog box displays, accept the settings and hit **OK**. The loop traverse will import.
13. Right click on **Networks** and choose **New....**
14. Name the new network **Chapter 11-b cross check**. Hit **OK**.
15. Expand **Networks**, right click on **Chapter 11-b cross check**, and select **Import Field Book...**; select **Chapter 11-b-crosscheck.fbk** from where your files are stored for the textbook
16. When the **Import Field Book** dialog box appears, hit **OK**.
17. A dialog box may appear that warns that Point 1 exists. Select **Ignore**.
18. The cross check traverse computes. Note that it is an open traverse.
19. Expand the Network called **Chapter 11-b**. Right click on the **Traverses** item and hit **New...**; name it **Chapter 11-b** and hit **OK**.
20. Click on the traverse called **Chapter 11-b** and fill in the following data: the Initial Station (1), Initial Backsight (4), Stations (2,3, 4,1), and Final Foresight (1).
21. Right click on the traverse **Chapter 11-b** and select **Traverse Analysis....**
22. When the **Traverse Analysis** dialog box appears, accept the defaults and hit **OK**.
23. You may get a dialog box that reads, **First Occupied Point does not have known coordinates**. Hit **OK**.
24. Then a dialog box may appear that indicates you have made changes and not applied them. Hit **Yes** to apply the changes now.
25. Then click **Chapter 11-b** and right click again. From the choices, select **Traverse Analysis....**
26. The Notepad files for the traverse analysis, a balanced angle analysis, and the Compass rule display, as shown in Figures 11-14, 11-15, and 11-16. Again note in the compass rule how Point 1 returns to 2000,2000 after balancing.

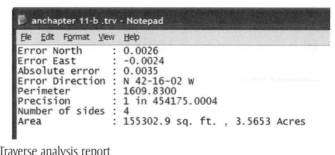

```
 anchapter 11-b .trv - Notepad
File  Edit  Format  View  Help
Error North     : 0.0026
Error East      : -0.0024
Absolute error  : 0.0035
Error Direction : N 42-16-02 W
Perimeter       : 1609.8300
Precision       : 1 in 454175.0004
Number of sides : 4
Area            : 155302.9 sq. ft. , 3.5653 Acres
```

Figure 11-14 Traverse analysis report

```
 bachapter 11-b .trv - Notepad
File  Edit  Format  View  Help
                  RAW    TRAVERSE              No RULE
        Point#         Coordinates          Coordinates            Delta
          1        N     2000.0000      N     2000.0000          0.0000
                   E     2000.0000      E     2000.0000
N 81-54-54 W  Dist:513.4300
          2        N     2072.2098      N     2072.2098          0.0000
                   E     1491.6732      E     1491.6732
N 29-50-13 E  Dist:440.5200
          3        N     2454.3366      N     2454.3366          0.0000
                   E     1710.8466      E     1710.8466
S 69-45-10 E  Dist:308.2000
          4        N     2347.6774      N     2347.6774          0.0000
                   E     2000.0024      E     2000.0024
S 00-00-00 W  Dist:347.6800
          1        N     1999.9974      N     1999.9974          0.0000
                   E     2000.0024      E     2000.0024
Error North     : 0.0026
Error East      : -0.0024
Absolute error  : 0.0035
Error Direction : N 42-16-02 W
Perimeter       : 1609.8300
Precision       : 1 in 454175.0004
Number of sides : 4
Area            : 155302.9 sq. ft. , 3.5653 Acres
```

Figure 11-15 Traverse analysis, balanced angles report

```
 chapter 11-b .lso - Notepad
File  Edit  Format  View  Help
                  RAW    TRAVERSE              COMPASS RULE
        Point#         Coordinates          Coordinates            Delta
          1        N     2000.0000      N     2000.0000          0.0000
                   E     2000.0000      E     2000.0000
N 81-54-54 W  Dist:513.4309
          2        N     2072.2098      N     2072.2107          0.0011
                   E     1491.6732      E     1491.6725
N 29-50-13 E  Dist:440.5203
          3        N     2454.3366      N     2454.3382          0.0021
                   E     1710.8466      E     1710.8452
S 69-45-10 E  Dist:308.1994
          4        N     2347.6774      N     2347.6794          0.0028
                   E     2000.0024      E     2000.0005
S 00-00-00 W  Dist:347.6794
          1        N     1999.9974      N     2000.0000          0.0035
                   E     2000.0024      E     2000.0000
Area             : 155303.0 sq. ft. , 3.5653 Acres
```

Figure 11-16 Traverse analysis report—Compass rule applied report

Reduce the cross check traverse.

27. Expand the network called **Chapter 11-b-crosscheck**. Right click on the **Traverses** item and hit **New...**; name it **Chapter 11-b-crosscheck** and hit **OK**.
28. Click on the traverse called **Chapter 11-b-crosscheck** and fill in the following data: the Initial Station (1), Initial Backsight (2), Stations (101, 102), and Final Foresight (4).

29. Right click on **Chapter 11-b-crosscheck** and choose **Traverse Analysis...**, hit **OK** for the dialog box that appears.
30. You may get a dialog box that reads, **First Occupied Point does not have known coordinates**. Hit **OK**.
31. Then a dialog box may appear that indicates you have made changes and not applied them. Hit **Yes** to apply the changes now.
32. Then click **Chapter 11-b-crosscheck** and right click again. From the choices, select **Traverse Analysis....**
33. The Notepad files for the traverse analysis, a balanced angle analysis, and the Compass rule display.

TIP When importing field books containing field collected data, the approach is exactly the same except, of course, that you may not reduce the traverse. You will simply see the data import into the Civil 3D drawing.

Exercise 11-6: Performing a Field Book Reduction in Civil 3D

You now execute a field book that will import an example of field-collected planimetric features. It will draw linework for an edge of pavement that was collected using a total station.

1. Open the drawing Chapter 11-1c.dwg.
2. In the **Toolspace**, click on the **Survey** tab.
3. Right click on the **Survey Databases** item; select **New local survey database...**
4. Name it **Chapter 11-c**, when the dialog box appears to name the database.
5. Right click on the **Chapter 11-c database** and select **Open database**. If the right-click menu has only **Close database**, then it is already open.
6. Right click on **Networks** and choose **New...**; when the dialog box for naming appears, type in **Chapter 11-c.**
7. Expand **Networks**, right click on **Chapter 11-c**, and select **Import Field Book...**; select **Chapter 11-c.fbk** from where your files are stored for the textbook. Accept any defaults that appear on executing.
8. The processing begins and a roadway edge of pavement and two associated buildings will be drawn.

Student Files

Note:
In the **Toolspace** within the **Survey** tab, there is a button at the top of the Prospector that displays the **Survey User Settings** dialog box shown in Figure 11-17.

You have just imported the field book and automatically drawn the points and linework collected in the field. The previous exercises illustrate how powerful this new addition to the Civil 3D software is. The entire survey department can now use Civil 3D to develop existing ground base information that the engineering design team can use directly. The surveyor will, of course, receive the data from the engineers at the back end of the project for checking and stakeout preparation or 3D/GPS data preparation for the contractor. The formatting of this data is covered in appendix A on data sharing.

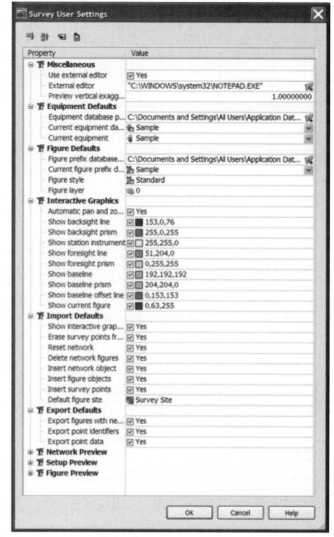

Figure 11-17   Survey User Settings dialog box

CHAPTER TEST QUESTIONS

Multiple Choice

1. The survey features added into this release include which of the following?

   a. Traverse entry, editing, and management
   b. Traverse balancing
   c. Importing filed data collector files
   d. All of the above

2. The interface for the surveying capabilities includes which of the following?

   a. Importing of SRTM satellite data
   b. Automatically pounding hubs and stakes into the ground
   c. Enabling the software to set up total stations so field crews are no longer needed
   d. None of the above

3. The survey functionality allows users to create the formats required for 3D/GPS Machine Control equipment. Which manufacturers are supported?

   a. Topcon, Trimble, Leica
   b. Caterpillar, Komatsu, VanNatta
   c. Mercedes, Range Rover, Ford
   d. All of the above

4. Survey functions include which of the following?

   a. Collecting planimetric and 3D data of existing conditions
   b. Quality control of engineering data and preparation of stakeout data for construction

c. Computing property boundaries

d. All of the above

5. The types of data surveyors prepare includes which of the following?

a. Proposed finished ground engineering, drainage improvements, stormwater retention facilities

b. Surface data, geometric data, and control data

c. Site improvements for dams, airport, or highways

d. Design of subdivisions, harbors, and treatment plants

6. Which of the following pertains to traverse data established by surveyors?

a. They are often tied into control monuments.

b. They are repeatedly occupied and used in surveying and perhaps construction and provide a constant location from which to compute additional point data.

c. They are usually corrected for errors caused by human, ambient, or instrument lapses.

d. All of the above.

7. When a traverse is corrected and balanced, what components are corrected?

a. The area within the closed traverse is corrected and rounded.

b. The control points governing the traverse are moved to new and more accurate locations.

c. The points or vertices of the traverse figure with its incorrect distances and angular errors are corrected.

d. All of the above

8. Surveyors also work in which of the following areas?

a. Aerial photogrammetry, GIS research, and GPS surveying

b. Hydraulic drainage computations for custom drainage structures

c. Designing beams and columns for standard building structures

d. None of the above

9. Why are VIOs (Visually Identifiable Objects) used in aerial photogrammetry?

a. They can be seen in the photographs.

b. They have horizontal and vertical control associated with them.

c. They assist in calibrating everything in the photo to a known coordinate.

d. All of the above.

10. What makes Imaginary Reference Stations an interesting technology for surveyors?

a. They eliminate the need for base stations for many surveyors.

b. They create virtual stations that interpolate errors and coorections.

c. They are being developed by both public and private concerns.

d. All of above.

True or False

1. True or False: Civil 3D can create, edit, and balance traverses; import field shot data; and draw linework and points.

2. True or False: 2D and 3D control points are often established randomly and are never tied to anything official.

3. True or False: In photogrammetry, a VIO is a Vertical Interference Obstacle that interferes with the airplane's flight path.

4. True or False: We use the Compass rule to find north when we are in the field.

5. True or False: A polygon with six sides should have the sum of the interior angles equal to 750 degrees.

6. True or False: There are several files produced when balancing a traverse, and they include analysis files, which display horizontal closure and angular error, a file showing raw and adjusted elevations; a file showing the balancing of the angular error; and the final adjusted coordinates.

7. True or False: The survey language can be customized so that the order of the data entry can be changed to the user's preferences.

8. True or False: The survey language can be customized so that the actual names of the commands (such as AD or BS) can be changed to what the user prefers.

9. True or False: An interface specifically developed for the surveyor was developed in Civil 3D 2007.

10. True or False: Civil 3D 2007 will perform the Compass rule only when balancing traverses.

STUDENT EXERCISES, ACTIVITIES, AND REVIEW QUESTIONS

1. What functions do surveyors perform on engineering projects? Do surveyors work in two dimensions or three?

2. Why is establishing project control important?

3. How many degrees does a 16-sided closed traverse have inside it?

4. How might the following table be adjusted to correct the angular error in this traverse?

Point	Measured Interior Angle	Average Correction (Degrees)	Adjusted Angle (Degrees)
A	120.5		
B	110.0		
C	110.5		
D	99.5		
E	100.0		
	————	————	————
Result:			

Adjustment of Angles

5. When speaking of traverse misclosure, what is the latitude and departure?

6. What is the relative precision for a 1000′ traverse that has a linear misclosure of 0.1?

7. If a traverse has a departure of 0.2 and a latitude of 0.3, what is the linear misclosure?

8. When using the survey language in Civil 3D, what do the following commands indicate: STA 10, BS 5?

9. When using the survey language in Civil 3D, what do the following commands indicate: NE 10 1000 1000, NE 5 1500 1000?

10. When shooting a traverse, why does it not close perfectly?

Data Sharing in Civil 3D

A

INTRODUCTION

You have worked with Civil 3D using AutoCAD drawings and have performed your design tasks using a single AutoCAD drawing each time. In a production atmosphere this may not always be the case. Sometimes engineers work in an environment in which they need to share data among teammates. In order to provide an understanding of how this could work, this appendix addresses data sharing for several purposes: (1) to share data with others on your team and (2) to share data with people external to Civil 3D users.

The 2007 version adds a concept called the Vault, which allows for securities and protections of the project data. The user is asked to log in with a password. If the password is lost or the employee who sets it leaves the organization without communicating to others the password, issues can arise. If the employee had administrator privilege, then only Autodesk can break the password. Contact Autodesk and fill out the required legal paperwork to accomplish this. Otherwise, if the employee did not have administrator access, then the system administrator can gain access to the Vault to clear the password.

NEW
to AutoCAD
2007

DATA SHARING

The exercises you have worked on in this text have included the use of point data, parcel data, surfaces, alignment data, profile views and profiles, corridors, and section data. The main method for data sharing has traditionally been through passing AutoCAD files back and forth. This allows others to see and manipulate the graphical data involved but does not provide strong access to the actual design data.

Civil 3D allows some very effective methods for sharing that design data. In fact there is the ability to share the data not only with other Civil 3D users but also with non–Civil 3D users. In fact data can be shared with others who do not even have AutoCAD as a foundation CAD platform.

Many larger projects have subconsultants working on them whereby each party contributes its own specialized expertise, or perhaps projects are multidisciplinary. In any event, data need to be shared between these groups in order to maximize efficiency, avoid errors, and minimize redundancies. The surveyor's base conditions need to be shared with the designers so they can develop the proposed conditions. When the proposed design conditions are finished, these must be sent back to the surveyor for quality control and preparation of stakeout or 3D/GPS Machine Control data. If machine control data are involved, they must be shared with the contractor.

This appendix covers Civil 3D's four main methods of data sharing. The first method is by simply copying the AutoCAD file and sending it to others who have Civil 3D for their use. In this case, the data are inherently available.

The second method is through the use of LandXML. The third method is performed through the use of data shortcuts. And the fourth method is accomplished through the use of Civil 3D's project abilities.

Method 1: Inserting or Copying a Drawing

As discussed, Civil 3D makes it easy to incorporate data into a drawing by inserting one drawing into another drawing or by simply saving as.... When the user wishes to obtain access to data within another drawing, all he or she needs to do is insert the source drawing into the active drawing or open up the copy. All of the object data such as terrain models, alignments, and so on are immediately available to the working session. Note that there is no live link to the source drawing when using this procedure.

Method 2: LandXML

Civil 3D includes a simple methodology called LandXML, which works similarly to the way it did in Land Desktop. This feature is a data migration tool developed to send design data across operating systems (say from Windows XP to Unix), across different versions of CAD systems (say between AutoCAD or Microstation), or between systems that have different CAD standards (say between a consulting firm and the state department of transportation). LandXML, which is being promoted by a nonprofit group, can be accessed via the Web at Landxml.org. It is supported by a group of software manufacturers interested in making their customers' data available to others in the interest of efficiency.

The ability to import or export LandXML makes use of a file format that is open, is text based, and is not software dependent. In other words, it uses a simple definition for each data type, and each software manufacturer can read it and reassemble its respective data into its software's definition. It is also an archival facility because the data are in a generic text-based format.

Let us look at some examples of the way data are stored in LandXML. The following example involves an alignment, and this is how it appears in a LandXML file.

```
<?xml version="1.0"?>
<LandXML xmlns="http://www.landxml.org/schema/LandXML-1.0"
xmlns:xsi="http://www.w3.org/2001/XMLSchema-instance" xsi:schemaLo-
cation="http://www.landxml.org/schema/LandXML-1.0
http://www.landxml.org/schema/LandXML-1.0/LandXML-1.0.xsd"
date="2005-06-26" time="10:42:00" version="1.0" language="English"
readOnly="false">
    <Units>
        <Imperial areaUnit="squareFoot" linearUnit="foot"
volumeUnit="cubicYard" temperatureUnit="fahrenheit"
pressureUnit="inchHG" angularUnit="decimal degrees"
directionUnit="decimal degrees"></Imperial>
    </Units>
        <Project name="C:\Admin\prentiss-hall-civil 3d\civil 3d-
applications\student files\appendix-a.dwg"></Project>
    <Application name="Autodesk Civil 3D" desc="Civil 3D"
manufacturer="Autodesk, Inc." version="2007"
manufacturerURL="www.autodesk.com" timeStamp="2005-06-
26T10:42:00"></Application>
    <Alignments name="Site 1">
        <Alignment name="Alignment - 1" length="2154.87367186"
staStart="0." desc="Chapter 6 alignment">
            <CoordGeom>
                <Line dir="354.45437497" length="200.">
                    <Start>18553.70716247
14599.21456424</Start>
```

```
                <End>18534.37948987 14798.27847627</End>
            </Line>
            <Curve rot="cw" chord="386.34076129"
crvType="arc" delta="30.03271185" dirEnd="324.42166312"
dirStart="354.45437497" external="26.35954132" length="390.79932237"
midOrd="25.45941315" radius="745.55877333" tangent="200.">
                <Start>18534.37948987
14798.27847627</Start>
                <Center>17792.31025954
14726.22889691</Center>
                <End>18398.68871707 15160.00654818</End>
                <PI>18515.05181728 14997.3423883</PI>
            </Curve>
            <Line dir="296.56505118" length="0.">
                <Start>18398.68871707
15160.00654818</Start>
                <End>18398.68871707 15160.00654818</End>
            </Line>
            <Curve rot="ccw" chord="380.90154719"
crvType="arc" delta="35.55331015" dirEnd="359.97497326"
dirStart="324.42166312" external="31.2774218" length="387.08193805"
midOrd="29.78404589" radius="623.80018303" tangent="200.">
                <Start>18398.68871707
15160.00654818</Start>
                <Center>18906.03838059
15522.94316422</Center>
                <End>18282.23825707 15522.67068898</End>
                <PI>18282.32561686 15322.67070806</PI>
            </Curve>
            <Line dir="359.97497326" length="200.">
                <Start>18282.23825707
                15522.67068898</Start>
                <End>18282.15089729 15722.6706699</End>
            </Line>
            <Curve rot="ccw" chord="383.19792514"
crvType="arc" delta="33.33121545" dirEnd="393.30618871"
dirStart="359.97497326" external="29.29379845" length="388.65516287"
midOrd="28.06330696" radius="668.09146379" tangent="200.">
                <Start>18282.15089729
15722.6706699</Start>
                <Center>18950.24229735
15722.96249153</Center>
                <End>18391.88615615 16089.82026175</End>
                <PI>18282.06353751 15922.67065082</PI>
            </Curve>
            <Line dir="33.30618879" length="0.00099774">
                <Start>18391.88615615
16089.82026175</Start>
                <End>18391.88670402 16089.82109561</End>
            </Line>
            <Curve rot="cw" chord="382.73122679"
crvType="arc" delta="33.79240661" dirEnd="359.5137823"
dirStart="33.30618891" external="29.70487123" length="388.3352508"
midOrd="28.42259646" radius="658.43108381" tangent="199.99900122">
                <Start>18391.88670402
16089.82109561</Start>
                <Center>17841.60420785
16451.37422662</Center>
                <End>18500.01158371 16456.96167135</End>
                <PI>18501.7087748 16256.96987144</PI>
            </Curve>
            <Line dir="359.5137823" length="200.00100003">
                <Start>18500.01158371
```

```
16456.96167135</Start>
                    <End>18498.31437567 16656.95546999</End>
            </Line>
          </CoordGeom>
        </Alignment>
      </Alignments>
</LandXML>
```

The next example is of a surface definition. It begins with header information and then defines the coordinates of each point involved with a reference
point number. It follows that with a triangle referencing those point numbers
so that the exact triangulation is honored.

```
<?xml version="1.0"?>
<LandXML xmlns="http://www.landxml.org/schema/LandXML-1.0"
xmlns:xsi="http://www.w3.org/2001/XMLSchema-instance"
xsi:schemaLocation="http://www.landxml.org/schema/LandXML-1.0
http://www.landxml.org/schema/LandXML-1.0/LandXML-1.0.xsd"
date="2005-06-26" time="10:48:52" version="1.0" language="English"
readOnly="false">
    <Units>
    <Imperial areaUnit="squareFoot" linearUnit="foot"
volumeUnit="cubicYard" temperatureUnit="fahrenheit"
pressureUnit="inchHG" angularUnit="decimal degrees"
directionUnit="decimal degrees"></Imperial>
    </Units>
    <Project name="C:\Admin\prentiss-hall-civil 3d\civil 3d-
applications\student files\appendix-a.dwg"></Project>
    <Application name="Autodesk Civil 3D" desc="Civil 3D"
manufacturer="Autodesk, Inc." version="2007"
manufacturerURL="www.autodesk.com" timeStamp="2005-06-
26T10:48:52"<</Application>
    <Surfaces>
<Surface name="Corridor - (1) Surface - (1)" desc="">
    <Definition surfType="TIN">
<Pnts>
    <P id="1">18492.21916265 14593.24452217 325.96527275</P>
    <P id="2">18517.37799852 14595.68726399 321.75240699</P>
    <P id="3">18520.3639572 14595.97717908 321.75240699</P>
    <P id="4">18532.30779192 14597.13683944 324.75240699</P>
    <P id="5">18533.30311148 14597.2334778 324.73240699</P>
    <P id="6">18537.28438973 14597.62003125 324.65240699</P>
    <P id="7">18539.27502885 14597.81330798 324.61240699</P>
    <P id="8">18539.68974533 14597.85357396 324.61240699</P>
    <P id="9">18539.77268863 14597.86162716 324.11240699</P>
    <P id="10">18541.76332775 14598.05490389 324.27907366</P>
    <P id="11">18553.70716247 14599.21456424 324.51907366</P>
    <P id="12">18565.65099719 14600.3742246 324.27907366</P>
    <P id="13">18567.64163631 14600.56750132 324.11240699</P>
    <P id="14">18567.72457961 14600.57555452 324.61240699</P>
    <P id="15">18568.13929609 14600.61582051 324.61240699</P>
    <P id="16">18570.12993521 14600.80909723 324.65240699</P>
    <P id="17">18574.11121345 14601.19565068 324.73240699</P>....
. . . </Pnts>
  <Faces>
    <F>679 117 116</F>
    <F>369 281 280</F>
    <F>735 73 734</F>
    <F>783 126 127</F>
    <F>693 131 130</F>
    <F>519 391 392</F>
    <F>693 3 2</F>
```

```
<F>688 125 124</F>
<F>690 5 4</F>
<F>690 6 5</F>
<F>689 7 6</F>
<F>737 70 71</F>
<F>687 8 7</F>
<F>229 174 175</F>
<F>685 122 684</F>
<F>528 403 404</F>
<F>739 68 69</F>
<F>741 65 66</F>
<F>317 227 228</F>
<F>683 119 682</F>
<F>742 64 65</F>
<F>681 118 117</F>
<F>675 674 21</F>
<F>230 175 176</F>
<F>707 34 706</F>
<F>716 56 715</F>
<F>682 13 12</F>
<F>463 357 356</F>....
```

By inspecting these data, one can see why this is a generic, simple, and text-based method for data sharing that can easily cross platform lines, CAD system lines, or CAD standards lines.

One company can send its LandXML data to another, and each organization's CAD system can invoke its respective CAD standards when the data are imported into the design software.

You can import the LandXML information using your own native CAD standards and be unaffected by the graphics used to generate the data because LandXML exports only design data and not the graphical appearance of it. The importation process then creates the data with the aesthetics or properties of the CAD system performing the import. The critical aspect of this process is maintaining the design data integrity.

Method 3: Data Shortcuts

Civil 3D provides a data collaboration tool called data shortcuts, which allow you to externally reference objects that are either inside or outside of a project. Data shortcut users are one of the following two types of people. Either they have the source object's drawing open and produce the data shortcut (which is an external XML file that contains the AutoCAD DWG file name, path, and object name), or they are the users of the data shortcut. If they are the users, they can create a reference to the object or read it and obtain ownership of the copy of the object. The referenced object initially has read-only geometry properties and takes up less file space when the drawing is saved, because it is only referenced.

Data shortcuts can be used for obtaining access to surfaces, alignments, and profiles. Alignment and profile data shortcuts are closely tied to each other because a profile data shortcut's source drawing must be the same as the profile's alignment data shortcut.

One last note is that data shortcuts are visible and accessible to the next method of Civil 3D projects for data sharing.

Method 4: Civil 3D Projects

A method provided in Civil 3D to allow sharing of data between Civil 3D users that is somewhat similar to that used by Land Desktop users is the project method. The person who has the source data available attaches the drawing to a project and makes data available to the project. Other users can then obtain access to these data through a Check Out/Check In system.

The project object cannot be modified directly. In order to modify it, the object has to be checked out. Changes are made and the object is then checked back in to the project. If a user tries to modify a local copy of the data without first checking it out, then it cannot be checked back in to the project. The modifications made will exist only in the drawing in which they were made.

The next task is to perform some exercises using methods 2 through 4.

Exercise A-1: Using LandXML

1. Open the file called Appendix-a.dwg. There are several data types in this file including points, a corridor, surfaces, and a pipe network.
2. From the **General** pull-down menu, select the **Export LandXML...** command. The command prompt asks you to: Select Filter or Pick. **Filter** invokes a dialog box so that you can check one or more of the items for export, whereas **Pick** allows you to select an object directly from the AutoCAD file. In this case, select **F** for **Filter** and the **Filter** dialog box displays.
3. Check the **All Points** toggle under the **Point Groups** category in Figure A-1. Check the **Alignment - 1** in the **Prospector** under **Site, Site 1, Alignments,** and all of the individual profiles check automatically. Hit **OK**.

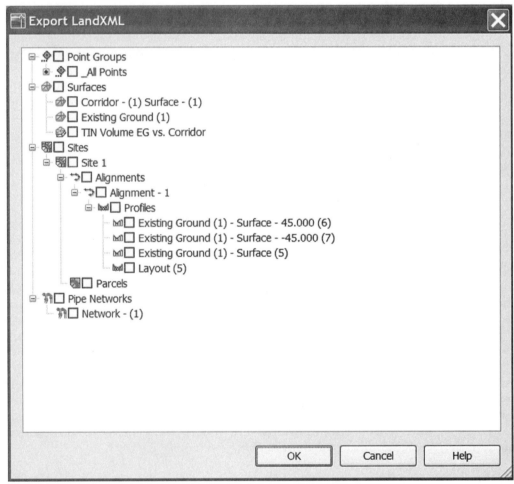

Figure A-1 LandXML

4. Provide a name for the file in a location that you can locate later, Hit **OK**.
5. When that process is complete, select **File** from the pull-down menu, choose **New**, and create a new Civil 3D file, using the Autodesk **Civil 3D Imperial By Layer.dwt** file as a template.

6. From the **General** pull-down menu, select the **Import LandXML...** command. The following **Import LandXML** dialog box in Figure A-2 displays.

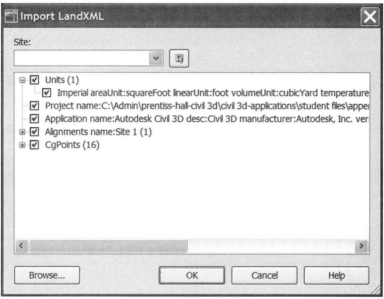

Figure A-2   Import LandXML

7. Accept that all of the data are checked for import, and hit **OK**.
8. **Zoom Extents** and you see the alignment and points in the new file.
9. Look in the **Prospector** and note that the alignment and points are visible here as well. See Figure A-3.
10. The only thing that is left to be done is to set the Styles for the objects imported.

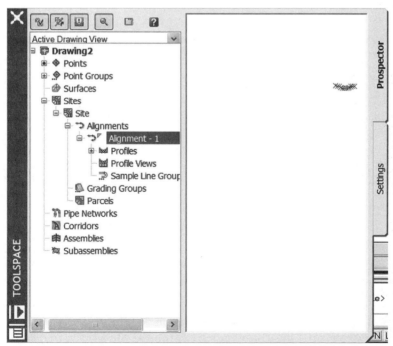

Figure A-3   Alignment in the Prospector

Exercise A-2: Using the Data Shortcuts Method

1. Open the file called Appendix-a.dwg. There are several data types in this file including points, a corridor, surfaces, and a pipe network.
2. From the Civil 3D pull-down menu, select **General >> Data Shortcuts >> Edit Data Shortcuts**.... The **Data Shortcuts** dialog box displays.

The icons across the top of the **Data Shortcuts** dialog box shown in Figure A-4 are **Import Data shortcut, Export Data shortcut, Validate shortcut, Create data shortcut by Selection, Create Reference, Open Source drawing, Expand,** and **Collapse all categories,** respectively.

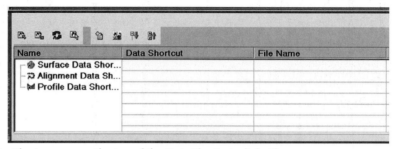

Figure A-4   The icons across the top of the Data Shortcuts dialog box

3. Click the fourth icon, **Create data shortcut by Selection,** and you are placed back into AutoCAD at the command prompt to select the object that you would like to make a data shortcut. Select the centerline alignment.
4. Repeat the command and select a red contour in the corridor. Note that the **Surface** and **Alignment** shortcuts in the dialog box now can be expanded. Expand them and they should appear as in Figure A-5.

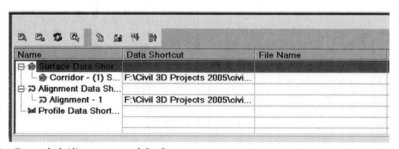

Figure A-5   Expanded Alignments and Surfaces

5. Now select the second icon in the dialog box, **Export Data shortcut,** to send it to the hard disk or the server. Provide a location and name for the file such as **My Data Shortcut**.
6. Close the file, there is no need to save it because no changes occurred.

You have created a data shortcut and saved it for use by others. Now let us explore how you or others who are on the project team can use these shortcuts.

7. From the Civil 3D pull-down menu, click **File** and then **New** and create a new drawing. Use the **Civil 3D Imperial By Layer** template to create the new file.

8. Save it as a properly named AutoCAD file of your choosing.
9. From the **Civil 3D** pull-down menu, select **General** >> **Data Shortcuts** >> **Edit Data Shortcuts**.... The **Data Shortcuts** dialog box displays.
10. Now select the first icon in the dialog box, **Import Data shortcut**. Select the location where you saved your shortcut earlier, and pick the file called **My Data Shortcut**.
11. The dialog box appears as in Figure A-5 with a **Corridor—(1) Surface** and an **Alignment—1**. Click on the **Corridor—(1) Surface**. Then click on the fifth icon from the left, **Create Reference**, and a **Create Surface Reference** dialog box appears, shown in Figure A-6. This dialog box is the same as a normal **Create Surface** dialog box except that this one has the keyword *reference* in its title. It is preparing a surface that is a reference. The user does not really "own" this surface but can use it and display it.

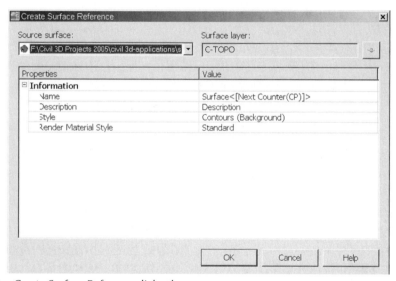

Figure A-6   Create Surface Reference dialog box

12. Say **OK** to the **Create Surface Reference**, taking the defaults.
13. Perform a **Zoom Extents** on the drawing, and you should see the contours for the corridor surface. If a **Panorama** appears saying the surface was created, close it.
14. Click on the **Alignment 1** item. Then click on the fifth icon from the left, **Create Reference**, and a **Create Alignment Reference** dialog box appears. Again, the user does not really "own" this alignment but can use it and display it. See Figure A-7.
15. Accept the defaults as shown in the figure, and hit **OK**. If a **Panorama** appears saying the surface was created, close it.
16. You see the alignment in the file now.

If you click on the alignment and right click, you see the typical commands to edit the alignment or the Style. If you select **Alignment Properties**, notice that nothing can really be modified because it is referenced. The Style, however, can be altered as needed.

There are two new commands in the right-click menu, **Synchronize** and **Promote**. **Synchronize** updates your reference should a change to the shortcut occur, and **Promote** turns the referenced shortcut into an individual and separate piece of data, now owned by the new drawing. This is similar to **Binding** and **External reference** in AutoCAD.

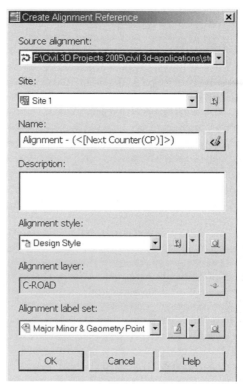

Figure A-7   Create Alignment Reference dialog box

17. While in the new file you created, in the **Data shortcuts** toolbar, select **Alignment 1**. Then click on the sixth icon, **Open Source drawing**. The original file from which the data shortcut was created opens.

18. Click on the **Alignment**, right click, and choose **Alignment Properties**. When the toolbar displays, click on the **Alignment Grid View** on the right side of the toolbar. The **Panorama** opens with the numeric alignment data in it.

19. Scroll to the right until you can see the radius for the first curve in the alignment. Change it to **700** and close the view. Hit <**Enter**> to terminate. The alignment radius is now 700′ for the first curve.

20. Save the file.

21. Using the **Window** pull-down menu, switch to the new drawing.

22. When you do, the **Data Shortcuts** toolbar as well as a pop-up window display, telling you that the data shortcut has been altered and you need to synchronize it.

23. Click on the blue hyperlink where it says **Synchronize**, shown in Figure A-8.

24. The command prompt notifies you that the Alignment - (1) is now up to date, and that the Surface Surface 1 is now up to date.

25. Close the **Panorama**. Close the **Data Shortcuts** toolbar.

If you wish to take complete control of the data shortcut, then you can promote it. This detaches it from its source.

26. Click on the **Alignment**, right click, and select **Promote**. The command prompt tells you that 1 Data References successfully promoted.

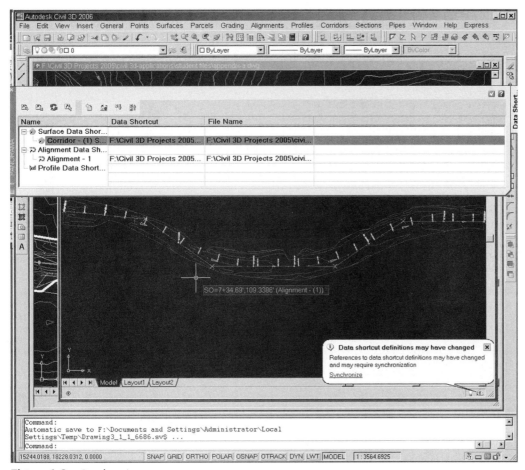

Figure A-8   Synchronize

27. Click on the **Alignment** now, and notice that grips appear, which allow you to edit the alignment, because you now "own" it.

This example shows how data shortcuts can be used. These shortcuts can be placed on a server for each member of the team to use in referencing the data that is in your drawing. They can be used to compute profiles, to change the appearances of data for different uses, or for report creation.

Exercise A-3: Using the Projects Method

1. Open Appendix-a-projects.dwg.
2. Check in the **Toolspace Prospector,** and notice that there is a **Projects** item when the **Master View** is showing. See Figure A-9.
3. Right click on **Projects** and you can **Manage Project Path.** This means that you can set the project path wherever you wish. Under the project path will reside the individual projects that you create.
4. The system defaults to **\Civil 3D Projects**, and you can switch that to a more desired location for this exercise if you like.
5. Then right click on the location you choose or **\Civil 3D Projects** if you left it alone. A menu displays allowing you to create a project, with **New Project...** Select this and a **Project Properties** dialog box displays.

Student Files

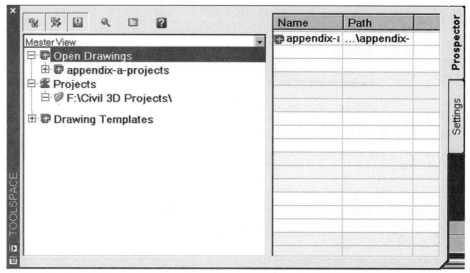

Figure A-9   A projects item in the Toolspace Prospector

6. Call it **My Project,** as shown in Figure A-10.

Figure A-10   Project Properties dialog box

7. Check in the **Toolspace Prospector,** and notice that there is a project called **My Project** under the **Civil 3D Projects** item.
8. Expand **My Project,** as shown in Figure A-11, and you see areas for **Drawings** where your drawings can be saved from now on, **Points, Point Groups, surfaces,** and **Data Shortcuts.**

Figure A-11   My Project expanded

9. Now in the **Prospector** right click on your drawing name, **Appendix-a-projects.dwg,** and notice another selection available called **Attach to Project**.

10. Choose **Attach to Project,** and select the **My Project** project, as shown in Figure A-12.

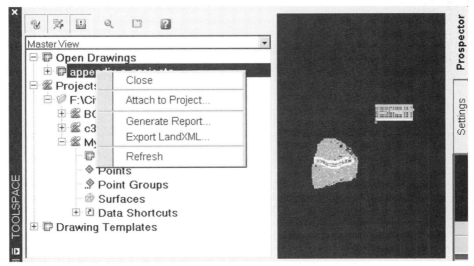

Figure A-12   Attach to Project

11. Next, expand the drawing, expand **surfaces** so you can see the **Corridor—(1) Surface—(1)**. Right click on the **Corridor—(1) Surface—(1)** and you see an option to **Attach to Project**. Select **Add to Project**. See Figure A-13.

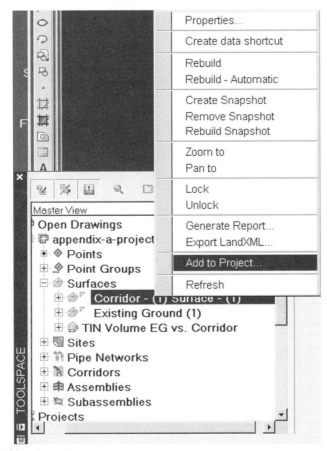

Figure A-13   Add to Project

12. An **Add to Project** dialog box displays, allowing you to attach the surface to the project, but it needs to know under what terms you are attaching it. The default is **Check in**. Click in the pop-down window where it says **Check in**. The other options are **Check in** and **Keep Checked in** and **Check in** and **Protect**. Leave the default. Hit **OK**. See Figure A-14.

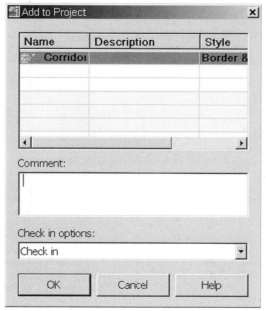

Figure A-14   Add to Project

13. You now see the **Corridor—(1) Surface—(1)** under the **Surfaces** item in the **Projects** area of the **Prospector**. It can now be used by others on the project team. Let us see how in the next part of the exercise.

Close the file you are in and there is no need to save it.

14. Click **File >> New** and create a new drawing. Use the **Civil 3D Imperial By Layer** template to create the new file.
15. Save it as a properly named AutoCAD file of your choosing.
16. In the **Prospector**, right click on your AutoCAD file, and select **Attach to Project**.
17. Choose the project called **My Project**.
18. In the **Prospector**, expand **Surfaces** and you will see the **Corridor—(1) surface—(1)**. Right click on the **Corridor—(1) surface—(1)** and you are able to **Check Out, Get from Project** or **Protect the surface**. Choose **Check Out**.
19. **Zoom Extents** and you see the corridor surface in the default style. You also see the **Surface** in the **Prospector** under your file's **Surfaces**.
20. After using it or making modifications to the surface, you can right click on the **Corridor—(1) Surface—(1)** in the **Prospector** under your file's **Surfaces** and choose **Check it In**.
21. Comments can be added and a log of who checked it in/out is maintained.

Check Out means that you can create an accessible read/write copy of a project object in your drawing. When you check an object out, it then overwrites equivalent objects in your file, thereby keeping the drawing updated with the project.

Get from Project means that you can create a read-only copy of the project object in your drawing.

Protect prevents, others from checking out and changing a project object.

Check In means that you can update the project with changes you made to that object while it was checked out by you. This feature locks the local copy of the object in the drawing. See Figure A-15.

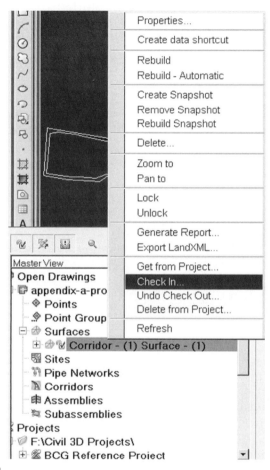

Figure A-15   Check In

Unprotect means that you allow others to check out and change a project object.

These are the main methods for sharing data in Civil 3D. As the use of the software matures, the need to share the data will likely arise, and these exercises will assist in accomplishing that.

SHARING DATA WITH CONTRACTORS FOR 3D/GPS MACHINE CONTROL

Three software add-ons assist designers using Civil 3D in developing their 3D data for contractors using robotic earthmoving devices, commonly known as machine control. This technology has been growing very rapidly and can decrease the amount of time a contractor needs to construct a project. As a result the contractor increasingly wants the data that engineers create, and he or she wants them in 3D!

Autodesk has recognized this need and offers three manufacturer's add-ons into the Civil 3D software: Carlson Connect, which addresses Topcon, Trimble Link from Trimble, and Leica X-change. Topcon, Trimble, and Leica are the three main suppliers of machine control equipment in the world. These add-ons are of no cost to the user.

Carlson Connect

Carlson Connect (Figure A-16) is another solution that offers a collection of routines for transferring and converting data between Autodesk Land Desktop or Autodesk Civil 3D and several popular data collectors. Carlson Connect runs inside Civil 3D and uses the current project data.

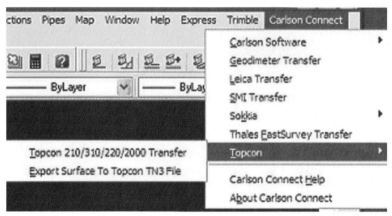

Figure A-16   Carlson Connect

Once an engineer or designer has created his or her data in 3D, say a roadway or a site grading project, there will be a surface involved. By using the **Carlson Connect** pull-down menu and choosing the **Topcon** command and then **Export Surface to Topcon TN3 File**, the surface can be formatted into the appropriate binary file for Topcon equipment. The TN3 file can be copied to a flashcard for insertion directly into the equipment on site.

Trimble Link

Trimble's equivalent routine would be in two parts. The first is **Export Surface** file for Sitevision for sending your Civil 3D surfaces. The second is for roadway data whereby you would use **Export Road** file for Sitevision. See Figure A-17.

Leica X-change

Leica Geosystems released the Leica X-change for Autodesk Civil 3D, which is a group of tools running inside Autodesk Civil 3D that allows the import of data from and the export of data to a Leica System 1200 job. Points, lines, and areas as measured in the field with System 1200 (TPS and GPS) can be directly imported into an Autodesk Civil 3D drawing. Conversely, points created in Civil 3D can be exported to a System 1200 job for later usage in the field. The benefit for the user is seamless data transfer between the System 1200 and Autodesk's Civil 3D software, thereby eliminating intermediate conversion steps, avoiding loss of data, and taking full advantage of the linework capabilities of the System 1200.

Figure A-17   Trimble Link

SHARING CIVIL 3D CONFIGURATION DATA

Before leaving this topic of sharing data, the last to discuss is sharing data among fellow project teammates. Civil 3D shares its CADD standards, Styles, and assemblies using AutoCAD files. The product configuration largely is performed in an AutoCAD file in which the environment variables and profiles can be established like normal.

Then the Styles and assemblies should be configured for Civil 3D. The Styles are stored in the AutoCAD template for dissemination to other users. The assemblies can be simply placed inside the AutoCAD file or dropped into **Tool Palettes**.

Labeling in Civil 3D 2007

This appendix briefly discusses some of the new labeling methods added to Civil 3D 2007. One of the main additions is the ability to label simple lines and curves. Up until this release, only defined objects could be labeled. This now can be achieved by choosing the **Add labels...** function available under several menu pull-downs. Under the **Feature:** option you can select **Line and curve**. Choose the label type of either **Single** or **Multiple** segment. Then select the style by which you want to label, and hit the **Add** button to perform the function.

Another addition to labeling is the new **Notes** function. This can be performed by choosing the **Add labels...** function available under several menu pull-downs. Under the **Feature:** option you can select **Note**. The label type is **Note**. Then select the style by which you want to label, and hit the **Add** button to perform the function. You will be asked to Pick label location: and then pick in the drawing where you wish the new label to be placed.

TIP

The Note is a free-form label whose main properties are that it will resize and rotate as described by the Style when plotted.

Exercise B-1: Expressions

Another important new labeling function is the ability to add expressions to the labeling. Let's do a quick example.

1. Expand the **Toolspace Settings**, and look under **Alignment, Label Styles and Curve**. You see a new option called **Expressions** in all of the labeling style areas. Expressions can be created for all of the labeling styles. They are then attached during Style creation.
2. Right click on **Expressions** and select **New....**
3. Provide a name for your expression, call it **Half-Delta**.
4. Just to the right of the **Expression** window is a button called **Insert Function**. Click it and select the function called **RAD2DEG**.
5. Then click in the **Expression** window to the right of the parenthesis. Now click on the **Insert Property** button to the right of the **Expression** window.
6. Select the **Delta Angle** property.
7. Then click in the **Expression** window and add another parenthesis to the right side of the {Delta Angle} term. Then type **/2** after the expression to divide the value in half.
8. It should say **RAD2DEG({Delta Angle})/2** on completion.
9. Hit **OK**.
10. Now right click on the **Curve style** called **Standard**. Choose the **Layout** tab.

11. Under the pop-down window called **Component name:** select **Delta**.
12. Now choose **Edit...** under the **Text Contents Value** field.
13. Click the button to the right of the field with the dots on it to enter the editing facility.
14. Click before the D= in the **Preview** window and hit **<Enter>** to open a blank line.
15. Move to the blank line and type **Half-Delta=**
16. Now click under the **Properties** pop-down window, and select the property called **Half-Delta** that you just created.
17. Hit the **blue arrow** to the right to send it into the **Preview** window. Hit **OK** and you see the half delta in the **Preview** window of the **Label Style Composer**. Hit **OK** again to leave.
18. Now in Civil 3D, draw an arc by typing **Arc** and selecting three points somewhere on the screen. Type **PE** for **PEdit** and convert the arc to a **Pline**.
19. Then choose the **Alignment** pull-down menu, select **Create from Polyline**, choose the arc, and accept the defaults when creating the alignment.
20. Then choose **Alignment >> Add labels....**
21. Select **Alignment** for the **Feature:**.
22. Select **Single Segment** for **Label type:**.
23. Click the **Add** button and choose the **arc**. A label appears showing the Length, Radius Delta, and Half-delta for the arc. Mission accomplished!

This example shows how to create an expression and use it for augmenting the labeling capabilities of the software.

Software and Hardware Versions

C

SOFTWARE VERSIONS AND SPECIFICATIONS

The software discussed in this text consists of Autodesk Civil 3D 2007. This version added significant labeling enhancements, corridor design improvements, new subassemblies, new piping tools, and an entire survey solution for data collectors and traverses. The Civil 3D 2006 version added the following enhancements to its feature set: corridor design enhancements, additional data sharing, and utility pipe drafting. The previous version, Civil 3D 2005, was the second iteration (but the first public release) of the Civil 3D solution and added Autodesk VIZ to the solution suite. The first version, a "prerelease" version called Civil 3D, was actually a preview of what was to come. The marketing philosophy of Autodesk was to release the software to its current clients in advance of it being made available to the public. In this fashion, Autodesk's goal was to build interest in the user population and persuade them to become acclimated to the revolutionary philosophies required to use the software successfully. It also wanted the customer to have time to consider the implementation expenses and compare them with the productivity improvements.

The Civil 3D software is provided to Autodesk Land Desktop users who have subscriptions for the Civil Design add-on as part of the subscription service. In 2005 it was formally introduced for sale to the general public.

HARDWARE SPECIFICATIONS

This section is included because very often an information technology department does the purchasing for the organization. These departments usually buy generic desktop PCs and laptops with little regard for the applications being used. CADD systems and visualization software are some of the most hardware-intensive applications available, and they deserve to be budgeted for accordingly. Hence three scenarios are described here. Scenario 1 describes the minimum requirements needed to run the software and these are per Autodesk. Scenario 2 consists of an arguable, optimal hardware solution for running the software for which cost is not restricted. Scenario 3 is a set of specifications that identifies a great system whereby budget is considered, costing a little over half of the optimal system.

Note:
Always check with the software manufacturer before making final decisions as specifications may change.

Scenario 1

These system specifications and *minimum requirements* are recommended for effective use of Civil 3D™ software:

- Intel® Pentium® IV processor, 3 GHz or higher
- Microsoft® Windows® XP Professional or Home Edition (SP2), Windows XP Tablet PC Edition (SP2), or Windows 2000 (SP4)

- 2 GB RAM
- 5 GB free disk space for installation
- 1280 × 1024 with true color
- OpenGL®-capable workstation class graphics card or DirectX® 9 support
- Mouse, trackball, or compatible pointing device
- Microsoft Internet Explorer 6.0 (SP1 or later)
- DVD drive

In order to run the software in an optimal environment, the following configuration is proposed. We have found, through our national seminars for organizations such as the Amercian Society of Civil Engineers (ASCE), that managers, engineers, and CADD gurus are often interested in what the "Perfect CADD Station" might consist of, assuming that money were no object. We often have fun with this because few of us find ourselves in this position, but as Alfred Lord Tennyson once said, "Dreams are true while they last, and do we not live in dreams?"

Because Civil 3D is a Windows-based solution, the perfect CADD Station is restricted to the Windows world and to Intel-based processors. Although I admit that some improvements in various processes can be achieved by using highly specialized adjunct solutions, sometimes keeping it simple, taking advantage of integrated design, and going with tried and true solutions is the best way to get the job done. So when the goal is to specify the Perfect CADD Station, also considered is that the station must be solid and stable enough to get an engineering project done without crashing, having to tinker with settings, or dealing with any one of many technical issues that could arise. A reliable and fast system is necessary.

First of all, the computer should be a workstation, not a PC. Hardware is relatively inexpensive these days, so to request a workstation for advanced users is not unthinkable. The productivity gains produce a significant return on investment for power users. Although debatable, the difference between a workstation and a typical desktop PC is defined as follows: A workstation is built to handle specific applications such as CADD and is manufacturer certified as such. The systems are often built to take advantage of top-of-the-line components that offer pure processing power, extremely high data throughput, high-end storage capacity, the fastest video, and reliability. A workstation also allows for Small Computer System Interface (SCSI) hard drives and dual CPUs.

Scenario 2

These system specifications and *optimal requirements* for effective use of Civil 3D software comprise the following workstation, which was specified at mid-year 2006 and runs around $10,000, not including the Civil 3D software.

CPU(s)

The CPU(s) would be an Intel® Xeon™ Processor 3.80 GHz, 2 MB L2 cache with a second Processor, also an Intel® Xeon™ Processor 3.80 GHz, 2 MB L2 cache. The Intel Xeon processor supports technologies such as the Intel NetBurst Micro-architecture, whose pipeline depth in the processor is doubled, allowing the processor to reach much higher core frequencies. It allows multiple branch prediction, speculative execution and data flow analysis to enhance processing and supports Hyper-Threading Technology and Extended Memory for large memory support. The Level 2 (L2) Cache is a collection of built-in memory chips that is faster than the main memory area and can help speed the operation of some applications.

The Operating System

The operating system would be Microsoft Windows XP Professional with Service Pack 2. The hard disk would be set up with an NTFS File System.

Random Access Memory (RAM)

The system's Random Access Memory (or RAM) would be 8 GB (if available, 4 GB otherwise) of DDR2 SDRAM Memory running at 400 MHz. It is Error Checking and Correction memory (or ECC) and is installed on Dual Inline Memory Modules (or DIMMS). The SDRAM is Synchronous Dynamic Random Access Memory and delivers data bursts at high speeds using a synchronous interface. DDR is an acronym for Double Data Rate, and DDR2 is the next version of this technology.

Keyboard and Mouse

The keyboard is subjective but should have the tactile response desired by the user. The mouse falls into the same category, although a 3D mouse such as a Motion Controller is also recommended. The SpaceBall 5000, 3D Motion Controller is designed to enhance 3D software applications. It has twelve programmable buttons and can keep function keys and other macros at the user's fingertips. It can simultaneously pan, zoom, and rotate 3D model scenes or manipulate the camera dolly with one hand, while the other hand selects, inspects, or edits with the mouse.

CD and DVD Drives

The system has both CD and DVD drives and is able to produce either media as Read-Write Devices. Something along the lines of a 16× DVD AND 16× DVD+/-RW with Sonic DM and DVDit! SE, CyberLink PowerDVD software associated with it.

Hard Drives

The hard drives are among the most important parts of the configuration and might include the following equipment. First the drives will be all SCSI drives. Two hard drives with the following specifications are recommended: 146GB Ultra 320 SCSI drives, running at 15,000 rpm. The hard drive internal controller will be a Hi-Performance U320 SCSI integrated controller used to connect internal hard drives.

Monitor

For the monitor, spare no expense: perhaps a Dell 24-inch UltraSharp™ 2405FP Widescreen, adjustable stand, supporting both VGA and DVI interfaces. Go for the second monitor for multitasking, and duplicate the monitor so you have two. The graphics card driving these monitors should be the best technology has to offer. This component is always in flux, but today's flavor is 512 MB PCIe ×16 nVidia Quadro FX 4500, Dual DVI or Dual VGA, or DVI + VGA.

Sound System

The sound system should be premium sound along the lines of a Dell AS500 Sound Bar with a Sound Blaster-Audigy 2 with Dolby Digital 5.1 and IEEE1394.

Installed Software

Installed software should include Microsoft Office Professional Edition and Norton Internet Security as basic components as well. Of course, the system also has the Civil 3D software on it.

Network and Internet Communications

Network and Internet communications should include a wireless Dell TrueMobile 1300 (802.11 b/g) WLAN USB 2.0 DT Adapter. Do not forget an 800 Watt battery backup and a strong 3-year, on-site support plan.

Printers

For printing, include two types of printers, one for small and medium-sized printing, but of very high photographic quality, and one for large-format printing. The first would be a printer similar to the Epson Stylus Pro 4000 Print Engine at about $1,795. It has a high-performance 1" wide print head that produces a resolution of 2880 × 1440 dpi and handles virtually any media type of either roll or cut sheet up to 17" wide. It has a large paper tray capable of handling cut-sheet media up to 17" × 22" as well as four different ways to load media. It also can handle media up to 1.5 mm posterboard using a front-loading straight-through path. The print head is an 8-channel head capable of handling eight separate ink cartridges simultaneously, which allows the handling of both Photo Black and Matte Black inks at the same time to maximize the black density on virtually any media type.

The second printer is a plotter for large format output. It would be similar to the HP Designjet 815 mfp, which has 2400 × 1200 dpi resolution enhanced for scanning or plotting and allows for scanning, copying, and printing as an all-in-one solution. It handles CAD output, GIS and graphics, large-format originals, and thick, rigid media. The integrated scan software produces electronic files and hard copies of color and black-and-white documents and can handle originals of up to 43" wide and up to 0.6" thick. It should produce the output on a large variety of media and in sizes up to 42" wide. It is around $20,000.

Scenario 3

The system specifications and *recommended requirements* for excellent use of the Civil 3D 2007 software by an advanced user comprises the following workstation, which runs around $7,000, not including the Civil 3D software.

- CPU(s): Intel® Xeon Processors 3.00 GHz, 2 MB L2 Cache
- 2nd Processor: Intel® Xeon Processor 3.00 GHz, 2 MB L2 cache
- Operating System: Microsoft® Windows® XP Professional, SP2 with NTFS file system
- Memory: 2 GB, DDR2 SDRAM Memory, 400 MHz, ECC (4 DIMMS)
- Keyboard and mouse with scroll
- CD-ROM, DVD: 48× CD-RW and 16x DVD+/-RW w/Sonic RecordNow! Deluxe, Sonic DVDit! SE, PowerDVD
- Hard Drive Configuration: All SCSI drives, Non-RAID
- First Hard Drive: 146 GB Ultra 320 SCSI, 15,000 rpm
- 2nd Hard Drive: 73 GB Ultra 320 SCSI, 15,000 rpm
- Hard Drive Internal Controller Options: U320 SCSI Integrated Controller
- External Storage Options: U320 SCSI Card—for external connectivity
- Floppy Drive Options: 3.5-inch floppy drive
- Monitor(s): 20 inch UltraSharp Flat Panel, adjustable stand, VGA/DVI
- 2nd Monitor: 17-inch flat panel

- Graphics Cards: 256 MB PCIe ×16 nVidia Quadro FX 3400, Dual DVI or Dual VGA/DVI+VGA
- Speakers: Dell AS500 Sound Bar
- Sound Card: Sound Blaster-Audigy 2 (D), w/Dolby Digital 5.1 and IEEE1394
- Productivity Software: Microsoft Office Professional Edition 2003 and Adobe Acrobat 6.0
- Hardware Support Services: 3-year Business Standard Plan
- Battery Backup: APC ES 500 Backup-UPS

Identify the Civil Engineering Industry

D

Now that the software used in this text as well as the suggested hardware have been identified, discussion follows on the work you will be doing, the business of civil engineering, and the kinds of projects that would be appropriate to apply Civil 3D toward.

The types of work falling into an industry that provides civil engineering design typically include services such as Planning, Preliminary Design, Surveying, Final Design, Value Engineering, Plans Preparation and Construction Document Preparation, and Site Data Preparation. The goal is to define which industries can make use of the Civil 3D product at this time.

Civil engineering projects are generally performed for two types of clients: public clients and private clients. Public clients include governmental agencies such as municipal, county, state, and federal agencies. Private clients include landowners, developers, and industrial, commercial, and legal organizations.

The projects performed for the **public sector** often include corridor design such as roads, tunnels, channels, and aqueducts; construction management; environmental engineering; transportation (which includes traffic flows, evacuation planning, and signage, signalization); airport engineering; bridge design; waste management; wetlands delineation and mitigation; water supply (which includes treatment, distribution, and pump stations); sewage networks (which includes collection and effluent treatment); and water resources.

Public Sector —Govenment agencies

The project types performed for the private sector may be comprised of design of subdivisions, roads, commercial sites, and industrial sites. They may also include design, analysis, and study of waste management, wetlands, water supply and distribution, sewage systems, and water resources. Additional work is performed as part of construction management on projects as well as the legal and real estate professions.

Civil 3D provides for some hydraulic tasks but does use object technology in its pipe networks. Significant control using the pipe style exists for laying out the pipe network and making modifications to it. The seed for formulas can be observed in some of the settings and portends the things to come.

Civil 3D does not provide tools that directly apply to bridge design. Neither does it support tools for some surveying tasks such as data collection and GPS computations.

Of the projects mentioned here, Civil 3D is a *primary solution* for corridors, airports, waste management, subdivision, road, commercial site, and industrial site design. Of the projects mentioned here, Civil 3D is a *strong supporting solution* for a variety of drafting and analyses functions for construction management, environmental engineering, water systems, sewage and storm networks, transportation, wetlands delineation and mitigation, water resources, and the legal and real estate fields.

IDENTIFY THE BREAKDOWN AND FLOW OF ENGINEERING TASKS AND HOW CIVIL 3D CONTRIBUTES TO THE WORK EFFORT

By recognizing the tasks performed and the project requirements for each project type identified, it is possible to distinguish which tasks within these civil engineering projects benefit most by applying Civil 3D functionality to them.

Although the Civil 3D software is a revolutionary product and performs many worthwhile and industry leading functions, it would be beneficial to point out where the software offers extraordinary assistance to the user. The first step in accomplishing this task is to break down the types of projects that an engineer might undertake. The purpose here is to identify immediately project areas that are out of the scope of this text; identify project types that are strong candidates for use of Civil 3D as the primary solution, and identify those areas that can use Civil 3D as a supporting tool.

The second step is to identify the functionality that Civil 3D offers in performing computations or drafting assistance of those engineering tasks. The tasks are identified as well as how Civil 3D can provide improved productivity and functionality to those who use it.

The listing that follows provides a step-by-step description of each project's requirements. **The tasks that benefit most from the use of Civil 3D are denoted in bold type.** <u>Those tasks that can make important use of Civil 3D as a supporting tool are underlined.</u> Those tasks in normal type are outside the current feature set of Civil 3D. The functionality that Civil 3D offers is shown using the following key:

- (1) Uses all Civil 3D functionality including alignments, parcels, corridor and grading tools, profiles, and sections
- (2) Uses Civil 3D's modification benefits where the changes ripple through the design
- (3) Uses alignments capabilities
- (4) Uses parcels capabilities
- (5) Uses corridor and/or profiles and sections capabilities
- (6) Uses terrain modeling, surfaces, and grading capabilities
- (7) Uses Civil 3D's drafting and labeling functions
- (8) Uses built-in rendering engine
- (9) Uses MAP 3D tools
- (10) Utility/storm sewer piping

For the Public Sector

The Public Sector has the following projects and can be DOTs, city, county, state agencies, public utility districts (PUDs), or municipal utility districts (MUDs). They are shown with the workflow milestones they involve.

Corridor Design

See Figure D-1.

- **Preliminary planning and alternative layout creation: (1), (2)**
- Communication with decision makers
- **Preliminary engineering: (1), (2)**
- Project funding and approval and notice to proceed
- <u>Surveying activities including property research, zoning issues, ownership, determination of existing site conditions: (7)</u>
- **Terrain modeling: (1), (2), (6)**
- <u>Determination of subsurface and geological conditions: (7)</u>
- <u>Wetlands determinations: (7)</u>

Figure D-1   Corridor design

- **R.O.W. determinations: (4)**
- **Curvilinear 2D geometry development, easements, alignments, parcels: (3), (4), (7)**
- **Drafting and labeling: (1)**
- **Corridor profile, section and corridor creation and analysis, roadway design, intersection ramp and gore development: (5)**
- Paving requirements: (7)
- **Intersection and cul-de-sac design: (5), (6), (7)**
- Utilities crossings, relocations, and conflict resolutions: (7)
- Drainage improvements, storm water piping, and retention: (10)
- Landscape design plans: (7)
- Irrigation design plans: (7)
- **Signage, electrical, and striping plans: (7)**
- **Erosion and siltation design plans: (7)**
- Submit review plans
- **Make modifications based on review comments: (2)**
- **Construction plan development: (7)**
- Quantity takeoffs and cost estimations: (6), (7)
- **QA/QC and survey stakeout: (6), (7)**
- **Public communication and public hearings with potential for visual renderings: (8)**

Construction Management

See Figure D-2.

- Obtain RFP, verifying plans, analyze design documents: (1)
- Quantities, earthworks takeoffs, and cost estimates: (6), (7)
- Proposal preparation: (8)
- Project awarded and notice to proceed
- Surveying activities including establishing project control, verify existing site conditions: (6), (7)
- **Making design changes based on change orders and for constructability: (2)**
- **Curvilinear 2D geometry development: (1)**
- Quantity takeoffs and cost estimations: (6), (7)

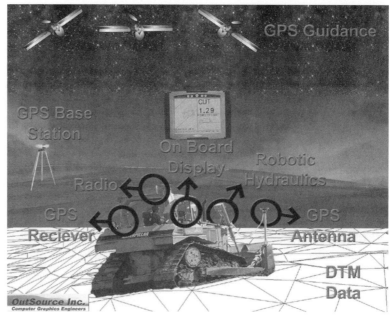

Figure D-2   3D/GPS Machine Control

- **Terrain modeling for existing and proposed conditions: (6)**
- **Earthworks computations: (6)**
- **Build/rebuild project data for road grading: (2), (5)**
- **Build/rebuild site grading data: (2), (5)**
- Build/rebuild utilities: (6), (7)
- **QA/QC of design plans and construction approaches: (2), (7)**
- **Generate stakeout information and/or GPS Machine Control: (1), (2)**
- **Format for machine control usage: (1), (2)**
- **Construction plan development: (1), (2)**
- Pavement design and analysis as-built surveying and plan generation: (6), (7)
- Project operations and maintenance: (6), (7)

Environmental Engineering
- **Curvilinear 2D geometry development for property lines, Planimetrics, etc.: (1), (2), (7)**
- **Terrain modeling and analysis: (3), (4), (6)**
- Proposed impact studies
- Data collection of above and below ground conditions, chemicals, species, etc.
- Subsurface water analysis, conditions, modeling: (6)
- Air quality conditions, analysis, modeling
- GIS development and mapping: (6), (7), (9)

Water Supply (Which Includes Treatment, Distribution, and Pump Stations)
- **Curvilinear 2D geometry development for waterline trunks, distribution systems: (3), (4)**
- Elevated storage: (6), (7)
- **Planimetrics, etc.: (7)**

- **Terrain development and analysis for waterline depth and conflict resolution: (1), (2), (6)**
- Water pressure analysis, demand loading, etc.
- <u>Location of elevated storage, lift stations, capacities, etc.: (6), (7)</u>
- <u>Water treatment plants and processes (see Industrial Projects): (6), (7), (10)</u>

Sewage Networks (Which Include Collection and Effluent Treatment)

- **Curvilinear 2D geometry development for trunkline, vacuum sewers and collection system, Planimetrics, etc.: (3), (4)**
- <u>Location and design of collection system, lift stations, capacities, etc.: (6), (7)</u>
- **Mapping and GIS tasks: (6), (8), (7)**
- **Drafting and labeling: (7)**
- **Terrain development and analysis of gravity flow systems: (1), (2), (6)**
- System capacity analysis
- <u>Location and design of collection systems from main trunks to treatment facilities: (6), (7)</u>
- <u>Wastewater treatment plants and processes (see Industrial Projects): (6), (7)</u>

Transportation (Which Includes Traffic Flows, Evacuation Planning, Signage, and Signalization)

See Figure D-3.

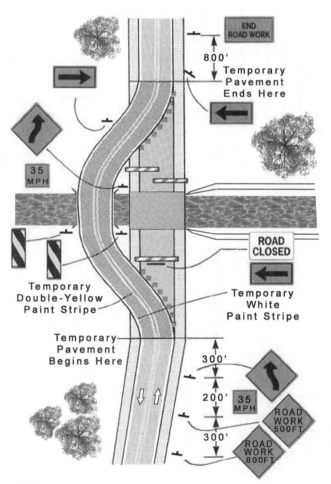

Figure D-3   Transportation

- **Geometry development: (1), (2)**
- **Planimetrics: (7)**
- Data collection, traffic counts
- Signalization
- HOV development plans and GIS studies: (6), (7)
- Data collection, traffic counts
- **Mapping and GIS tasks: (6), (7), (9)**
- Pavement design and analysis: (6), (7)
- **Drafting and labeling: (7)**

Wetlands

- Delineation, classification, and mitigation: (7)
- **Curvilinear 2D geometry development for wetlands and other boundaries, property lines, treelines, Planimetrics, etc.: (1), (2)**
- **Mapping and GIS tasks: (6), (7), (9)**
- **Drafting and labeling: (7)**
- Data collection and reduction
- RTK/GPS and traditional surveying

Water Resources

See Figure D-4.

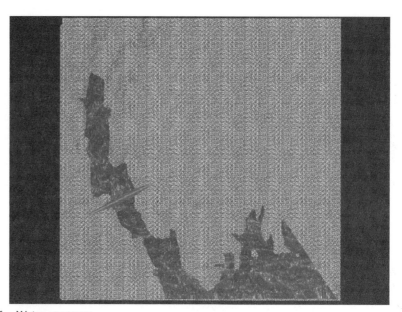

Figure D-4 Water resources

- **Preliminary planning: (1), (2)**
- Project funding and approval and notice to proceed
- Surveying activities including property research, ownership, determination of existing site conditions: (6), (7)
- Determination of subsurface and geological conditions, flows, and magnitudes: (6), (7)
- Surface and groundwater water analysis
- Design/analysis of wells, water transfer systems, irrigation

- Hydrology and water supply
- River conditions and forecasts
- Water quality
- Weather forecasts
- Flood emergency information
- Water use, conservation, and planning
- Design and analysis of dams
- **Geometry development for Planimetrics: (1), (7)**
- Rainfall determinations
- Flood studies
- <u>Wetlands determinations: (6), (7)</u>
- <u>Stormwater analysis and piping</u>
- **Erosion and siltation design: (6), (7)**
- <u>Landscape analysis: (7)</u>
- Chemical analysis
- **Development of maps: (6), (7), (9)**
- <u>Report generation</u>
- **QA/QC: (2)**
- **Public hearings with potential for visual renderings: (8)**

Airport Engineering

See Figure D-5.

Figure D-5   Airport engineering

- Obtain FAA and local grants
- **Preliminary planning and alternative layout creation: (1), (2)**
- Communication with decision makers
- **Preliminary engineering: (1), (2)**
- Project funding and approval and notice to proceed
- <u>Surveying activities including property research, zoning issues, ownership, determination of existing site conditions</u>

- **Terrain modeling: (1), (2), (6)**
- Determination of sub-surface and geological conditions: (6), (7)
- **Curvilinear 2D geometry development, runway/taxiways, FAA obstruction/clearance plans, parking lots, service industry access roads: (1), (7)**
- Security considerations: (6), (7)
- **Planimetrics, etc.: (1), (7)**
- FAA and state regulations compliance
- **FAA height contours, clearance analysis, etc.: (1), (7)**
- Pavement design and analysis: (6), (7)
- Runway and taxiway profiles and sections or corridors: (1), (2), (6), (7)
- **Drafting and labeling: (6), (7)**
- **Drainage improvements: (10)**
- Storm water retention: (6)
- Underground retention facilities: (6), (7)
- **Rough grading, finished grade development: (2), (6)**
- **Erosion and siltation design: (2), (6)**
- **Final design engineering, pad siting for hangars and services: (1), (2), (6)**
- Landscape design: (6), (7)
- Irrigation design: (6), (7), (10)
- **Easement determinations: (4)**
- **Value engineering and review: (2)**
- **Engineering modifications: (2)**
- **Construction plan generation: (1)**
- **Quantities and cost estimates: (6), (7)**
- **QA/QC and survey stakeout: (1), (2), (7)**
- Obtain FAA and local construction grants
- **Public hearings with potential for visual renderings: (8)**

The Legal Field

See Figure D-6.

Figure D-6 The legal field

- <u>Rezoning issues: (4)</u>
- <u>Study, research, analysis: (1)</u>
- <u>Legal, forensic graphics: (8)</u>
- Expert witness

Bridge Design

See Figure D-7.

Figure D-7   Bridge design

- **Preliminary planning and alternative layout creation: (1), (2)**
- Communication with decision makers
- **Preliminary engineering: (1), (2)**
- Project funding and approval and notice to proceed
- <u>Surveying activities including property research, zoning issues, owner-ship, determination of existing site conditions: (6), (7)</u>
- **Terrain modeling: (1), (2), (6)**
- <u>Determination of subsurface, underwater, and geological conditions: (6), (7)</u>
- <u>Wetlands determinations: (6), (7)</u>
- **ROW determinations: (4), (7)**
- **Curvilinear 2D geometry development, easements, alignments, parcels: (3), (4), (7)**
- **Drafting and labeling: (1)**
- **Approach design, corridor profile, section and corridor creation and analysis, roadway design, intersection ramp and gore development: (5)**
- Structural requirements for beams, bents, columns, decks, etc.
- <u>Paving requirements: (6), (7)</u>
- **Intersection and cul-de-sac design: (5), (6), (7)**
- <u>Utilities crossings, relocations, and conflict resolutions: (7)</u>
- <u>Drainage improvements: (10)</u>

- <u>Storm water retention: (6)</u>
- <u>Noise abatement design plans: (7)</u>
- <u>Landscape design plans: (7)</u>
- <u>Irrigation design plans: (7)</u>
- **Signage, electrical, and striping plans: (7)**
- **Erosion and siltation design plans: (7)**
- Submit review plans
- **Make modifications based on review comments: (2)**
- **Construction plan development: (7)**
- <u>Quantity takeoffs and cost estimations: (6), (7)</u>
- **QA/QC and survey stakeout: (6), (7)**
- **Public communication and public hearings with potential for visual renderings: (8)**

For the Private Sector

Subdivision Design

See Figure D-8.

Figure D-8 Subdivision design

- **Preliminary planning and alternative layout creation: (1), (2)**
- Communication with decision makers
- **Preliminary engineering: (1), (2)**
- Project funding and approval and notice to proceed

- Surveying activities including property research, zoning issues, ownership, determination of existing site conditions: (7)
- **Terrain modeling: (1), (2), (6)**
- Determination of subsurface and geological conditions: (7)
- Wetlands determinations: (7)
- **Refined site layout, park, and greenspace requirements**
- **Curvilinear 2D geometry development, easements, alignments, parcels: (3), (4), (7)**
- **Roadway design: (3), (5)**
- Paving requirements: (6), (7)
- **Intersection and cul-de-sac design: (5), (6), (7)**
- **Corridor profile, section, and corridor creation and analysis: (5), (6), (7)**
- **Drafting and labeling: (7)**
- **Rough grading, overlot, and finished grade development: (6)**
- **Drainage improvements: (10)**
- Storm water retention: (6)
- Sanitary sewer design plans and profiles: (5), (6), (7), (10)
- Waterline design plans and profiles: (5), (6), (7)
- Landscape design plans: (6), (7)
- Irrigation design plans: (6), (7)
- **Erosion and siltation design plans: (2), (6)**
- Submit review plans
- **Make modifications based on review comments: (2)**
- **Construction plan development: (7)**
- Quantity takeoffs and cost estimations: (6), (7)
- **QA/QC and survey stakeout: (6), (7)**
- **Public communication and public hearings with potential for visual renderings: (8)**

Environmental Engineering

- **Curvilinear 2D geometry development for property lines, Planimetrics, etc.: (1), (2), (7)**
- **Terrain modeling and analysis analysis: (3), (4), (6)**
- Proposed impact studies
- Data collection of above and below ground conditions, chemicals, species, and so on
- Subsurface water analysis, conditions, modeling: (6)
- Air quality conditions, analysis, modeling
- **GIS development and mapping: (6), (7), (9)**

Water Supply (Which Includes Reticulation Distribution and Pump Stations)

- **Curvilinear 2D geometry development for waterline trunks, distribution systems: (3), (4)**
- Elevated storage: (6), (7)
- **Planimetrics: (7)**
- **Terrain development and analysis for waterline depth and conflict resolution: (1), (2), (6)**

- Water pressure analysis, demand loading, etc.
- <u>Location of elevated storage, lift stations, capacities, etc.: (6), (7)</u>
- <u>Water treatment plants and processes (see Industrial Projects): (6), (7), (10)</u>

Sewage Networks (Which Include Collection, not Treatment)

- **Curvilinear 2D geometry development for trunkline, vacuum sewers and collection system, Planimetrics, etc.: (3), (4)**
- **Mapping and GIS tasks: (6), (8), (7)**
- **Drafting and labeling: (7)**
- **Terrain development and analysis of gravity flow systems: (1), (2), (6)**
- System capacity analysis
- <u>Location and design of collection systems up to main trunks or lift stations: (6), (7)</u>

Transportation (Which Includes Traffic Flows for Commercial and Large Subdivision Impact)

- **Geometry development: (1), (2)**
- **Planimetrics: (7)**
- Data collection, traffic counts
- **Mapping and GIS tasks: (6), (7), (9)**
- <u>Pavement design and analysis: (6), (7)</u>
- **Drafting and labeling: (7)**

Wetlands Delineation and Mitigation

- <u>Delineation, classification, and mitigation: (7)</u>
- **Curvilinear 2D geometry development for wetlands and other boundaries, property lines, treelines, Planimetrics, etc.: (1), (2)**
- **Mapping and GIS tasks: (6), (7), (9)**
- **Drafting and labeling: (7)**
- Data collection and reduction
- RTK/GPS and traditional surveying

Commercial Design

See Figure D-9.

- **Preliminary planning and alternative layout creation: (1), (2)**
- Communication with decision makers
- **Preliminary engineering: (1), (2)**
- Project funding and approval and notice to proceed
- <u>Surveying activities including property research, zoning issues, ownership, determination of existing site conditions: (7)</u>
- **Terrain modeling: (1), (2), (6)**
- <u>Determination of subsurface and geological conditions: (7)</u>
- **Access requirements: (4), (7)**
- **Site layout, parking space requirements, curvilinear 2D geometry development, easements, BRLs, parking lots, parcels, Planimetrics: (4), (7)**

Figure D-9   Commercial design

- <u>Paving requirements: (7)</u>
- **Drafting and labeling: (7)**
- **Drainage improvements: (10)**
- **Stormwater retention: (6)**
- <u>Underground retention facilities: (7)</u>
- **Rough grading, finished grade development: (6)**
- **Erosion and siltation design: (7)**
- **Final design engineering, pad siting: (6)**
- <u>Environmental considerations, oil/water separators, etc.: (7)</u>
- <u>Pavement design and analysis: (7)</u>
- <u>Water and sanitary plans and profiles: (7), (10)</u>
- **Erosion and siltation plans: (7)**
- <u>Landscape design: (7)</u>
- <u>Irrigation design: (7)</u>
- **Easement determinations: (4)**
- **Value engineering and review: (2)**
- **Engineering modifications: (2)**
- **Construction plan generation: (1)**

- **Quantities and cost estimates: (6), (7)**
- **QA/QC and survey stakeout: (1), (2), (7)**
- **Public hearings with potential for visual renderings: (8)**

Industrial (and Waste Management) Site Layout

See Figure D-10.

Figure D-10   Industrial (and waste management) site layout

- **Preliminary planning and alternative layout creation: (1), (2)**
- Communication with decision makers
- **Preliminary engineering: (1), (2)**
- Project funding and approval and notice to proceed
- <u>Surveying activities including property research, zoning issues, ownership, determination of existing site conditions: (7)</u>
- **Terrain modeling: (1), (2), (6)**
- <u>Determination of subsurface and geological conditions: (7)</u>
- **Public and staff access requirements: (7)**
- **Site layout, parking space requirements, curvilinear 2d geometry development, easements, BRLs, parking lots, parcels, Planimetrics: (3), (4), (7)**
- **Drafting and labeling: (1)**
- Environmental engineering analyses and considerations
- **Drainage improvements: (10)**
- <u>Storm water retention, underground retention facilities: (6)</u>
- **Rough grading, finished grade development: (2), (6)**
- **Erosion and siltation design: (2), (6)**
- **Final design engineering, pad siting: (1), (2), (6)**
- <u>Environmental considerations, oil/water separators, etc.: (7)</u>
- <u>Pavement design and analysis: (6), (7)</u>

- Water and sanitary plans and profiles: (1), (2), (6), (7), (10)
- **Erosion and siltation plans: (6), (7)**
- Landscape design: (7)
- Irrigation design yard piping design for plant activities: (7)
- EPA requirements for spill and containment: (7)
- Disaster and evacuation plans: (7)
- **Easement determinations: (4), (7)**
- **Value engineering and review: (1)**
- **Engineering modifications: (2)**
- **Construction plan generation: (1)**
- Quantity takeoffs and cost estimations: (6), (7)
- **QA/QC and survey stakeout: (6), (7),**
- **Public hearings with potential for visual renderings: (8)**

Hydrology and Water Resources: Surface Water Analysis, Groundwater Analysis, Wells, Water Transfer, Irrigation

See Figure D-11.

Figure D-11   Surface water analysis

- **Preliminary planning, project funding and approval, and notice to proceed: (1), (2)**
- Surveying activities including property research, ownership, determination of existing site conditions: (7)
- Determination of subsurface and geological conditions, flows, and magnitudes: (7)
- **Geometry development for Planimetrics: (4), (7)**
- Rainfall determinations
- Flood studies
- Wetlands determinations: (7)
- Storm water analysis
- **Erosion and siltation design: (2), (6)**
- Landscape analysis: (6), (7)

- Chemical analysis
- **Development of maps: (6), (7), (9)**
- <u>Report generation: (1)</u>
- **QA/QC: (1), (2), (7)**
- **Public hearings with potential for Visual Renderings: (8)**

The Legal Field

See Figure D-12.

Figure D-12   Rezoning issues, shadow studies

- <u>Rezoning issues: (4)</u>
- <u>Study, research, analysis: (1)</u>
- <u>Legal, forensic graphics: (8)</u>
- Expert witness consulting

Supplement to Chapter 3: The Simple, but Time Honored Point

E

This supplement is included to provide the reader and instructor with additional routines and procedures for using the extensive set of Point-related commands in the **Point Creation** toolbar. The commands in this supplement are no less powerful nor less useful than those in Chapter 3 and should be given equal respect in learning and use in production.

CREATING "THINKING POINTS"

The next segment delves into procedures for placing Interpolation-related points into a project drawing. Refer to Figure E-1. These commands allow the ability to generate "thinking points." Thinking points are points that may or may not be used in design, but they will assuredly be used to develop other data.

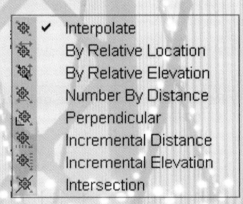

- ✓ Interpolate
- By Relative Location
- By Relative Elevation
- Number By Distance
- Perpendicular
- Incremental Distance
- Incremental Elevation
- Intersection

Figure E-1 "Thinking points"

TIP Consider that you know the elevation of a flow line of a curb and are wondering what the elevation would be if you went in a specified direction for a specified distance at a specific grade or slope. And then, depending on where it landed, you would make a decision on what would happen next. Decide whether the point is too high or too low or erase it and provide different criteria. These are thinking points because they allow you to see the results of design criteria rapidly and then either proceed forward or step back and try again with little to no cost.

Exercise E-1: Interpolation—Interpolate Command

The next command, **Interpolate,** can be used to set intermediate points in 3D. The software uses the next available point numbers and prompts for other information as it goes.

> This command might be used to densify a data set for DTM triangulation.

1. Continue in the same drawing, or open the drawing called Chapter 3-a. Dwg.
2. Using the **V** for **View** command, highlight the view called **Points-I,** hit **Set Current,** and then hit **OK.** You see two points already set, Points 1001 and 1002. You will densify these by adding three points in between them at prorated elevations based on the two points' elevations.
3. Using the drop-down arrow to the right of the fifth icon from the left, choose **Interpolate.**
4. `Pick a first point object:` Select **Point 1001** with your mouse; no Osnap is needed.
5. `Pick a second point object:` Select Point 1002 with your mouse; no Osnap is needed.
6. `Enter number of points <0>:` **3**
7. `Specify an offset <0.000>:` **0,** this will create the new points on line with the 1001 and 1002.
8. `Enter a point description <.>:` **Spot**
9. `Enter a point description <Spot>:`**<Enter>**
10. `Enter a point description <Spot>:`**<Enter>**
11. Hit **<ESC>** to terminate.

Notice that three new points were created in between Points 1001 and 1002.

Exercise E-2: Interpolation—By Relative Location Command

The next command, **By Relative Location,** can be used to set points in 3D by using Grade, Slope, Elevational, or Difference based criteria.

> This command can be used to create proposed ground points based on the engineer's desired slope, grade, or elevation. These thinking points can be used to test what elevations, slopes, and grades might result based on the data you are providing in your design.

1. Continue in the same drawing, or open the drawing called Chapter 3-a. Dwg.
2. Using the **V** for **View** command, highlight the view called **Points-I,** hit **Set Current,** then hit **OK.** You see two points already set, Points 1001 and 1002. You will set a point from one of the contour endpoints toward Point 1001 at a grade of 2%.
3. Using the drop-down arrow to the right of the fifth icon from the left, choose **By Relative Location.**
4. `Specify first point or [Entity]:` Pick a point at an endpoint of a contour with the Endpoint Osnap.

5. `Specify an elevation for the first control point <356.000'>:` **\<Enter\>**
6. `Specify second point:` **Pick a point on Point 1001 using the Node Osnap.**
7. `Specify an elevation for the second control point or [Difference/Slope/Grade] <356.911'>:` **g**

TIP

Note that a comment is warranted here to explain the other options available. Difference allows the user to enter a change in height from the first elevation and can be either plus or minus. So if the first elevation is 360.00, a difference can be applied of, say, 5' so that the second elevation becomes 365.00. Slope is similar to grade except that it refers to a ratio instead of a percentage. Slopes are in values such as 2:1, 3:1, and so forth, where the 2:1 means 2' horizontally for every 1' vertically. Of course, you are using the Grade option in the example where 2% is 2' vertically for every 100' horizontally.

8. `Grade (percent) or [Slope] <0.00>:` **2**
9. `First Elev: 356.000', Second Elev: 358.617', Elevation Difference: 2.617 Horiz Dist: 130.850, Grade: (percent): 2.00, Slope: (run:rise): 50.00:1`
10. `Distance:` **100**
11. `Specify an offset <0.000>:` **\<Enter\>**
12. `Enter a point description <.>:` **Spot**
13. Hit **\<ESC\>** to terminate.

Notice a new point showed up 100' from the contour endpoint you selected. The elevation is established at a 2% grade from the contour elevation in a direction toward the Point 1001.

Exercise E-3: Interpolation—By Relative Elevation Command

The next command, **By Relative Elevation**, can be used to set points in 3D by using Grade, Slope, Difference, or Elevational criteria. This is different from the previous command in that it establishes the grade to be met but then requests the final elevation to be set. It then sets a point at the desired elevation using the grade computed at whatever horizontal distance it takes to meet the elevation prescribed.

TIP

Again, these can establish thinking points that allow you to consider the results of moving ahead with your actions. They can be used to find out where a point falls running at, say, −2% until it hits a critical hydraulic grade elevation of, say, 101.33. If the point falls into the next county, then it likely is not feasible. If it falls near the ditch flowline, then maybe it makes sense.

1. Continue in the same drawing, or open the drawing called Chapter 3-a. Dwg.
2. Using the **V** for **View** command, highlight the view called **Points-I**, hit **Set Current**, and then hit **OK**. You see two points already set, Points 1001 and 1002. You will set a point starting at 1002, heading toward 1001, and hold the grade between them such that it keeps traveling until it meets the elevation specified.

Student Files

3. Using the drop-down arrow to the right of the fifth icon from the left, choose **By Relative Elevation**.
4. `Specify first point or [Entity]:` Pick a point on Point 1002 with the Node Osnap.
5. `Specify an elevation for the first control point <360.000'>:` **<Enter>**
6. `Specify second point:` Pick a point on Point 1001 using the Node Osnap.
7. `Specify an elevation for the second control point or [Difference/Slope/Grade]`
   `<356.911'>:` **<Enter>** the system then reports the following data to you for inspection.
8. `First Elev: 360.000¢, Second Elev: 356.911¢, Elevation Difference: -3.089`
   `Horiz Dist: 187.628, Grade: (percent): -1.65, Slope: (run:rise): -60.75:1, Elevation <0.000'>:` **356** Type this in to see where a point of elevation 356 would land if it held the slope between Point 1002 and Point 1001.
9. `Specify an offset <0.000>:` **<Enter>**
10. `Enter a point description <.>:` **<Enter>**
11. It then allows you to continue setting new points at new specified elevations, so hit **<ESC>** to terminate. Notice that a point was set to the right of Point 1001 at exactly the elevation of 356.00.

Exercise E-4: Interpolation—Number By Distance Command

The next command, **Number By Distance**, can be used to set interpolated points in 3D by using the Grade between two other 3D objects.

TIP   This command can be used to densify your data set. This can be important when you have too little data where the TIN activity can cause ambiguous results.

1. Continue in the same drawing, or open the drawing called Chapter 3-a. Dwg.
2. Using the **V** for **View** command, highlight the view called **Points-I**, hit **Set Current**, and then hit **OK**. You see two points already set, Points 1001 and 1002.
3. Using the drop-down arrow to the right of the fifth icon from the left, choose **Number By Distance**.
4. `Specify first point or [Entity]:` Pick a point on a contour vertex using the Endpoint Osnap
5. `Specify an elevation for the first control point <356.000'>:` **<Enter>**
6. `Specify second point:` Pick a point on Point 1002 with the Node Osnap.
7. `Specify an elevation for the second control point or [Difference/Slope/Grade]`
   `<356.911'>:` **<Enter>** the system then reports the following:
   `First Elev: 356.000', Second Elev: 356.911', Elevation Difference: 0.911, Horiz Dist: 130.850, Grade: (percent): 0.70 Slope: (run:rise): 143.57:1`
8. `Enter number of points <0>:` **3** so you set three intermediate points between these two objects.
9. `Specify an offset <0.000>:` **<Enter>**
10. `Enter a point description <.>:` **<Enter>**
11. `Enter a point description <.>:` **<Enter>**
12. `Enter a point description <.>:` **<Enter>**
13. `Enter number of points <0>:` **<ESC>** to terminate.

Exercise E-5: Interpolation—Perpendicular Command

The next command, **Perpendicular,** can be used to set points in 3D at a perpendicular location from a point to a vector.

> On a project you often need to identify the elevation of a centerline for a driveway as it hits the flowline of a curb. This command can be used to accomplish that or similar 3D perpendicular tasks.

1. Continue in the same drawing, or open the drawing called Chapter 3-a. Dwg.
2. Using the **V** for **View** command, highlight the view called **Points-I**, hit **Set Current**, and then hit **OK.** You see two points already set, Points 1001 and 1002.
3. Using the drop-down arrow to the right of the fifth icon from the left, choose **Perpendicular.**
4. `Specify first point or [Entity]:` Pick a point on Point 1002 with the Node Osnap.
5. `Specify an elevation for the first control point <360.000'>:` **<Enter>**
6. `Specify second point:` Pick a point on Point 1001 with the Node Osnap.
7. `Specify an elevation for the second control point or [Difference/ Slope/Grade] <356.911'>:` **<Enter>**

The software responds with data: `First Elev: 360.000', Second Elev: 356.911', Elevation Difference: -3.089, Horiz Dist: 187.628, Grade: (percent): -1.65, Slope: (run:rise): -60.75:1`

8. `Please specify a location that is perpendicular to the interpolated line:` Pick a point on the circle in the view with the Center Osnap.
9. `Specify an offset <0.000>:` **<Enter>**
10. `Enter a point description <.>:` **<Enter>**
11. Hit **<ESC>** to terminate.

Exercise E-6: Interpolation—Incremental Distance Command

The next command, **Incremental Distance**, can be used to set points in 3D by using Grade, Slope, Elevational, or Difference based criteria. The variation of this command is that it sets points at specified intervals, say every 25'.

> This routine might be used to create a centerline of driveway elevations at the flowline of a curb section every 75' based on the grade of the curb section.

1. Continue in the same drawing, or open the drawing called Chapter 3-a. Dwg.
2. Using the **V** for **View** command, highlight the view called **Points-I**, hit **Set Current**, and then hit **OK.** You see two points already set, Points 1001 and 1002.

3. Using the drop-down arrow to the right of the fifth icon from the left, choose **Incremental Distance**.

4. `Specify first point or [Entity]:` Pick a point on a contour vertex using the Endpoint Osnap.

5. `Specify an elevation for the first control point <356.000'>:` **<Enter>**

6. `Specify second point:` Pick a point on Point 1001 with the Node Osnap.

7. `Specify an elevation for the second control point or [Difference/ Slope/Grade]`

8. `<356.911'>:` **<Enter>**

The software responds with data: `First Elev: 356.000', Second Elev: 360.000', Elevation Difference: 4.000, Horiz Dist: 97.374, Grade: (percent): 4.11, Slope: (run:rise): 24.34:1`

9. `Distance between points <10.0000>:` **<Enter>**
10. `Specify an offset <0.000>:` **<Enter>**
11. `Enter a point description <.>:` **<Enter>**
12. `Enter a point description <.>:` **<Enter>**
13. `Enter a point description <.>:` **<Enter>**
14. `Enter a point description <.>:` **<Enter>**
15. `Enter a point description <.>:` **<Enter>**
16. `Distance between points <10.0000>:` **<ESC>** to terminate.

Exercise E-7: Interpolation—Incremental Elevation Command

The next command, **Incremental Elevation**, can be used to set points in 3D by using Grade, Slope, Elevational, or Difference based criteria. The variation of this command is that it will set points at specified *Elevational* intervals, say every 1'.

TIP This routine might be used to create points that have only even elevations, such as 360.0, 361.0, 362.0, 363.0. This could be helpful in a planning exercise in which the planner is trying to obtain a rough feel for the design possibilities.

1. Continue in the same drawing, or open the drawing called Chapter 3-a. Dwg.

2. Using the **V** for **View** command, highlight the view called **Points-J**, hit **Set Current**, then hit **OK**. You see two contours that you will work with that have white circles identifying them.

3. Using the drop-down arrow to the right of the fifth icon from the left, choose **Incremental Elevation**.

4. `Specify first point or [Entity]:` Pick a point on a contour vertex inside the circle on the green contour using the Endpoint Osnap.

5. `Specify an elevation for the first control point <360.000'>:` **<Enter>**

6. `Specify second point:` Pick a point on a contour vertex inside the circle on the brown contour using the Endpoint Osnap.

7. `Specify an elevation for the second control point or [Difference/ Slope/Grade] <366.000'>:` **<Enter>**

The software responds with data: `First Elev: 360.000', Second Elev: 366.000', Elevation Difference: 6.000, Horiz Dist: 247.076, Grade: (percent): 2.43, Slope: (run:rise): 41.18:1`

8. `Elevation Difference <0.000'>:`**1**
9. `Specify an offset <0.000>:` **<Enter>**
10. `Enter a point description <.>:` **<Enter>**
11. `Enter a point description <.>:` **<Enter>**
12. `Enter a point description <.>:` **<Enter>**
13. `Enter a point description <.>:` **<Enter>**
14. `Enter a point description <.>:` **<Enter>**
15. `Elevation Difference <1.000'>:` **<ESC>** to terminate.

Exercise E-8: Interpolation—Intersection Command

The next command, **Intersection**, can be used to set points in 3D by using the slope of a vector defined via two points or a 3D object with another object that crosses the first object's path.

 TIP This command could be helpful in a determining the elevation at the edge of pavement where an underground utility crosses the edge of pavement, or where a roadway crown might intersect another roadway centerline. This occurs in 3D.

Student Files

1. Continue in the same drawing, or open the drawing called Chapter 3-a. Dwg.
2. Using the **V** for **View** command, highlight the view called **Points-J**, hit **Set Current**, then hit **OK**. You see two contours that you will work with that have white circles identifying them. You also see a white line that crosses the path if the two circles were connected.
3. Using the drop-down arrow to the right of the fifth icon from the left, choose **Intersection**.
4. `Specify first point or [Entity]:` **Pick a point on a contour vertex inside the circle on the green contour using the Endpoint Osnap.**
5. `Specify an elevation for the first control point <360.000'>:` **<Enter>**
6. `Specify second point:` **Pick a point on a contour vertex inside the circle on the brown contour using the Endpoint Osnap.**
7. `Specify an elevation for the second control point or [Difference/Slope/Grade] <366.000'>:` **<Enter>**

The software responds with data: `First Elev: 360.000', Second Elev: 366.000', Elevation Difference: 6.000, Horiz Dist: 247.076, Grade: (percent): 2.43, Slope: (run:rise): 41.18:1`

8. `Specify an offset <0.000>:` **<Enter>**
9. `Specify first point or [Entity]:` **e**
10. `Select an arc, line, lot line, feature line, or polyline or [Points]:` **Select the line crossing the path between the two circles.**
11. `Specify offset from entity <0.000>:` **<Enter>**
12. `Enter a point description <.>:` **<Enter>**
13. **<ESC>** to terminate.

This concludes this segment on placing Interpolation-based points into your project.

PERFORMING SLOPE-BASED COMPUTATIONS

This next segment delves into procedures for placing Slope-based points in the file. See Figure E-2. These additional 3D commands can be used as thinking points or directly within the design as 3D data points. These commands are interesting because they solve real-life problems that engineers encounter. For instance, the **Slope—High/Low Point** command can take a known slope in one direction, intersect it with a known slope in another direction, and set the intersection point in 3D.

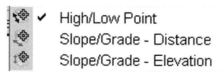

Figure E-2   Placing Slope-based points

Exercise E-9: Slope—High/Low Point Command

The next command, **High/Low Point,** can be used to set points in 3D by using the slope of one vector and intersecting it with the slope of another vector. This could be helpful in the following situation.

This can be solved for many design conditions, but one example in particular might be in a subdivision, while grading between two houses. There is a drainage envelope around each house, assuring that drainage occurs away from the house. If one house is higher than the other and has a differing drainage slope, you may want to locate a center of swale point where the two envelopes intersect so you can guide the drainage toward either the front or the back of the houses.

1. Continue in the same drawing, or open the drawing called Chapter 3-a. Dwg.
2. Using the **V** for **View** command, highlight the view called **Points-K**, hit **Set Current**, then hit **OK**. You see two house templates with drainage envelopes around them. You will set a high/low point between the two houses, which are at different finish floor elevations and have 10′ drainage envelopes.
3. Using the drop-down arrow to the right of the sixth icon from the left, choose **High/Low Point**. The prompts are as follows.
4. `Specify start point:` Select a point on the lower right drainage envelope corner of the left house using the Endpoint Osnap.
5. `Specify second point:` Select a point on the lower left drainage envelope corner of the right house using the Endpoint Osnap.
6. `First Slope (run:rise) or [Grade] <Horizontal>:` **g** to indicate that you will give it a grade (in %).
7. `Grade (percent) or [Slope] <0.00>:` **−5**, this is a 5% downward grade. (Remember that grades are in % whereas slopes are in ratios.) The system responds with data: `Grade: (percent): −5.00, Slope: (run:rise): -20.00:1`
8. `Second Grade (percent) or [Slope] <−5.00>:` **−.05**, to indicate that the second grade will be ½ % downward. Therefore, the first grade is coming down faster because the house is at a 2′ higher elevation while the second grade is coming down slower because that house is lower. This allows an intersection point to be computed between the two drainage envelopes that can be used as a center of swale point. The swale can guide drainage between the houses to either the front or the back of the property. Again, the system will respond with data: `Grade: (percent): −0.05, Slope:`

(run:rise): −2000.00:1, New point: X: 15876.145, Y: 19171.564, Z: 348.997. Note that a tick mark displays where the new computed point will be placed if you proceed with the command. If the point will fall in an unsatisfactory location, say **No** to the next prompt and redo the command with different, more satisfactory data.

9. `Add point [Yes/No] <Yes>:` **y**
10. `Enter a point description <.>:` **<Enter>**
11. `Specify start point:` **<ESC>** to terminate.
12. Save this drawing as you will use your results to perform the next two exercises.

Exercise E-10: Slope—Slope/Grade—Distance Command

The next command, **Slope/Grade—Distance,** can be used to set points in 3D by using the elevation of an object and providing it with Slope, Grade, and Distance criteria to set another point in 3D.

 This could be helpful in designing a swale between two houses once a low point has been constructed as was accomplished in the previous exercise. A designer can extend the swale between the houses for, say, 50′ at a grade of −2%, thereby ensuring drainage away from the houses toward the front or back of the property.

1. Continue in the same drawing.
2. Using the **V** for **View** command, highlight the view called **Points-K**, hit **Set Current**, and then hit **OK**. You see two house templates with drainage envelopes around them. You will use the high/low point set in the last exercise to design a swale to convey drainage between the two houses.
3. Using the drop-down arrow to the right of the sixth icon from the left, choose **Slope/Grade—Distance**. The prompts are as follows.
4. `Specify start point:` Select the point placed using the **High/Low Point** command in the last example using the Node Osnap.
5. The system will respond: `Elevation <348.997′>`**<Enter>**
6. `Specify a point to define the direction of the intermediate points:` Pick a point in the front yard of the houses (in other words, to the north and between the two houses).
7. `Slope (run:rise) or [Grade] <Horizontal>:` **g**
8. `Grade (percent) or [Slope] <0.00>:` **−1**
9. `Grade: (percent): −1.00, Slope: (run:rise): −100.00:1`
10. `Distance:` **100**, you will set points 100′ feet from the first point you picked.
11. `Enter the number of intermediate points <0>:` **4**, you will set four intermediate points along the way with one last point at 100′.
12. `Specify an offset <0.000>:` **<Enter>**, this allows for offset points if desired; you will say none.
13. `Add ending point [Yes/No] <Yes>:` **<Enter>**
14. `Enter a point description <.>:` **<Enter>**
15. `Enter a point description <.>:` **<Enter>**
16. `Enter a point description <.>:` **<Enter>**
17. `Enter a point description <.>:` **<Enter>**
18. `Enter a point description <.>:` **<Enter>**

19. `Elevation <348.997'>`**<Enter>**
20. `Specify a point to define the direction of the intermediate points:` <ESC> to terminate.

You should see several points developed between the two houses in Figure E-3.

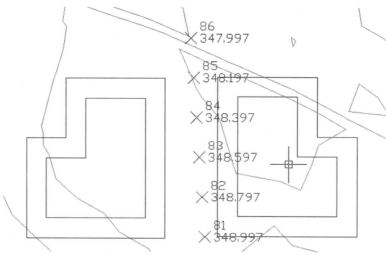

Figure E-3   Points set between two houses

Exercise E-11: Slope—Slope/Grade—Elevation Command

The next command, **Slope/Grade—Elevation,** can be used to set points in 3D by using the elevation of an object and providing it with Slope, Grade, and Distance criteria to set another point in 3D.

This differs from the previous command in that it asks for the ending elevation instead of a distance. You can use it to lay some points out for the front yard of the left house in the example.

1. Continue in the same drawing.
2. Using the **V** for **View** command, highlight the view called **Points-K**, hit **Set Current**, and then hit **OK**. You see two house templates with drainage envelopes around them. The points you placed between the two houses should be there from the previous exercise. Use the last point placed in the front yard to set some new points. Select the last point in the front yard, provide a direction to the west (or left), and travel at a grade of positive 4% until you hit an elevation of 351.5.
3. Using the drop-down arrow to the right of the sixth icon from the left, choose **Slope/Grade—Elevation**. The prompts are as follows.
4. `Specify start point:` Select the last point placed in the front yard from the last exercise using the Node Osnap.
5. The system will respond: `Elevation <347.997'>`**<Enter>**
6. `Specify a point to define the direction of the intermediate points:` Pick a point in the front yard of the left house (in other words, to the left, but somewhere in the front yard of the left house).
7. `Slope (run:rise) or [Grade] <Horizontal>:` **g**

8. Grade (percent) or [Slope] <0.00>: **4**
9. Grade: (percent): 4.00, Slope: (run:rise): 25.00:1 Ending Elevation: 351.5
10. Enter the number of intermediate points <0>: **3**
11. Specify an offset <0.000>: **<Enter>**
12. Add ending point [Yes/No] <Yes>: **<Enter>**
13. Enter a point description <.>: **<Enter>**
14. Enter a point description <.>: **<Enter>**
15. Enter a point description <.>: **<Enter>**
16. Enter a point description <.>: **<Enter>**
17. Enter a point description <.>: **<Enter>**
18. Elevation <347.997'>**<Enter>**
19. Specify a point to define the direction of the intermediate points: **<ESC>** to terminate.

Notice that the last point placed has an elevation of 351.500, as shown in Figure E-4, which is what was requested. The points are placed such that their grade is 4% from the first point chosen.

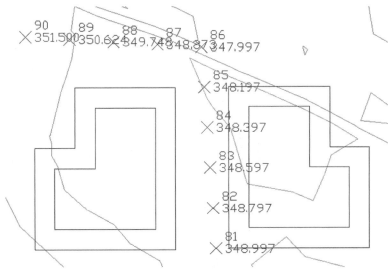

Figure E-4   The last point placed is at elevation 351.500

INDEX

Emboldened page numbers denote a definition. Figures and tables, indicated respectively with an f or t trailing the page number, are cited only when they appear outside the related text discussion. Citations for material in the appendices are denoted with a capital letter preceding the page number(s).